Blockchain Technology for IoE

This book explores opportunities and challenges in the field of Internet of Everything (IoE) security and privacy under the umbrella of distributed ledger technologies and blockchain technology, including distributed consensus mechanisms, crypto-sensors, encryption algorithms, and fault tolerance mechanisms for devices and systems. It focusses on the applicability of blockchain technology, including architectures and platforms for blockchain and IoE, authentication and encryption algorithms for IoE, malicious transactions detection, blockchain for forensics, and so forth.

- Outlines the major benefits as well as challenges associated with integration of blockchain with IoE;
- Describes detailed framework to provide security in IoE using blockchain technology;
- Reviews various issues while using distributed ledger technologies for IoE;
- Provides comprehensive coverage of blockchain for IoE in securing information including encryption schemes, authentication, security issues, and challenges;
- Includes case studies in realistic situations like healthcare informatics, smart industry, and smart transportation.

This book is aimed at researchers and graduate students in computing, cryptography, IoT, computer engineering, and networks.

Computational Intelligence Techniques

Series Editor: Vishal Jain

The objective of this series is to provide researchers with a platform to present state-of-the-art innovations, research, and design and implement methodological and algorithmic solutions to data processing problems, designing and analyzing evolving trends in health informatics and computer-aided diagnosis. This series provides support and aid to researchers involved in designing decision-support systems that will permit societal acceptance of ambient intelligence. The overall goal of this series is to present the latest snapshot of ongoing research as well as to shed further light on future directions in this space. The series presents novel technical studies as well as position and vision papers comprising hypothetical/speculative scenarios. The book series seeks to compile all aspects of computational intelligence techniques from fundamental principles to current advanced concepts. For this series, we invite researchers, academicians, and professionals to contribute, expressing their ideas and research in the application of intelligent techniques to the field of engineering in handbook, reference, or monograph volumes.

Computational Intelligence Techniques and Their Applications to Software Engineering Problems
Ankita Bansal, Abha Jain, Sarika Jain, Vishal Jain and Ankur Choudhary

Smart Computational Intelligence in Biomedical and Health Informatics
Amit Kumar Manocha, Mandeep Singh, Shruti Jain and Vishal Jain

Data Driven Decision Making Using Analytics
Parul Gandhi, Surbhi Bhatia and Kapal Dev

Smart Computing and Self-Adaptive Systems
Simar Preet Singh, Arun Solanki, Anju Sharma, Zdzislaw Polkowski and Rajesh Kumar

Advancing Computational Intelligence Techniques for Security Systems Design
Uzzal Sharma, Parmanand Astya, Anupam Baliyan, Salah-ddine Krit, Vishal Jain and Mohammad Zubair Kha

Graph Learning and Network Science for Natural Language Processing
Edited by Muskan Garg, Amit Kumar Gupta and Rajesh Prasad

Computational Intelligence in Medical Decision Making and Diagnosis: Techniques and Applications
Edited by Sitendra Tamrakar, Shruti Bhargava Choubey and Abhishek Choubey

Applications of 5G and Beyond in Smart Cities
Edited by Ambar Bajpai and Arun Balodi

Healthcare Industry 4.0: Computer Vision-Aided Data Analytics
Edited by P. Karthikeyan, Polinpapilinho F. Katina and R. Rajaagopal

Blockchain Technology for IoE: Security and Privacy Perspectives
Edited by Arun Solanki, Anuj Kumar Singh, and Sudeep Tanwar

For more information about this series, please visit: www.routledge.com/Computational-Intelligence-Techniques/book-series/CIT

Blockchain Technology for IoE

Security and Privacy Perspectives

Edited by
Arun Solanki, Anuj Kumar Singh,
and Sudeep Tanwar

CRC Press is an imprint of the
Taylor & Francis Group, an **informa** business

Designed cover image: © Shutterstock

First edition published 2024
by CRC Press
6000 Broken Sound Parkway NW, Suite 300, Boca Raton, FL 33487-2742

and by CRC Press
4 Park Square, Milton Park, Abingdon, Oxon, OX14 4RN

CRC Press is an imprint of Taylor & Francis Group, LLC

Library of Congress Cataloging-in-Publication Data
Names: Solanki, Arun, 1985– editor. | Singh, Anuj Kumar, editor. | Tanwar, Sudeep, editor.
Title: Blockchain technology for IoE : security and privacy perspectives /
edited by Arun Solanki, Anuj Kumar Singh, and Sudeep Tanwar.
Description: First edition. | Boca Raton, FL : CRC Press, [2024] |
Series: Computational intelligence techniques, 2831–6150 |
Includes bibliographical references and index.
Identifiers: LCCN 2023012494 (print) | LCCN 2023012495 (ebook) |
ISBN 9781032431741 (hbk) | ISBN 9781032431758 (pbk) | ISBN 9781003366010 (ebk)
Subjects: LCSH: Internet of things. | Internet–Security measures. | Blockchains (Databases)
Classification: LCC TK5105.8857 .B5829 2024 (print) | LCC TK5105.8857 (ebook) |
DDC 005.8–dc23/eng/20230727
LC record available at https://lccn.loc.gov/2023012494
LC ebook record available at https://lccn.loc.gov/2023012495

ISBN: 978-1-032-43174-1 (hbk)
ISBN: 978-1-032-43175-8 (pbk)
ISBN: 978-1-003-36601-0 (ebk)

DOI: 10.1201/9781003366010

Typeset in Times
by Newgen Publishing UK

Contents

Preface

Blockchain Technology, beyond the bitcoin, has created an exceptional revolution in the reliability and security of many Internet amenities, services, and applications. IoT cannot be exempted from this transition where the blockchain technology permits it to securely store restrained data and use this data to decide the compliance with the restrictions of a smart contract. However, blockchain technology only guarantees the immutability of information once it reaches the blockchain, whereas the data or sensors themselves can be forged or emulated and, thus, added security and privacy procedures must be enforced to the sensors at the edge level. The IoE is a multi-domain environment. IoE contains a huge number of devices and services integrated for exchange of information. The security areas of IoE include user privacy, data protection, authentication, identity management, trust management and policy integration, authorization, access control, end-to-end security, availability, and resistance against dynamic attacks. Blockchain Technology has been proven to be very successful in securing this kind of environment with increased trust and transparency.

This book aims to explore new frontiers, opportunities, and challenges in the field of IoE security and privacy under the umbrella of distributed ledger technologies and blockchain technology, at the same time also including the mentioned distributed consensus mechanisms, crypto-sensors, encryption algorithms, fault tolerance mechanisms for devices and systems in IoE, and so on. This book tends to focus on the primary fundamentals of the applicability of blockchain technology in securing IoE environments like architectures and platforms for blockchain and IoE, authentication and encryption algorithms for IoE, IoE malicious transactions detection, blockchain for forensics in IoE, crypto-elements for the security in IoE, fog/edge computing and sidechains for IoE security, and blockchain for forensics in IoE.

The book has been organized in three parts where, Part I consists of three chapters introducing the security of IoE and blockchain. Part II presents blockchain-based security mechanisms for IoE and includes five chapters. Part III comprises of four chapters focusing on application areas integrating blockchain and IoE.

Chapter 1 of the book introduces IoE and aims to describe the modern era of the future technologies along with the concepts of privacy, safety, and security that can be easily implemented with the help of the combination of blockchain and IoE. A case study to identify the attack surface for vulnerability finding has also been presented in this chapter. Chapter 2 reveals the need for security in the context of IoE and explores the security requirements, issues, and challenges in IoE. The different security issues including confidentiality, integrity, privacy, availability, authentication, authorization, and access control in context to IoE have been discussed here. Chapter 3 enlightens the importance and impact of IoE and blockchain convergence to enhance security in financial and commercial sectors. The study presented in this chapter reveals various features of blockchain technology and its modern-day applications to make the digital transactions more secured.

Chapter 4 of the book presents the Artificial Intelligence (AI)-based blockchain-enabled access control framework for the mitigation and detection of malicious attacks to enhance security of IoE environment. How the security of the messages transmitted by the IoE devices to the blockchain can be enhanced by using a novel timestamp and Ed25519 (Edward Curve) based signature algorithms has been demonstrated in Chapter 5. Chapter 6 focuses on the architecture and security issues of IoE. Machine Learning (ML) and Blockchain are implemented together to increase the security in IoE. This chapter also talks about how the industrial IoE fits into the picture of such an application. How even after adding network and things, how people, data and process come together and help in solving the real-world problems. Chapter 7 addresses the issue of single authentication systems by discussing blockchain-based two-factor authentication. In this chapter, the OTP-SMS framework has been introduced as a two-factor authentication mechanism utilizing blockchain technology to improve security and overcome login problems. A high data priority endorsement and profile overhaul using blockchain against remapping attacks in MANET-IoT has been presented in Chapter 8 of the book. In the proposed scheme the novel hash function blocks the attacker's misbehavior by identifying the new update. The effectiveness of a network has been evaluated based on its performance metrics, which may include routing traffic, throughput, and other similar indicators.

Blockchain applications and case studies for IoE including smart airports, environmental surveillance, smart cities, energy sector, and smart water management have been explained in Chapter 9. Chapter 10 presents an easy-to-implement anti-counterfeiting architecture that will provide authenticity and traceability for expensive medicines using QR codes, web technology, and blockchain. In Chapter 11, a thorough analysis along with the systematic review of current technologies and medicine business logistics management using blockchain technology to handle secure drug logistics records have been discussed. To ensure the confidentiality and soundness of the novel structure, an apt agreement has been developed in combination with permissioned blockchain architecture. Chapter 12 explains the proposed design followed by implementation of a new voice-controlled home automation system for Industry 4.0 that uses block chain-enabled 5C protocols of the Cyber Physical System (CPS). In this study, the common CPS procedures are made possible by block chains and categorized according to how time-sensitive and throughput-intensive they are. Several characteristics of block chains, such as security, privacy, fault tolerance, interoperability, automation, data/service exchange, and trust, are categorized according to the varied degrees to which they assist CPS.

The book presents the recent research trends in the area of blockchain and IoE to its readers and covers every significant aspect of it. The approach of this book is novel and one of its kind, presenting the fusion of two latest technologies, that is IoE and Blockchain.

Editors
Arun Solanki
Anuj Kumar Singh
Sudeep Tanwar

Overview of the Book

The continuing dispersal of Internet of Things (IoT) technologies is opening new opportunities, and the numerous amazing applications are evolving which include smart cities, smart industry, smart homes, smart agriculture, smart healthcare, and many others. The application of IoT in many diverse areas is a primary cause of generating a huge amount of data originating from different things. This raises a lot of privacy and security vulnerabilities since most of the data is produced by wireless and less computationally capable devices like sensors, RFID, smart cards, and others. But with the evolution of IoE which provides connectivity not only among things, but also among people, data, and processes, there is an increased requirement for secure communication and protected storage of information. The IoE enhances connectivity and intelligence to about every entity in the system including things, data, processes, and the people, giving it special functions.

Blockchain Technology, beyond the bitcoin, has created an exceptional revolution in the reliability and security of many Internet amenities, services, and applications. IoT cannot be exempted from this transition where the blockchain technology permits it to securely store restrained data and use this data to decide the compliance with the restrictions of a smart contract. However, blockchain technology only guarantees the immutability of information once it reaches the blockchain, whereas the data or sensors themselves can be forged or emulated and, thus, added security and privacy procedures must be enforced to the sensors at the edge level.

The IoE is a multi-domain environment. IoE contain a huge number of devices and services integrated for exchange of information. The security areas of IoE include user privacy, data protection, authentication, identity management, trust management and policy integration, authorization, access control, end-to-end security, availability, and resistance against dynamic attacks. Blockchain Technology has been proven to be very successful in securing this kind of environment with increased trust and transparency. This book aims to explore new frontiers, opportunities, and challenges in the field of IoE security and privacy under the umbrella of distributed ledger technologies and blockchain technology, at the same time also including the mentioned distributed consensus mechanisms, crypto-sensors, encryption algorithms, fault tolerance mechanisms for devices and systems in IoE, and so on. This book tends to focus on the primary fundamentals of the applicability of blockchain technology in securing IoE environments like architectures and platforms for blockchain and IoE, authentication and encryption algorithms for IoE, IoE malicious transactions detection, blockchain for forensics in IoE, crypto-elements for the security in IoE, fog/edge computing and sidechains for IoE security, and blockchain for forensics in IoE.

About the Editors

Arun Solanki is working as Assistant Professor in the Department of Computer Science and Engineering, Gautam Buddha University, Greater Noida, India where he has been working since 2009. He has served as the Time Table Coordinator and as a member of the Examination, Admission, Sports Council, Digital Information Cell, and other university teams. He received his M.Tech. degree in Computer Engineering from YMCA University, Faridabad, Haryana, India. He received his Ph.D. in Computer Science and Engineering from Gautam Buddha University in 2014. He has supervised more than 6o M.Tech. dissertations under his guidance. His research interests span expert system, machine learning, and search engines. He has published many research articles in SCI/ Scopus indexed international journals/conferences. He has participated in many international conferences. He has been a technical and advisory committee member of many conferences. He has organized several Faculty Development Programs (FDPs), conferences, workshops, and seminars. He has chaired many sessions at international conferences. Arun Solanki works as Associate Editor of the *International Journal of Web-Based Learning and Teaching Technologies* (IJWLTT). He has been working as Guest Editor for special issues in *Recent Patents on Computer Science*, Bentham Science Publishers. Arun Solanki is the editor of many books with reputed publishers. He is working as a reviewer for many other reputed publishers of journals.

Anuj Kumar Singh is Associate Professor in the Department of Computer Science and Engineering at Adani University, Ahmedabad, India. He has more than 18 years of teaching experience in technical education. He holds a Ph.D in Computer Science and Engineering from Dr. A.P.J. Abdul Kalam Technical University, Lucknow. He passed his M.Tech degree with First Distinction from Panjab University, Chandigarh and his B.Tech degree with First Honors from U.P.T.U Lucknow in Computer Science and Engineering. In addition, he has also achieved the national qualification UGC NET. Having published more than 35 research papers in journals and conferences including SCIE and Scopus, he has also authored one book and edited three. He has also filed five design patents. He has also supervised more than ten dissertations at PG level. His areas of specialization include intelligent systems, cryptography, network and information security, blockchain technology, and algorithm design.

Sudeep Tanwar (Senior Member, IEEE) is currently Professor with the Computer Science and Engineering Department, Institute of Technology, Nirma University, India. He is also a Visiting Professor at Jan Wyzykowski University, Polkowice, Poland, and the University of Pitesti in Pitesti, Romania. He received his B. Tech in 2002 from Kurukshetra University, India, his M. Tech (Honor's) in 2009 from Guru Gobind Singh Indraprastha University, Delhi, India, and his Ph.D. in 2016 with specialization in Wireless Sensor Network. He has authored two books and edited 13 books and more than 250 technical articles in top journals and for top conferences. He initiated the research field of blockchain technology adoption in various verticals, in

2017. His H-index is 44. He actively serves his research communities in various roles. His research interests include blockchain technology, wireless sensor networks, fog computing, smart grid, and the IoT. He is a Final Voting Member of the IEEE ComSoc Tactile Internet Committee, in 2020. He is a Senior Member of IEEE, Member of CSI, IAENG, ISTE, and CSTA, and a member of the Technical Committee on Tactile Internet of IEEE Communication Society. He has been awarded the Best Research Paper Awards. He has served many international conferences as a member of the Organizing Committee, such as the Publication Chair for FTNCT-2020, ICCIC 2020, and WiMob 2019, a member of the Advisory Board for ICACCT-2021 and ICACI 2020, a Workshop Co-Chair for CIS 2021, and a General Chair for IC4S 2019, 2020, and ICCSDF 2020. He also serves on the editorial boards of *Computer Communications*, *International Journal of Communication Systems*, and *Security and Privacy* and is leading the ST Research Laboratory, where group members are working on the latest cutting-edge technologies.

Contributors

Avishake Adhikary
Amity Institute of Information Technology, Amity University, Kolkata, India

Akshat Agrawal
Department of Computer Science & Engineering, Amity School of Engineering and Technology, Amity University Haryana, Gurugram, India

Md. Nafees Imtiaz Ahsan
University of Asia Pacific, Bangladesh

Atul Bandyopadhyay
Department of Physics, University of Gour Banga, Malda, India

Qazi Basheer
Department of IT, Nawab Shah Alam Khan College of Engineering, Hyderabad, India

Arindam Biswas
Department of Computer Science Engineering, Kazi Nazrul University, Asansol, West Bengal, India

Preeti Chandrakar
National Institute of Technology, Raipur, India

Kamlesh Chandrawanshi
Department of Computing Science Engineering VIT Bhopal University, Bhopal, MP, India

Yash Chawla
Department of Operations Research and Business Intelligence, Wroclaw University of Science and Technology, Wroclaw, Poland

Soma Debnath
Amity Institute of Information Technology, Amity University, Kolkata, India

Narendra Kumar Dewangan
National Institute of Technology, Raipur, India

P. Gayathiri
Nirmala College for Women, Coimbatore, India

Swati Gupta
Centre of Excellence, Department of Computer Science and Engineering, School of Engineering and Technology, K.R. Mangalam University, Gurugram, India

Sk. Yeasin Kabir Joy
University of Asia Pacific, Bangladesh

Anil George K
Department of Computer Science, St. Thomas's College, Thrissur, Kerala, India

Asha K
Department of Advanced Computing, St. Joseph's University, Bangalore, Karnataka, India

Gurudas Mandal
Department of Metallurgical Engineering, Kazi Nazrul University, Asansol, West Bengal, India

Neha Mathur
Amity University, Gurugram, India

Abdullah Al Omar
University of Alberta, Canada

Nuzhat Tabassum Progga
University of Asia Pacific, Bangladesh

Shaik Qadeer
Department of EED, Muffakhamjah College of Engineering and Technology, Hyderabad, India

Roheen Qamar
Department of Computer Science, Quaid-e-Awam University of Engineering, Science and Technology, Nawabshah, Pakistan

Mohammad Sanaullah Qaseem
Depart. of CSE, Nawab Shah Alam Khan College of Engineering, Hyderabad, India

Mohammad Shahriar Rahman
United International University, Bangladesh

Aloke Kumar Saha
University of Asia Pacific, Bangladesh

Sandeep Sahu
Department of Computing Science Engineering VIT Bhopal University, Bhopal, MP, India

Rahul Samanta
Department of Metallurgical Engineering, Kazi Nazrul University, Asansol, West Bengal, India

Devdas Saraswat
IES University, Bhopal, India

Dhrubasish Sarkar
Amity Institute of Information Technology, Amity University, Kolkata, India

Shweta Sinha
Amity University, Gurugram, India

Gaurav Soni
Department of Computing Science Engineering VIT Bhopal University, Bhopal, MP, India

Milind Udbhav
Centre of Excellence, Department of Computer Science and Engineering, School of Engineering and Technology, K.R. Mangalam University, Gurugram, India

Ravi Verma
Department of Computing Science Engineering VIT Bhopal University, Bhopal, MP, India

Meenu Vijarania
Centre of Excellence, Department of Computer Science and Engineering, School of Engineering and Technology, K.R. Mangalam University, Gurugram, India

Baqar Ali Zardari
Department of Information Technology, Quaid-e-Awam University of Engineering, Science and Technology, Nawabshah, Pakistan

Part I

Introduction to Security of IoE and Blockchain

1 Internet of Everything and Blockchain

An Introduction

Milind Udbhav, Meenu Vijarania, Swati Gupta, Akshat Agrawal, and Yash Chawla

CONTENTS

1.1 INTRODUCTION TO THE BLOCKCHAIN AND INTERNET OF THINGS

Blockchain potential can be considerably enhanced for future applications and security purposes when combined with the IoE. The Internet of Everything (IoE) is a technology that can enable a digital ecosystem consisting of natural and non-natural things resulting in a smart and connected Earth.

In the near future, every data element will be connected using artificial intelligence (AI) and machine learning (ML) systems, resulting in a smart and integrated approach toward sustainable development and advancement. IoE functioning is not feasible at the current 'low' Internet speed and bandwidth. Instead, we need to use the technology of limitless memory to create a massive pool of constant Internet connectivity that will be managed by AI and will replace the manual technology of the human generation and can efficiently perform things that were previously thought to be fiction or impossible.

DOI: 10.1201/9781003366010-2

1.1.1 Blockchain

Blockchains are extremely popular in today's world. The name blockchain itself alludes to how they work: a series of blocks that contain data. Blockchain was discovered by a group of researchers, and the technique was originally intended to timestamp digital documents, making it impossible to backdate. Satoshi Nakamoto, a pseudonym, initially adopted the technology in 2009 to create Bitcoin, a digital cryptocurrency. Blockchain is a type of distributed ledger that anyone can access and is composed of blocks. Each block contains data, hash, and the previous block's hash. The data stored within the block is determined by the type of blockchain in use. The Bitcoin blockchain, for example, stores transaction details such as sender data, receiver data, and the number of coins. The hash function acts as a fingerprint and provides a unique authorization identity. When a block is created, its purpose is to verify the hash calculation, which is required to detect changes in a block. The hash of the previous block is the final element within each block. The previous block's hash is used to create a chain of blocks, and the technique is used to secure the blockchain. Blockchain employs a method known as Proof of Work (PoW) to authenticate the creation of new blocks. In the case of Bitcoin, calculating the required PoW and adding a new block to the chain takes ten minutes. It is tough to tamper with blocks when using the mechanism. If someone tampers with one block, the entire process of recalculation must be repeated for the following block to function as the proof defense mechanism is activated. Blockchain also employs a distribution mechanism for security purposes. Instead of relying on a central entity to manage the chain, blockchain uses a peer-to-peer network that anyone can join. Users who join the network receive a complete copy of the blockchain, which the node can use to verify whether everything is still in order. The functional diagram of the blockchain for finding a link with IoT is shown in Figure 1.1.

1.1.2 Limitations of Blockchain

Blockchain technology has numerous advantages. However, it's also essential to look at the limitations as well. One of the first limitations of blockchain technology is its scalability. Due to its consensus mechanism, centralized companies like Visa can process thousands of transactions per second. However, due to the high time taken, only ten transactions can be done on the Bitcoin blockchain. Another limitation of

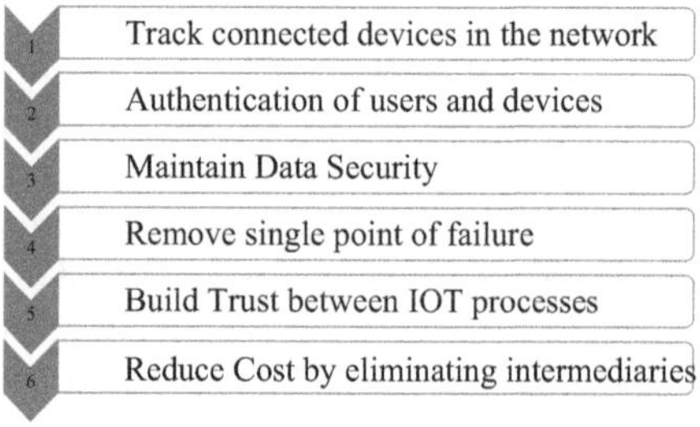

1 Track connected devices in the network
2 Authentication of users and devices
3 Maintain Data Security
4 Remove single point of failure
5 Build Trust between IOT processes
6 Reduce Cost by eliminating intermediaries

FIGURE 1.1 Blockchain finding the link between security and IoT.

the blockchain is that it uses excessive energy. The miners in the Bitcoin blockchain use tremendous amounts of energy to find the hash for the next block. At the end of every ten minutes, one block is added, that is, only one miner finds that correct hash. So, the work done by the other miners is not very fruitful, which can be a costly deal in terms of wasted energy. Also, energy generation is a climate issue in today's world, and according to the survey, the Bitcoin blockchain alone uses the equivalent of Switzerland's annual energy usage. Another drawback of the technology is immutability, and users cannot make changes in the block. This makes it perfect for someone who wants to ensure that no one ever tampers with any of the records. However, it's not the best choice in certain cases. For example, in the case of traditional banks, if the sender accidentally sent a large amount of money to someone, then the user can request the bank to cancel that transaction, and the money would be refunded back to the user's account. In the case of blockchain, once the entry is validated and logged onto the block, there is no central authority to whom the user can request a refund. Another limitation of blockchain technology is that it still has an unavoidable security flaw. If more than half of the computers acting as nodes to service the network lie, validation is false. The truth becomes true when transactions are modified and rendered to their public key. This type of attack is known as a 51% attack. Finally, the community monitors Bitcoin mining pools to ensure no one gains such network influence. Each user has a public key and a private key. The blockchain is decentralized, and no one can access the key information.

Some things need to be clarified about blockchain as well. The first misconception about blockchain is that blockchain and Bitcoin are the same; people tend to use Bitcoin and blockchain interchangeably and think both are the same thing. However, that is not the case. Bitcoin uses blockchain, the leading underlying technology that enables Bitcoin to work. The second misconception about blockchain is that cryptocurrency is its only application, and that is the most popular or the only application of blockchain. People forget that the Internet is a large electronic system in which people can build multiple applications; here, cryptocurrency is just one of them. Blockchain is used in various fields like healthcare, voting businesses, logistics, insurance, and many more than usually perceived. The third misconception about blockchain is that blockchain activity is public, but most transactions are public and can be viewed by anyone worldwide. Last, people think there is only one blockchain and all the transactions from everywhere are stored on this one. It's essential to know that there are multiple blockchains, each serving a specific purpose. Different blockchains include public blockchains, private blockchains and hybrid blockchains. Public blockchains are open to everyone; anyone can view them and make transactions on them, for example, the Bitcoin and Ethereum blockchain. Private blockchains are used by businesses and tend to be more centralized. At the same time, hybrids contain the property of both types of blockchain.

1.1.3 Internet of Things

To begin, let's discuss what the Internet of Things is. It is a system of interconnected computing devices, mechanical, and digital mechanics, each with a unique identifier, capable of transferring data over and at work without requiring human-to-human or human-to-computer interaction. Consider the home's components or devices, for instance the refrigerator. Connecting it to the Internet would provide much information in real-time, such as its operating status, temperature, and so on. This process helps monitor or control devices or appliances in the user's home by using direct communication. The process can also be done by using a proxy to transfer the data between the user and the appliances. For the whole process, there are two main approaches – for instance, IoT devices or home appliances such as a refrigerator, toaster, or microwave can be directly connected to the Internet, which will help transfer the data by connecting directly to an Ethernet cable or Wi-Fi. The approach followed in the whole process is called the gateway approach. The IoT devices may be used as a proxy or the calling process. An IoT gateway uses the mechanism to communicate with the Internet. The second approach is the gateway; here, for instance, users have many ultrasound devices, X-ray stations or monitors, and many more. All of them are connected to an IoT gateway, which can be considered a router or a device that might relate to the router. Those devices will be connected to the IoT gateway through different mediums, for instance, Wi-Fi or many other wireless technologies. The IoT gateway might be connected to a router through Wi-Fi; for example, a cable connected to the router will be connected to the Internet for a data storage server. The user will connect to the Internet through an app which will fetch the data sent by the IoT devices to the data storage server and get information about the IoT devices. For the whole process, an assumption can be made. For instance, an X-ray station will be connected to an output gateway, and an IoT gateway will be connected to the Internet and consist of stations responsible for transferring data in IoT devices. To access the information, there is a need for a username and password to fetch the data from the data storage server. There are many wireless technologies between the IoT devices and the gateway, for instance Bluetooth low energy or Wi-Fi or radio frequency identification, called the RFID (radio frequency identification). So the whole process totally depends on the device architecture, and it's up to the user to use the medium for its purpose. Many devices nowadays can be found on the Internet and come with an Internet connection that helps serve the data and fetch the data from the required place. The functional diagram of the IoE is given below in Figure 1.2.

1.1.4 Internet of Everything

Around 15 to 20 years from now, every person on the globe will be connected, as well as many animals, both domesticated and wild, as well as many inanimate and natural objects, which will be linked to technology via the Internet to create an IoE. The technology has the potential to become a fully electronic ecosystem, making our planet and its surroundings smart, and connecting every aspect of life, including health data, to one another via the IoE. AI will manage a massive network in the future, and the future appears to be in some electronic cosmos. The IoE would not function the same way as today's Internet speed and capacity do. Limitations will be a relic from the history of the early semi-mechanical Internet. The future will consist of a planet and a region in near space that will be a massive pool of constant connectivity. The entire connectivity will be managed by AI, and the cost of the IoE for human privacy will

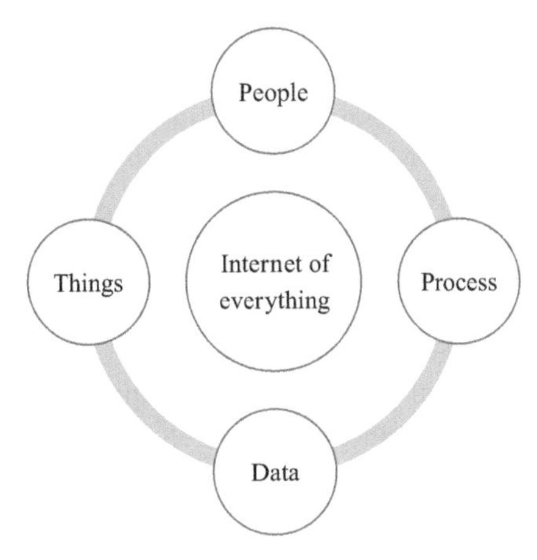

FIGURE 1.2 Internet of Everything working scientific diagram.

be far beyond human comprehension. Even in an anonymous pool, the self-driving car will pick people up and apply algorithms to learn about their travel plans, just as healthcare monitors, security monitors, and so on lead to information that will be considered private in the future. The security monitors will all need information that is now regarded as private. The IoE will soon produce such massive and rapid data flows that only AI systems can manage. The future is when humans know everything that exists on the planet and near space. The IoE will become the successor to the nation-state in which AI will make and administer laws and will govern the electronic rules for a better future which is part of a fused electronic cosmos that might one day become sentiment. IoT will play a significant role in developing the future IoE, which will play a major role in the security and privacy of the system and use technologies like AI and ML to improve and increase its accuracy.

The organization of the remainder of this chapter is as follows. Section 2.2 presents state-of-the-art technologies and mechanisms for enhanced security for IoE. Section 2.3 describes the IoE and blockchain technology protection using reverse engineering and Malware Analysis. Section 2.4 explains the case study of Vulnerability Finding to figure out the attack surface. Finally, the conclusion and future work are presented in Section 2.5.

1.2 LITERATURE REVIEW

Over the last few years, the global IoT market has grown significantly. According to projections, the number of connected devices will reach 75.44 billion by 2025 (Alzoubi et al., 2022). This means that many devices will be connected to the Internet and will play an important role in driving IoT systems, which are all about making business decisions based on data collected by smart devices (Hazra et al., 2022). A slew of smart devices will collect data or be attached to physical assets (Na and Park, 2021). The collected data will be transferred to the back-end systems before storage and processing. Data will be processed further after that. Customers can decide to do whatever the owner wants based on the data (Sarac et al., 2021). In practice, the system is usually realized using a cloud-centric approach, which means various IoT devices. Some of them are very small, such as sensors and actuators (Evans, 2011). The IoT gateway will send the data to the cloud, allowing IP-capable IoT devices to communicate directly with the Outback end. So, all that data will first be sent to the cloud's connected gateway, and the cloud service will host various types of services to assist patients (Cisco, 2012). Based on those services, the user device management system subscripts central can build a different hotel application and provide insight for users. Vehicles along the systems with various types of entities create the security and safety mechanism for securing IoT applications (Weissberger, 2014). The common challenge focuses on the security of the various stages. For example, Windows considers secure hardware that runs a secure OS by loading trusted firmware into smart devices that can establish a secure communication channel with the cloud back-end on the cloud side. Several security services will need to be hosted. It is critical in IoT systems to ensure data trustworthiness by assuring the confidentiality and integrity of data based on those security services (Evans, 2012). There is a need to realize the Internet's vision or trust things in general to consider security throughout the data life cycle. Internet trust items include data collection, data in transit, and data retention (Kitchenham, 2012). There is a need to ensure that security has been properly implemented across the entire data lifecycle and that blockchain can be used to realize this end-to-end security for the proper functioning and usage of the Internet of Things (Jula et al., 2014). The main thing that sets us apart from other species is not intelligence. It is the way people connect and the way to manage larger groups of people. It refers to the ability of people to collaboratively work toward shared objectives and visions when groups of people grew bigger in earlier times. The larger the flow of information and the greater the goals in the treatments, it's time for the entire human race to collaborate and leave more differences behind. The time is to organize in a way which serves the entire planet, leaving no one behind blockchain; it's like the backbone as the Internet was meant to be. It serves as a decentralized web that allows people to connect securely to transfer value to do business directly with each other without the need for unnecessary things. It is mesmerizing to see the leaps of progress in things like robotics and nanotechnology, biotechnology, or AI blockchain, which is the one technology connecting fields seamlessly, raising them into new spheres with incredible potential. AI will finally have a platform (Asaduzzaman et al., 2020) to exchange value in the form of trained data sets. Because of its trustless nature, blockchain is a language that AI only understands but will be able to speak fluently, which will help to create entire libraries before anyone can even turn a single page. It's not a race of people versus technology. As such, technology evolved from the seed that was first planted, and has enabled the fundamental development of humans and aids them in creating technology, and thus people cannot separate themselves from it. The technique also enables people to merge with the connected super-brains, who frequently dread change because they fear the unknown, despite the fact that reality's most important constant is change (Hossain et al., 2020). Everything changes all the time. It's the most natural law, and nobody can stop the process. So, it's more important to let evolution happen and take its course in the most decentralized and transparent way possible. Technology is more than just Bitcoin (Weiss et al., 2016).

Blockchain technology is still in its infancy, but please don't underestimate the exponential times of accelerating technological advancement. In a prosperous world, a global technological web, a global superbrain, the IoE, not only are humans merging with it, but everything which is present around humans such as trees, cars, animals, everything, algorithms will steer and govern the dynamics of the new web and will let them do it, because the new technology is just so much better at its free-flowing data (Ramos et al., 2016). The key to optimizing data processing and growing the intelligence of the new web blockchains will bring the necessary security and trust that is needed to let this valuable data flow as freely as it needs to; the world has never been more vibrant and exciting than it is today (Susi et al., 2013). The process will bring the fourth industrial revolution, which is the revolution of humanity. This is the awakening which is a new paradigm (Muralidharan et al., 2014).

1.3 IOE AND BLOCKCHAIN TECHNOLOGY PROTECTION USING REVERSE ENGINEERING AND MALWARE ANALYSIS

Anyone can understand deductive reasoning by using reverse engineering. This concept also explains how a previously created wise process, system, or piece of software accomplishes any task with ease and demonstrates how it works.

1.3.1 Introduction to Reverse Engineering

The reverse engineering method is commonly used in computer science engineering, mechanical engineering, design electronic engineering, and biology. The complex nature of anatomy requires people to understand each part of the body. With the help of this technology, people can redo and reproduce origami by unfolding it at first and repeating the process until proper output is achieved. The car uses the concept of reverse engineering to understand its major and minor mechanical parts, and its purpose is to reduce problems for better working and experience. In the software security industry, reverse engineering is commonly used for day-to-day tasks. The concepts also help to safeguard from trojan and malware attacks. The methodology of reverse engineering also talks about the high-level code and review process, which is useful in sorting the quality or verification for better implementation. The aim of the test activities performed is to ensure that the vulnerabilities are found and fixed. The diagram of reverse engineering is given below in Figure 1.3.

1.3.2 Malware Analysis

Before delving into the concept of malware analysis, it is necessary to first define malware.

Malware is a piece of code that performs malicious actions in the form of an executable script. Malware is used by attackers to steal sensitive information and spy on the victim's system, which can then be used to gain control and use it for illegal purposes. Malware enters the system without authorization and is used for illegal purposes. Malware analysis is a widely used concept for stopping malware attacks and reducing their impact. Malware analysis is typically used to provide the information required

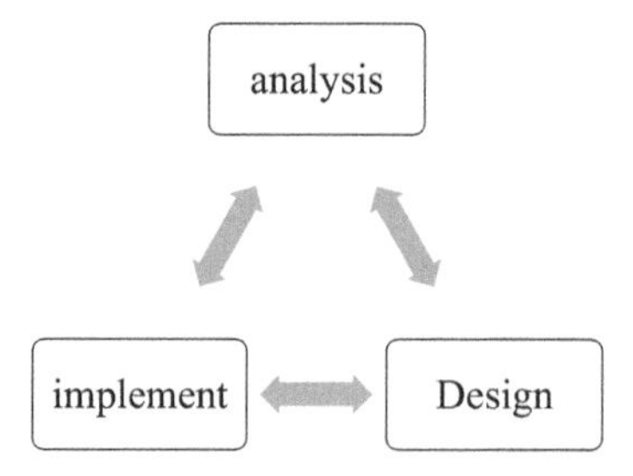

FIGURE 1.3 Working diagram of reverse engineering.

to respond to a network intrusion. To achieve the goal, it is necessary to determine what exactly occurred, as well as to locate an infected device and its files to analyze the suspected malware. When the malware has been identified and analyzed, it is time to create a signature that can detect malware infections on the network. Malware analysis, developed, host-based, and network signatures are commonly used. Malicious code is detected on a victim's computer using host-based and network signatures. The indicators frequently identify files created or modified by malware or specific changes detected. Malware indicators concentrate on what malware does to a system rather than on the malware's characteristics. Using these techniques increases the effectiveness factors, making it easier to detect malware that has changed its original form or has been deleted from the hacked disc. Network signatures are used by network traffic monitoring systems to identify malicious code and provide a higher detection rate. With the help of network techniques, the understanding of the working of malware and its advantage, as well as disadvantages, can be easily understood. Malware analysis plays a major role when integrated with blockchain and IoT, providing a better solution to deal with new types of cyber-attacks and online frauds. Malware analysis has two types of approaches, one is static approach and the other is a dynamic approach. Static analysis involves the mining techniques for finding malware without testing it, whereas dynamic analysis runs the malware particles for categorizing in basic or advanced forms. Using these techniques, the division of different techniques for different types of malware can be easily achieved. By using the techniques of malware analysis, gathering difficult information can be easily achieved and helps to reduce and stop the chance of getting infected by malware attacks, as given in Figure 1.4.

1.3.3 Important Aspect of IoT: Firmware and Firmware Analysis

With the aid of the operating system, the configuration, and the filesystem, the data and programmes that are stored together in memory improve persistence performance and the user's overall experience. Multiple Process Architecture (MIPS, ARM, AND INTEL ATOM) depends on the Central Processing Unit (CPU). In today's time, most of the firmware is based on Linux. Firmware contains a bootloader, os kernel, file system, configuration data, and busy box (Xuan et al., 2010). The busy box is a container that is responsible for containing a stripped-down version of important utilities like passwords. This ensures optimization from the space perspective and acts as a trainer who guides better output. The firmware serves as a starting point for any IoT-based

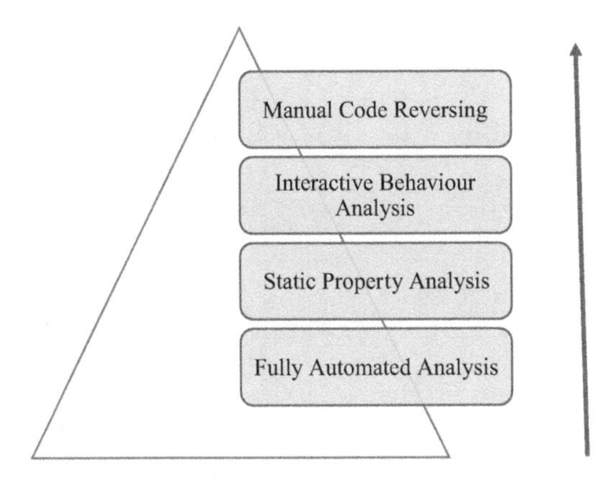

FIGURE 1.4 Working strategy of malware analysis.

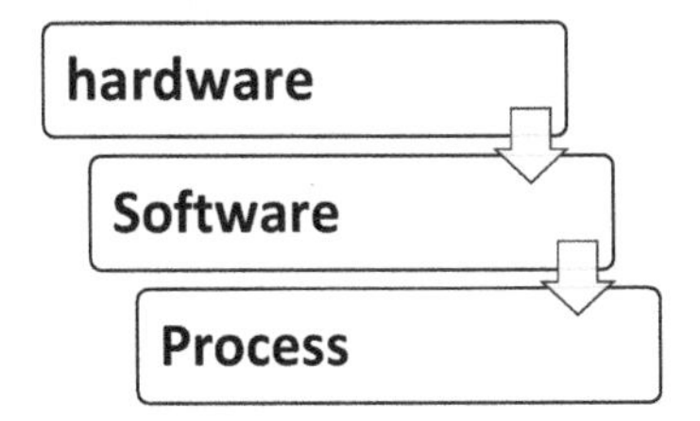

FIGURE 1.5 Firmware analysis approach.

security research, which can unpack a firmware-based image and uses reverse engineering for binaries, an inspection of web sources, and review configurations. Firmware is cheap and easily available. Firmware can be obtained by the use of a USB key/cd from a manufacturer, downloaded using a manufacturing website, a serial connection from the device, and many other techniques are present for the process, as shown in Figure 1.5. The firmware analysis is an important part of recognizing the usefulness of firmware. The firmware analysis contains three approaches: hardware, software, and process. The most focused perspective is the software approach which uses a method of downloading software and testing its use cases to know the advantages and limitations of the software. When we talk about the hardware approach we tend to focus on the hardware of the device, like analyzing IoT devices, the CPU, and the architecture of the motherboard. At last comes the process approach, which tends to focus on data transformation inside a particular IoT device. The major tools that are generally used in software approaches are binwalk. Binwalk is a tool that helps to identify the known file types inside a binary file and is used to extract those files into separate files.

1.4 CASE STUDY 1

Vulnerability Finding
Aim: To figure out the attack surface
Software: Binwalk

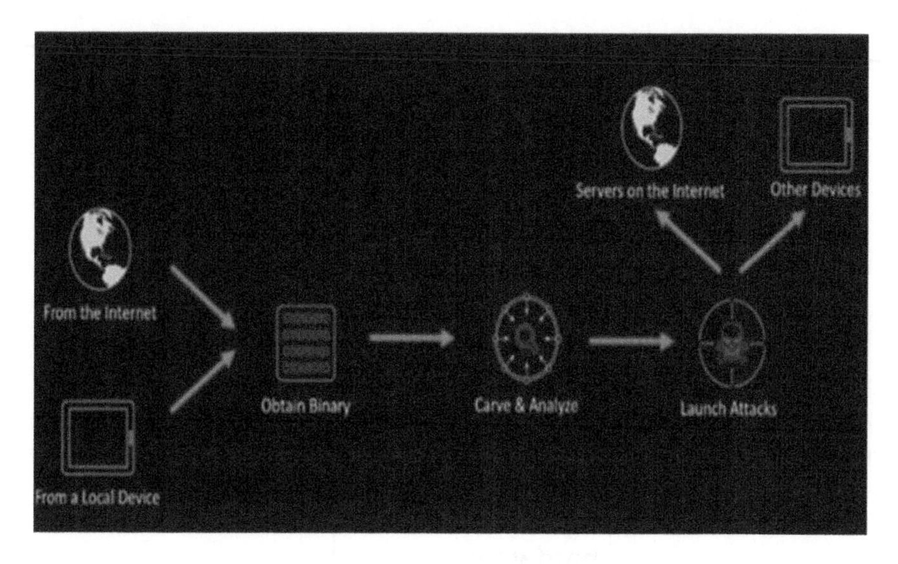

FIGURE 1.6 Diagram of firmware surface attacks.

1.4.1 Working

There is a need to figure out the attack on a surface that the whole operation is going to carry out. Before explaining the practical side, there is a need to obtain the firmware by visiting the manufacturer's website and then downloading it on the desktop. It is simply a binary file that uses binwalk tools to extract the firmware and analyze the file sub-files inside it. After that process, it looks for some vulnerabilities to launch the attack. When finding the vulnerabilities, the first example will be this spiffy-looking device that is actually just a wireless bridge and access points at the same time (Saha et al., 2015). Usually, you don't need physical access to the devices to find them. After that, the files that were successfully modified to the firmware at that time must be put back together and uploaded to the IoT device with the modified firmware. You don't need physical access to these devices to find the vulnerabilities: all you need is to do is download their firmware from the official websites and then perform the analysis on some emulators to emulate the modified version of the firmware on the emulator to see what changes are made to the firmware (Liu et al., 2017). To fulfill this, there is an obstacle that needs to be faced while running across and that is encryption. All the firmware might be encrypted, which will prevent running any analysis on it; however, in most cases, encryption is often hard for some manufacturers to get rid of so it gets left out, especially on home-grade devices (Kang and Han, 2015). So let's see the first example of how potential vulnerabilities can be found inside it, and an antivirus will be used to carry out the process. Before carrying out the steps, you need to download the firmware of the D-Link on the OEM website. Dealing with the practical side, you need to go to the terminal where the checking process is done using a tool in Linux called coal-fired, which can drag and drop the firmware to the file. To learn more about it, just look for the bin file, which is nothing but a row of data (Chen et al., 2018). So first things first, you need to check

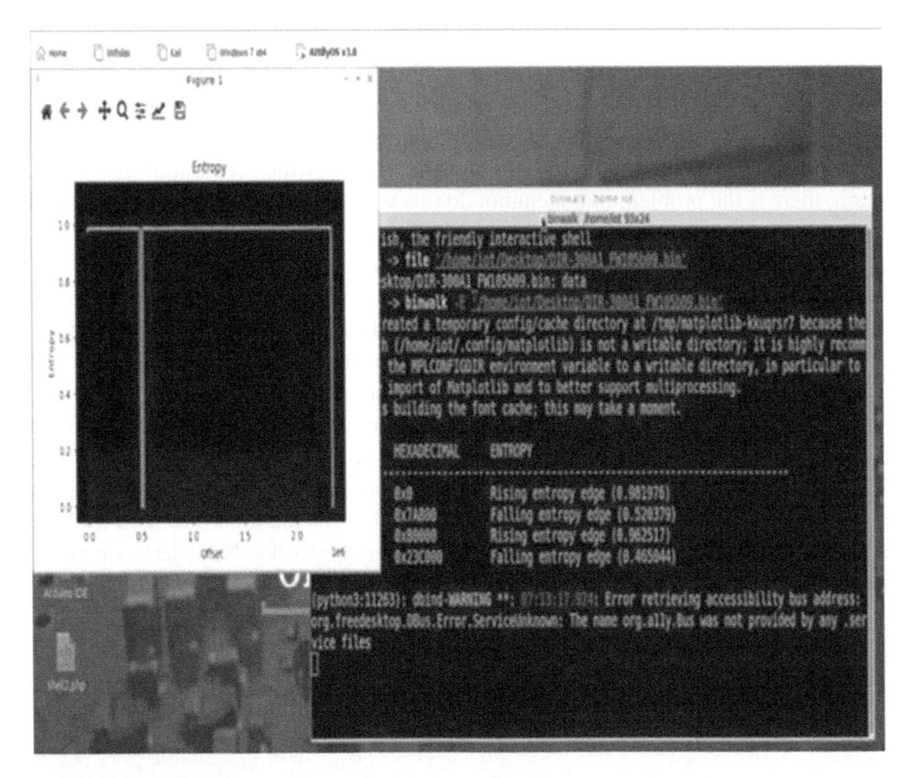

FIGURE 1.7 Shows the working of encryption software in Binwalk.

whether this firmware is encrypted or not by using dashi and then drag and drop the firmware, as shown below in Figure 1.7.

Figure 1.7 shows a straight line is present that indicates the firmware is not encrypted perfectly, so the next thing we will need to do is extract the firmware by running along with hash software this time so that the file has been extracted on our desktop so let's go and see what's inside the file: for that, go inside it and type LS2 to list the files in the folder. We have the located FS, which again is the file system of the firmware. Also, there is a need to navigate to this fire to see what's inside it. As expected, there is a bin file that contains the busy box, which is the container. All of these tools are stripped versions of the same Linux store tools spot with minimum functionalities, which can be used as a very important tool to look for some vulnerabilities which act as a grep and by using the "grep-ir" is used to look for instance telnet which is unsecured protocol usually used by the network administrator to access the network devices remotely (Poh and Poh, 2017). The problem of telnet is it doesn't know incorrect username and password while establishing the connection (Yan et al., 2017). The process never uses any type of encryption using the communication process. The alternative solution for telnet is SSH, so usually, when the finding of telnet, it's an indication that this device is using an unsecured protocol which can look for the default username. The password of the telnet that connects to it next time is provided using the credentials to do scripting to the telnet and to manage to download the firmware on the desktop (Fan et al., 2014). Just imagine the physical device which is the access point so now one can modify this firmware and re-upload it back to the device so that it can then be distributed to the customer (Zhou et al., 2014). Thus it could be used to spy on and look for the telnet to which credentials it uses. But again, there is the problem that it cannot do this without obtaining written permission from the owner, so there is a need to acquire the property.

Now in the command, look for grep – ir to search for the word telnet recursively in or in all the firmware files and press enter. A script is noticed, which contains the telnet D dash L and so on. So now check what is inside the script by simply running the cat tool and using scripts to ask the telnet so that it can see a bash script code and try to understand what's inside the code. In the command, a tonette dash is seen, which is there for the activation of code (Hyuga et al., 2015). A username which is alpha networks, along with some variables, is present, which is nothing but an image sign and variable (Malik et al., 2020). If the previous step is visited, going back to the beginning of the code we will see that the variable uses the toolkit to read this path and pass it to telnet along with the username alpha networks. Now, let's see what's inside the variable, which is the image sign which uses the cat tool and uses configuration to train the image.

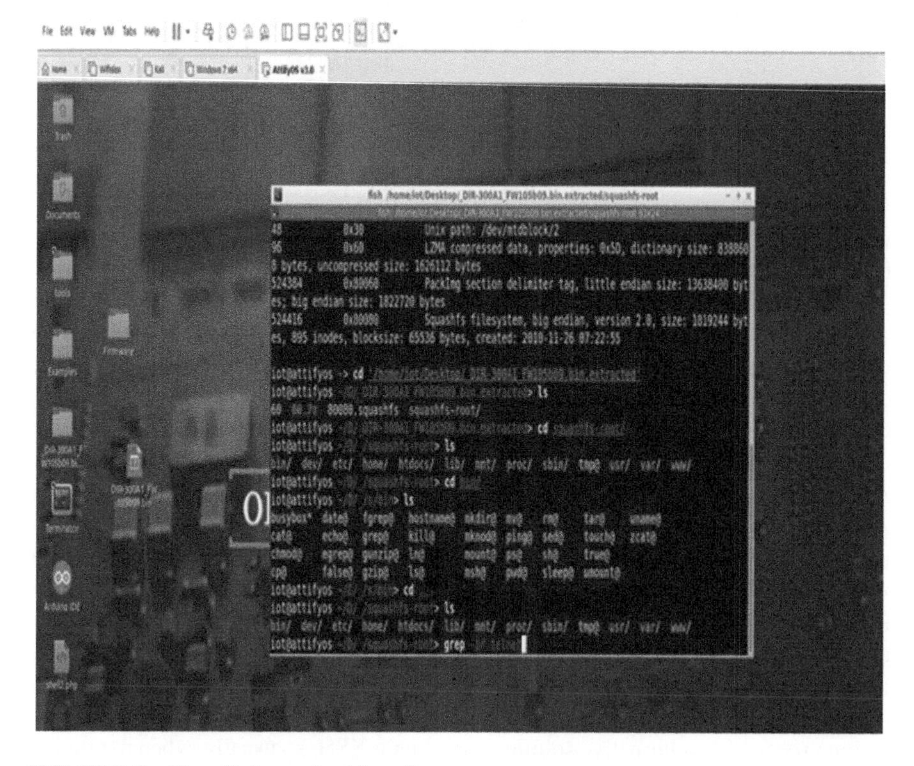

FIGURE 1.8 Use of telnet using binwalk.

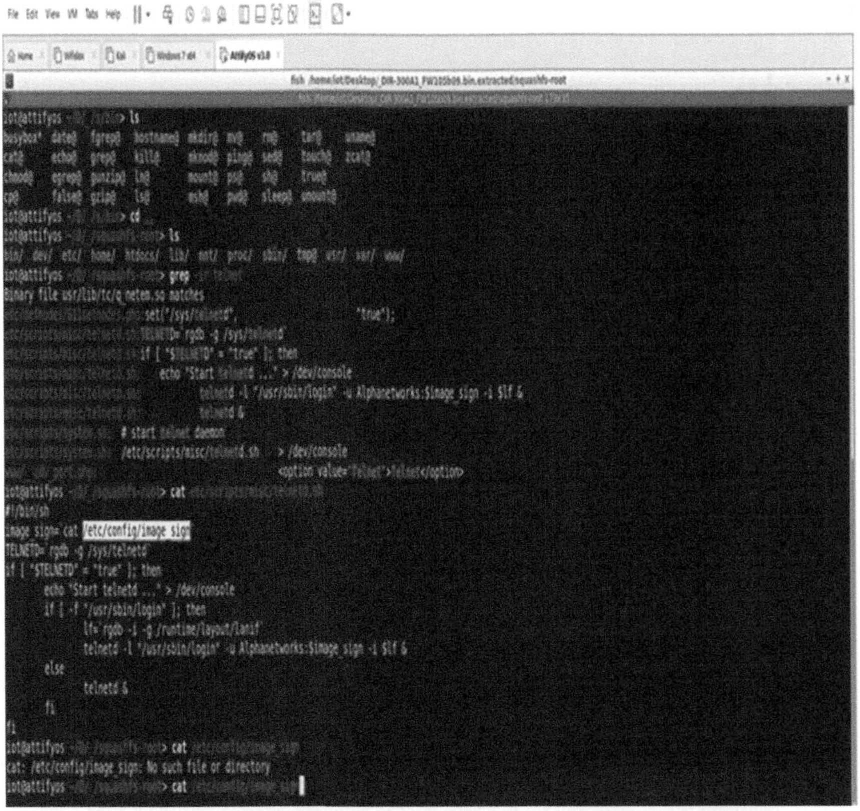

FIGURE 1.9 Use of cat tool for checking configuration and training purpose.

Given below is the final image showing the Ip address and other crucial information of the device found using telnet and this shows the process of the virtual hacking process. By using such types of methods, hackers try to get unauthorized details of user computers and try to access the sensitive data of the users without their permission and use it for malware attacks or to steal sensitive information, which could lead to cybercrime, online fraud, and data leaks. Through Case Study 1, the aim is to highlight how surface attacks are done with the help of vulnerabilities inside any computer. Clicking on untrusted sources and unauthorized access can also lead to infection of the system, which can be further used for cybercrime.

By the power of technology, the special IoE and blockchain combined can completely help to stop malware attacks and act as a protector to deal with such serious security breaches. The main goal of the case study is to highlight the functionality of vulnerabilities, as shown in Figure 1.10, malware analysis, and how cyber-attacks are conducted. With the help of the case study, users will know the working of online attacks.

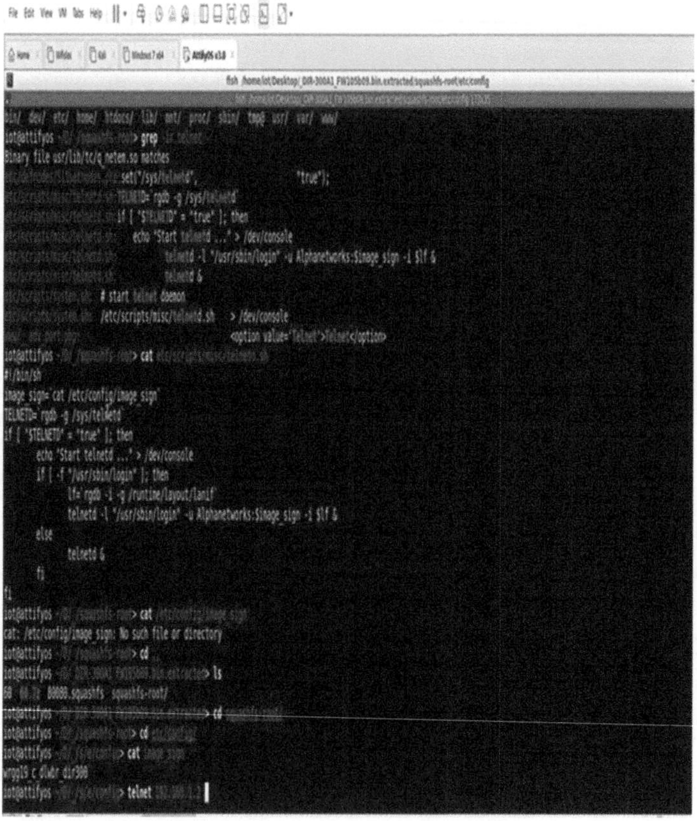

FIGURE 1.10 Final image showing IP address.

1.5 CONCLUSION

The most important part which needs to be considered is that in most IoT, IoE, and security-based products manufacturers solely focus on functionality, leaving out security and neglecting it partially or completely on their devices because the main focus is on selling and, of course, taking care of the functionality of the device itself. Also, there are many attacks which can be carried out against IoT, ranging from attacking the wireless infrastructure regardless of what kind of wireless, like Wi-Fi and Bluetooth devices, or even RFID, so, each of these devices can be attached with an individual through the wireless medium as well. Again, the majority of the IoT device firmware is corrupted, making it simple for trackers to extract, examine, and even insert backdoors into it. Firmware plays a significant role in the interaction of hardware and software. The hackers try to take advantage of firmware to breach the security system of software and access sensitive information without the user's permission. Due to firmware backdoor issues, the security system is badly affected, so there is a need to focus on firmware and add a security layer which acts as a bodyguard to secure private data and doesn't allow unauthorized access. For increasing security and making good security with functional features, the blockchain can be used to integrate with IoT and IoE systems for a secure future. A blockchain is a collection of transactional data blocks linked together—bringing the preceding block into balance. As each transaction is sequential, removing something would invalidate the entire record or ledger as new transactions. These transactions rely on the history of previous transactions; the same is true for each block. Because of the process, it is nearly impossible for thieves to alter the records. Some may believe that any user can simply change a copy of their blockchain database, allowing them to spend their coins once more. However, that person would have to change the databases of numerous entities with the exact copy that the attackers require. Massive computing power is needed to change and maintain the modified blockchain, which becomes more complex as more people use cryptocurrency, including those who run nodes, wallets, and businesses. As a result, virtual currencies are frequently referred to as cryptocurrencies, which a user creates for a cryptocurrency wallet generated by the software. The entire procedure includes a public and private key, which is referred to as a unique digital signature. The process allows them to sign transactions much like when a user signs a physical check, which are a few ways to validate transactions and generate new coins on the blockchain. Bitcoin employs a technique known as PoW. Miners are given cryptographic puzzles to solve as the difficulty of the puzzles increases. The method increases with the effects of time and the difficulty of solving problems exponentially. The miners are responsible for solving those equations and verifying account transactions. The incentive provided to miners is the possibility of being rewarded with new Bitcoin security, which gives blockchains the trust required to function according to the underlying design. Cryptocurrencies that use such methods are exciting and open the door to incredible innovation in the future. By combining the IoE with blockchain, a secure system can be created which monitors unauthorized access and prevents malware attacks for a better future. By the use of such technologies, people can be prevented from experiencing data leaks or online fraud. Using such technologies, an online bodyguard can be deployed, which protects and safeguards from misuse and the disadvantages of the Internet world, creating a new gen future technology. The chapter helps both to describe the importance of security systems and to explain the importance of privacy and encryption. With the help of blockchain and the IoE, the security system for protection in the Internet world can be achieved and it also helps in discovering new futuristic technologies, and linking with other technologies can be easily achieved, and the process can be made more accessible and safe for all, leading to more development and technological advancement.

REFERENCES

Alzoubi, Y.I., Al-Ahmad, A., Kahtan, H., and Jaradat, A. Internet of Things and Blockchain Integration: Security, Privacy, Technical, and Design Challenges. *Future Internet*, 14(7), 216, 2022.

Asaduzzaman, Md., Hasib, F., and Hafiz, Z. Bin. Towards Using Blockchain Technology for Microcredit Industry in Bangladesh. Proceedings of the 2020 23rd International Conference on Computer and Information Technology (ICCIT), pp. 1–6, 2020, Published By IEEE.

Chen, J., Zhang, Y., and Xue, W. Unsupervised Indoor Localization Based on Smartphone Sensors, iBeacon, and Wi-Fi. *Sensors*, 18(5), 2018.

Cisco. How Can Service Providers Face IPv4? A Review of Service Provider IPv4-IPv6 Coexistence Techniques. Cisco Internet Business Solutions Group (IBSG), Cisco Systems, Inc., San Jose, CA, White Paper 2012.

Evans, D. The Internet of Things: How the Next Evolution of the Internet Is Changing Everything. Cisco Internet Business Solutions Group (IBSG), Cisco Systems, Inc., San Jose, CA, White Paper 2011.

Evans, D. The Internet of Everything: How More Relevant and Valuable Connections Will Change the World. Cisco Internet Business Solutions Group (IBSG), Cisco Systems, Inc., San Jose, CA, White Paper 2012.

Fan, J., Han, F., and Liu, H. Challenges of Big Data Analysis. *National Science Review*, 1(2), 293–314, 2014.

Hazra, A., Alkhayyat, A., and Adhikari, M. Blockchain-aided Integrated Edge Framework of Cybersecurity for Internet of Things. *IEEE Consumer Electronics Magazine*, 2022.

Hossain, Md. P., Khaled, Md., Saju, S.A. Shanto Roy, Milon Biswas, Rahaman, M.A. Vehicle Registration and Information Management Using Blockchain-Based Distributed Ledger from Bangladesh Perspective. 2020 IEEE Region 10 Symposium (TENSYMP), 2020, pp. 900–903, 2020, IEEE.

Hyuga, S., Ito, M., Iwai, M., and Sezaki, K. Estimate a User's Location Using Smartphone's Barometer on a Subway. MELT 15 Proceedings of the 5th International Workshop on Mobile Entity Localization and Tracking in GPS-less Environments, pp. 2:1–2:4, Seattle, WA, November 2015.

Jula, A., Sundararajan, E., and Othmana, Z. Cloud Computing Service Composition: A Systematic Literature Review. *Expert Systems with Applications*, 41(8), 3809–3824, 2014.

Kang, W. and Han, Y. SmartPDR: Smartphone-Based Pedestrian Dead Reckoning for Indoor Localization. *IEEE Sensors Journal*, 15(5), 2906–2916, 2015.

Kitchenham, B.A. Systematic Review in Software, Re-Engineering: Where We Are and Where We Should Be Going. Proceedings of the 2nd international workshop on Evidential Assessment of software technologies, pp. 1–2, 2012, Published by ACM.

Liu, Z., Zhang, L., Liu, Q., Yin, Y., Cheng, L., and Zimmermann, R. Fusion of Magnetic and Visual Sensors for Indoor Localization: Infrastructure-Free and More Effective. *IEEE Transactions on Multimedia*, 19(4), 874–888, 2017.

Malik, N., Jaglan, V., Vijarania, M., Arora, N., and Gupta, S. Supernova Type Ia Diversity: A Study Using DBSCAN Algorithm. *International Journal*, 9(3), 2020.

Muralidharan, K., Khan, A.J., Misra, A., Balan, R.K., and Agarwal, S. Barometric Phone Sensors: More Hype than Hope! HotMobile '14 Proceedings of the 15th Workshop on Mobile Computing Systems and Applications, pp. 12:1–12:6, Santa Barbara, CA, February 2014.

Na, D. and Park, S. Fusion Chain: A Decentralized Lightweight Blockchain for IoT Security and Privacy. *Electronics*, 10(4), 391, 2021.

Poh, M.-Z. and Poh, Y.C. Validation of a Standalone Smartphone Application for Measuring Heart Rate Using Imaging Photoplethysmography. *Telemedicine and e-Health*, 23(8), 678–683, 2017.

Ramos, F.B.A., Lorayne, A., Costa, A.A.M., de Sousa, R.R., Almeida, H.O., and Perkusich, A. Combining Smartphone and Smartwatch Sensor Data in Activity Recognition Approaches an Experimental Evaluation. Proceedings of the 28th International Conference on Software Engineering and Knowledge Engineering, pp. 267–272, San Francisco Bay, CA, July 2016.

Saha, S., Chatterjee, S., Gupta, A.K., Bhattacharya, I., and Mondal, T. TrackMe – A Low Power Location Tracking System Using Smartphone Sensors. 2015 International Conference on Computing and Network Communications (CoCoNet), pp. 457–464, India, December 2015.

Sarac, M., Pavlovic, N., Bacanin, N., Al-Turjman, F., and Adamović, S. Increasing Privacy and Security by Integrating a Blockchain Secure Interface into an IoT Device Security Gateway Architecture. *Energy Reports*, 7, 8075–8082, 2021.

Susi, M., Renaudin, V., and Lachapelle, G. Motion Mode Recognition and Step Detection Algorithms for Mobile Phone Users. *Sensors*, 13(2), 1539–1562, 2013.

Weiss, G.M., Timko, J.L., Gallagher, C.M., Yoneda, K., and Schreiber, A.J. Smartwatch-Based Activity Recognition: A Machine Learning Approach. 2016 IEEE-EMBS International Conference on Biomedical and Health Informatics (BHI), pp. 426–429, Las Vegas, NV, February 2016.

Weissberger, A. TiECon 2014 Summary-Part 1: Qualcomm Keynote & IoT Track Overview. IEEE ComSoc, May 2014.

Xuan, Y., Sengupta, R., and Fallah, Y. Making Indoor Maps with Portable Accelerometer and Magnetometer. 2010 Ubiquitous Positioning Indoor Navigation and Location-Based Service, pp. 1–7, Kirkkonummi, Finland, October 2010.

Yan, B.P., Chan, C.K. Li, C.K. et al. Resting and Postexercise Heart Rate Detection from the Fingertip and Facial Photoplethysmography Using a Smartphone Camera: A Validation Study. *JMIR mHealth and uHealth*, 5(3), e33, 2017.

Zhou, P., Li, M., and Shen, G. Use it Free: Instantly Knowing Your Phone Attitude. MobiCom '14 Proceedings of the 20th annual international conference on Mobile computing and networking, pp. 605–616, Maui, HI, September 2014.

2 Security Requirements, Issues, and Challenges in IoE

Ravi Verma, Gaurav Soni, Sandeep Sahu, and Kamlesh Chandrawanshi

CONTENTS

DOI: 10.1201/9781003366010-3

2.1 INTRODUCTION TO THE INTERNET OF EVERYTHING (IOE)

The Internet of everything is one of the growing technologies of the present and future. It is also one of the most crucial and dynamic technologies of the future. In this way we can say that the Internet of Everything is considering multiple innovations in unique technology advancements that will offer various networking services. The conversion of information into specific action always creates new capabilities and opportunities, and it will create a big era of opportunity for businesses as well as individuals [1].

Technically, the Internet of Things considers millions of computational devices and client products that enable Internet connectivity to bring an intelligent network platform along with the most advanced and expanded computational features.

This is a particular type of technology that technically considers and compromises with various types of wireless devices, applications, and many other direct or indirect things which all take part in the process of communication so that these things can be globally applicable for these services. Currently, Internet connectivity is limited to mobile phones, tablets, desktops, and many other handful computing devices but when we discuss the technology working behind the Internet of Things, shortly machines will be more intelligent, more interactive and will work for cognitive science by using huge datasets along with expanding its network capacity and opportunities [2].

On the other hand, we can say the Internet of Everything is a type of technology that can smartly connect multiple peoples, processes, information, and things that will transmit to the local world in such a wonderful manner where we find billions of interconnected devices which are capable enough through sensor technology to detect, recognize and access as per their role, as we can see in Figure 2.1.

Understanding the difference between the Internet of Things versus the Internet of Everything comes from intelligent digital technology through smart connection. When we talk about the Internet of Things, we consider various physical devices and operational objects through which communication can be possible, but the Internet of everything extends the services of the Internet of Things by having a more intelligent network connection made possible by having smart sensor-based devices which work to requirement. In this manner we can say intelligent network connection and devices work together to advance a cohesive system, which we call the Internet of Everything.

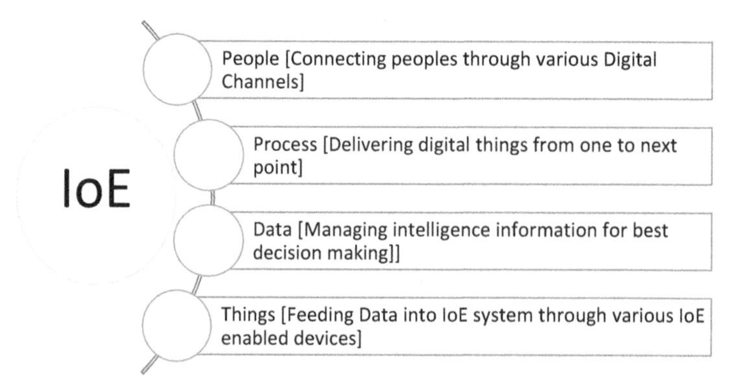

FIGURE 2.1 Essential elements of IoE.

IoT technology is dedicatedly available for machine-to-machine interaction, therefore the Internet of Things considers machines at the fundamental level, while on the other hand the Internet of Everything will come together along with people, data, and many other things , hence the Internet of Everything is not only a tablet or desktop computer but it will also integrate the people itself In this manner we can say that wristband wireless devices, and even coffee pots, can be the part of the IoE as an intelligent node of any communication network. Besides all such things we can say that the IoE is not only a machine-to-machine interaction system but it offers the machine to people and a technologically specific people-to-people communication system [3].

The Internet of Everything follows four major aspects of daily life's needs.

- **People**
  Now, in the era of the Internet of Everything, people can connect to the Internet service platforms in various ways, not only through PCs, laptops, or desktops but they can make connections by having multiple sensors placed on their skin or inside their clothes. This makes peoples' lives more comfortable in accessing the Internet without social media in this sense and now they themselves behave like a node of a network in the Internet of Everything.

  For example, we can understand the concept of the Internet of Everything with the help of the latest fitness band which can track the entire activity of any individual. In various time sessions the fitness band will keep track of their heart beat, blood pressure, and many other health-related parameters continuously, around the clock, as per their requirements [4].
- **Things**
  The things can play their roles in the form of sensors, wireless devices, consumer products, and in many other ways, they can directly connect to the Internet and interact across multiple things where they can find the information themselves from different multiple sources and exchange the same with others and then store it in some location so that after the analysis the system can make better intelligent decisions by comparing multiple information captured by sensor-enabled devices from their surrounding environment. According to various research studies, we found that in 1984 only 1,000 devices could interact with the Internet but in1992 it had expanded to one million devices connecting to the Internet. According to a study by CISCO, in 2010 the device count will be 10 billion and will reach 50 billion-plus after 2025. In this manner, we can say that, in the near future, there will be more devices involved in the process of interaction which brings better services to the clients and offers the easiest way to use Internet services [5].
- **Data**
  The intermediate IoE devices are sensor-enabled and can collect data as well as manage the entire network activity through each other; moreover, they can revert this data to the server so that analysis should be easy and helpful for the process of evaluation and analysis to take intelligent decisions. Here, all the components of intelligent devices have the capability to flow the process using valuable information and an action plan. This is why the devices will not deal with the random chunks of data sheets that are not valuable enough

for the process. Since the data comes from several sources, we call this big data or data science and it is always necessary to extract valuable information from a large set of data sets using intelligent analysis and decision-making techniques.

- **Process**
  Interference with the IoE processes is always recommended to ensure that the information is flowing to the right person at the right in a systematic way.

  The present and future decision-making of any business relies on information through which we can find better advancements in the technology for future perspectives. This is why the present and future growth of the business is completely based on data analysis which ensures and boosts the process of taking faster decisions.

  Many business leaders believe that Internet of Everything will contribute 15 trillion dollars to the domestic business economy, which is why it is proposed that the Internet of Things will contribute 19 trillion dollars to global domestic products, generating savings and profits for numerous multinational companies. The growing technology and involvement of sensor-enabled connected devices on the Internet and its growing numbers and day-to-day operation will put security at high risk; this would be the bigger challenge in front of multiple software developers but it is also common to say that the data we store should be secured and prevented from the harmful actions of attackers.

2.2 WHAT IS THE INTERNET OF EVERYTHING (IOE)

According to Cisco, the Internet of Everything is considered as a group of intelligently connected pills, processes, data, and many other things. In IoE everything is going through the process of M2M communication therefore both the terms IoE and M2M are a similar type of things but when we talk about the Internet of Everything we can also consider the terminology is like the communication between multiple advanced technologies which can assist the interaction between various peoples.

If we talk about the Internet of Things we can say it considers a broad category, it is a domain that includes various types of objects, physical or virtual objects to make communication and which allows bringing communication capacity among the different objects of the network, or in other words we can define the process of IoE in which we pursue the concept of exchanging the data without human interference that should be considered by devices. So as a conclusion we can define IoE as the process of automation of multiple devices to configure and perform as per the requirement of the user [6].

In this way, we can say that the Internet of Everything is one of the most creative ventures of future computing technology projects, products, and advancements everywhere in computing work. Digitalization is evolving so that it is highly demanded in the current business scenario to integrate automated and smarter approaches to improve product security and efficiency in the real-time environment. In having IoE now, we cannot imagine the type of network which may have been achieved in previous traditional technology because the conversion of information in an automated manner is very supportive now for business applications from the point of view of improving capabilities, the rich experience of operation and the efficient measurement of economic opportunities which are relevant to every part of the business like its people, economy and the complexity of various business operations.

In the future, the system will become more intelligent and smarter because IoE allows the entire networking system to make better decisions and to bring machines into a more intelligent and efficient state of operations, and that's why Internet connection is not limited to laptop-tablets or any mobile device because the Internet of Everything is considered the part of communication where everyone can have a way to access data and it will increase capacity and opportunity for the future computing world.

IoE considers various aspects of the entire process of communication in a prominent manner. The following diagram defines how the IoE is related to a different aspect of life, meaning what part of communication is encapsulated during the deployment of IoE within the system [7].

2.3 DEFINITION

The Internet of Everything is the concept of extension of the Internet of Things which consist of batter way in the process to make communication between machine to machine. By having the Internet of everything, we can easily perform the real-time task it offers various day to day services and also provides a smarter way of learning and configuring the devices through intelligence instructions. Therefore, the future of computing technology is completely based on the concept of Internet of everything.

2.3.1 Essential Security Requirements in IoE

Since Internet technology is growing, multiple intelligent smart devices have been involved in business operations. The growth defines the trend of today's Internet. People are relying on the system even though they are more comfortable with the interactive devices which make their daily life needs more comfortable. Nowadays the constant growth defines the status of a trend which is driven by all the connected sources of sensor-enabled devices with the Internet, including where there is needful entertainment operation, organizational equipment, unique vehicles, and other needful devices like medical equipment. Such patterns define the conversion of Internet technology into what we called today the Internet of Everything (IoE) [8].

The Internet of Everything enables us to deal with the huge amount of data that comes from numerous natural land and unnatural sources which we call artificial sources that have to be converted into some suitable form of information that can be utilized for better decision-making, understanding, and analyzing. We can say that the Internet of Everything has the potential to make our daily lives more comfortable, efficient and secure. Figure 2.2 shows the essential security requirements in IoE.

For example, we can use information which is collected from equipment-based industry for getting a better prediction of whenever maintenance is required, therefore reducing the point of failure and in a similar way we can also choose those services to check for traffic jams in busy metro cities to ensure a comfortable journey.

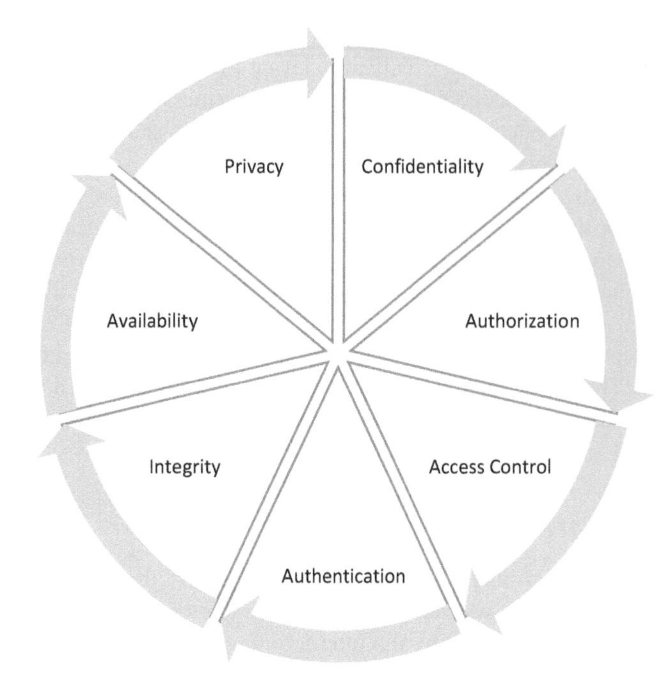

FIGURE 2.2 Essential security requirements in IoE.

Similarly, various household sensor-based devices are also offering home-based services to their users to save energy and use electricity and its devices in a more user-friendly and efficient manner. In healthcare, multiple sensors enable technology to capture the entire body of individuals whereby someone can make the treatment decision and can also observe their diabetes and blood pressure (BP) level at any point in time.

Moreover, IoE demands trust and belief since Internet technology does not come into the category of a secure place so IoE demands the security consideration at all parts of the operation [9].

Many times, the user realizes that Internet services are not available round the clock [5]; similarly, we heard huge amounts of sensitive data can be leaked [10], which will demand a big investment of cost every year to help prevent this in the future. Since Internet services, platforms and users are growing day by day, the outside eavesdroppers and hacking activities also increase their operational areas; they miss using the Internet services and devices for their own financial or other benefits, which increases cybercrime and criminal activities, breaking the trust of Internet users from various digital platforms.

The Internet of Everything offers network security in the following two ways: first, multiple outside attackers will target the IoE system and the associated gadgets that interact with the physical network. Therefore, multiple cyber-attacks can be possible according to the physical consequences, which range from destroying physical equipment to the unavailability of services and it might lead to the loss of a life. It may also be possible that some cyber-attacks affect the energy distribution concept which can damage expensive electrical devices and waste power resources which is not good for the entire networking environment [11].

In addition to this, the wastage of energy resources will unexpectedly stop the major services of the network, like emergency departments and surgical operations in the hospitals, and various military cooperations. Similarly, a cyber-attack may also be responsible for fatal poisoning in oil production units, and food production units will also face the poisonous impact due to cyber-attacks. Second, the direction through which the services of the Internet of everything are relevant to cyber security where outside attackers can break the entire security of a network or find sensitive information, affecting the entire networking service [12].

On the other hand, sometimes we find there are multiple options to ensure security at the infrastructure level in the Internet of Everything, where we can say that the multiple IoE devices communicate with each other through a network which brings a more convenient environment to manage the security consideration and also elevates the problem of network-based attack which we called botnets: these could be the biggest problem we face in the near future. The IoE will handle the botnet in a way that preserves the history of Distributed Denial of Service (DDoS) attacks throughout the network's operational history [13].

In this way, we can say that security is one of the most important considerations of the Internet of Everything so the network must manage the security measurements during the entire process of communication so every time whenever we are working on any application of the Internet of Everything, we always need to decide the security level first. Using the following technologies, we can easily compare and examine the level of satisfaction based on the security requirement; considering the different domains of networking and how the system will work for the security at the system level and application level are essential to manage and need to be prevented from the outsiders [14].

The most common security parameters that are required are confidentiality, integrity, and availability; when we talk about the traditional information systems this will also define the importance of these measurements. In this section we have tried to consider all the security-based considerations: one by one, we look at all the security measurements and parameters that one needs to keep in mind while working with any part of IoE.

2.3.2 Confidentiality

This feature will take into account the protection of sensitive data from unauthorized access; however, since there are many different types of information available, we discovered that not all of it needs to be kept secure, necessitating the need to protect sensitive and confidential information from a variety of types of access. When we talk about the traditional information technological system, multiple applications generate a high volume of data that needs to be managed with protection, especially data relevant to business plans, technical specifications, and financial specifications like the salaries of individuals, mailing communication with customers and administrators, and so on. All the inside information regarding the internal activities of an organization should always be considered a piece of sensitive information because no one wants to reveal its successful business planning to market competitors, therefore data should always be kept secure from unauthorized access along with various authentication tools like account credentials because through these parameters one can easily access sensitive information. Therefore, the network offers various types of cryptographic techniques to manage confidentiality.

The security always comes from different application domains and generally data comes from the sensors we did not manage secretly so anyone can access the data and obtain it on their own. But when we talk about the application area for home appliance, healthcare's few units of information need to be kept secret in accordance with the need and requirements list [15]. When it comes to surveillance video systems, we discovered that this information is crucial, so we need to manage confidentiality in accordance with the needs of users. Likewise, when it comes to healthcare, the best application through which systems collect patient data is crucial. Determining the level of secrecy is reserved by the information and in order to handle it, confidentiality is required.

When discussing data security, safety, and privacy, it is important to consider both storage components and transmission media as sources of information. This way, anyone attempting to compromise the security of these two crucial communication phases can either directly access information from storage or attempt to compromise the security of the media site. However, the use of encryption and decryption-based technology will keep the information secure. [16].

2.3.3 Integrity

We always favour IoE combined with data integrity features because protecting information from unauthorized access over data alteration is one of the most demanding security requirements. The sensors used in IoE will continuously generate the data and this set of data needs to keep track of and observe physical operations throughout the process so that we can manage the accuracy and consistency of the data during the entire communication process. Similar to how we observe when communication is taking place through the Internet of Everything, data has been captured by an outsider but it should not since modification will make the bigger change. If an attacker wants to change a set of data, through tracking one can find an inaccurate and wrong result at the end of the process. If a particular control command is changed by an attacker, that control will reflect a corrupted value. Therefore, maintaining integrity is one of IoE's top priorities for all application domains, but this is especially true for sectors like the healthcare industry, automation, and many others that will not tolerate any type of data breach under any circumstances

2.3.4 Availability

The requirement of availability is a top priority for all applications and you need to manage the control parameters, like in transport applications, the healthcare industry, and in the application process domain; it is another crucial security feature in any

Internet of Everything system. It means that the entire state of information is always available but in inaccessible mode for all the associates like entities. Sometimes availability plays a crucial role in any industry because if the required information is not available on time, it may damage the industry financially, and sometimes this could be a very serious situation [17].

Apart from confidentiality, data integrity, and information availability, two more essential security features are needed by the network. First consider the operation of authenticity and nonrepudiation and both of these additional features will be relevant to security considerations. Further, both designs are decided according to the scale of the organization in digital data. In the following section we try to understand both of these in detail.

2.3.5 Authenticity

Authenticity means every time the data is generated, we need to examine the source which should be verified by the receiver carrying out a site audit. This will then show that at the end we can examine the accuracy of the actual sender and receiver since sometimes we realize that the source of information is fake. Through this one can take the advantage of it, and therefore sometimes it is very important to find the validity of the sender similar to data integrity features in the world of Internet of Everything. In the absence of authenticity any attacker can take the advantage and send the data to the receiver system in the form of malicious data as a virus or irrelevant data, and in the world of Internet of Everything it is always recommended to update your application software only from a trusted and authentic site [18].

2.3.6 Non-repudiation

Non-repudiation is also a similar security feature to authentication. In this concept we need to recognize both the sender and the intermediate third party to find and recognize the actual entities and to ensure the accuracy of data that is coming from the trusted source. In traditional information systems, we typically find that data has been exchanged and there is not enough space in the memory of the system to make it available for a longer period of time. As a third party in the community, when we talk about the Internet of Everything, this feature is especially required in transport and healthcare applications, and that's why it is called non-reputation; it is also not always required as part of communication at all.

Features are necessary for managing information security using the Internet of Everything. But, in order to protect the intermediary device that will be a part of the interaction in the Internet of Everything, information security is managed with a variety of cryptographic algorithms and techniques for privacy. Cryptographic tools are not applicable for the devices. In order to keep things under control, the network must be made sufficiently secure. In addition, context-level requests must be taken into account. The authentication and authorization processes must also be improved before they affect the system.

2.3.7 Access Control and Authorization

Access control and authorization are two of the most important things in the Internet of Everything technologies. When we talk about the hospital domain, security is a major consideration. In all contexts access control and authorization are mechanisms so that one can easily access the functioning and permissions required to get operate the application, so if one wants to do something they have to get the accessing authorities in three ways in advance otherwise they will not have the permission to access this in context of the Internet of Everything. It doesn't mean that everything is available all the time, so one needs to take care of it and have permission for admission to get access to the specific object of the applications.

When we talk about the healthcare industry it is essential to have permission to get access to the restricted properties of the information system since there are multiple hierarchical positions available in the healthcare sector so different types of accessing privileges need to be assigned to different users according to their role. When we talk about the transport or other manufacturing unit it could be possible that all the domain will available for public use so we don't need to go for the authentication process otherwise the remaining process apart from the public domain will need to follow the process of getting access privileges to secure access to a particular set of information [19].

2.3.8 Trustworthy Computing

This is a very general requirement which means that all the services offered by the industry or system should be as per the expectation of its user all around the clock. Trustworthy computing is also a basic requirement for every individual because if any organization does not consider this feature it means they will also not consider the authentication, verification, and authorization process as well. This means the system and service level security authentication policies will not be run systematically along with the protocols in any system, and in this case, the system uses compromised non-security issues that would not be good for quality communication and productivity, the result being it will lose all the trust of its client. Moreover, if the system is secure enough then they can gain the trust of their people and never compromise with their attackers. It will resolve the problem of outsider attacks so that the Internet of Everything will be protected against all the attackers in an open environment along with all the IoE devices. Conclusively, we can say that trustworthiness is the story of security, by which we can say that system should never compromise with the outside elements, and the system has been completely verified based on some standards of the healthcare system.

2.3.9 Denial-of-Service Protection

One of the most common categories of attack is called Denial-of-service in which the attacker will either make the server system underperform during the dynamic process or will temporarily disable the server services and the networking resources. To manage the information more securely in an open environment, communication can be accomplished through the Internet of Everything technologies and then we can protect it from various types of attacks.

Since a DoS attack will make the server services unavailable for the other users, we need to protect the entire system along with the information itself. Information is one of the most important considerations during the process of safety in prevention so it should be available for its authorized users during the service period [20].

2.3.10 Privacy

Privacy is one of the most important essential security elements of IoE. It is how a user can protect their private information in the operation of the Internet of Everything. This could be the application environment where we deal with the private information of the user: this information could be sensitive information which we need to keep private for all those applications dealing with sensitive information, for example, as well as household and office application areas which will generate data that should not be revealed to normal users. Similarly, the healthcare system keeps security information relevant to location, background or other aspect of life, and in this manner, a person can readily use one set of data to identify another appropriate version that would not immediately indicate the new date to them. Because the Internet of Things is specifically intended to make life easier by incorporating background intelligence into ordinary life occurrences, we can define new things in this process by incorporating human and other object qualities. There are multiple privacy-preserving mechanisms available now that will help to maintain private security services and all these mechanisms are incorporated by the Internet of Everything to deal with sensitive information without losing privacy.

2.4 IOE SECURE FRAMEWORK

We require a scalable and flexible security framework in order to meet many additional security concerns as well as any demand for a large and diverse IoE environment domain. Generally, Figure 2.3 shows the key elements of any secure Internet infrastructure [21].

1. The process of Authentication
2. The process of Authorization and Access control
3. Network enforced scheme

2.4.1 The Process of Authentication

There is a method by which we can recognize machines with embedded sensors and other machine-to-machine components. An individual can be identified in an organization by using their ID and password, which can be used as a username and credential to access the system's services. The Internet of Everything may also be equipped with a similar type of authentication process, in which we may find that it occasionally uses a fingerprint and eye scanner system to reduce the need for human interaction.

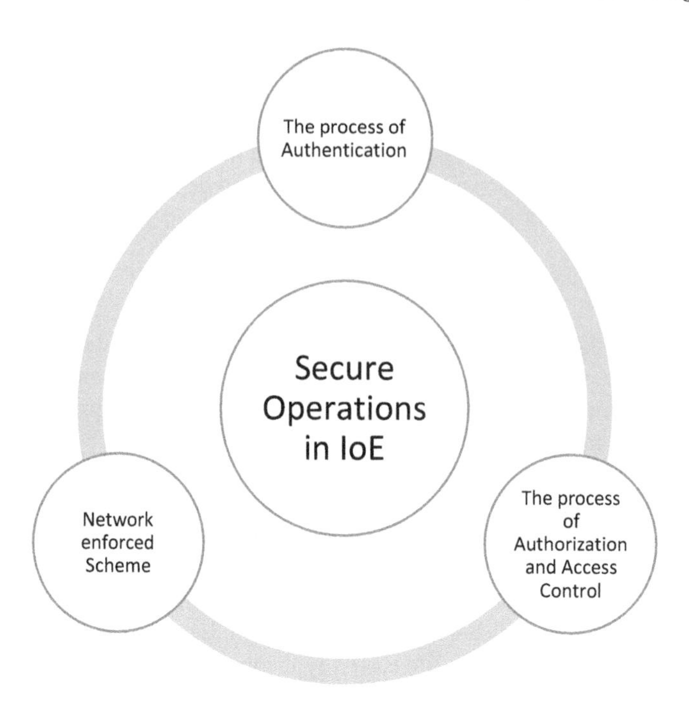

FIGURE 2.3 IoE secure framework.

In the world of the Internet of Everything, which includes devices that can be optimized with different type of operating systems, central processing units, different type of memory for scanning and foot printing, and various other wireless devices, all of these devices are low-cost, functional gadgets with network access that measures items like blood pressure or temperature. Due to geographical limitations, it may be possible that sometimes we find these devices have been located in a non-accessible location which is beyond the coverage. In this way there are multiple new challenges such as IP connectivity issues and various others related to the installation and configuration setting so that the Internet of Things will not be compromised with any type of configuration issues. In this way, we can say that the Internet of Everything is a fundamentally very different system, and nowadays connected devices will use IP connectivity at the initial level.

One should discuss the current authentication process, which is entirely dependent on the binding of acknowledged free communicated and shared passwords or any other automatically generated value. For instance, in the RSA algorithm, the key pairs are found to be the combination of the protocol with an associated X.509 certificate, which can be an OTP password-based system.

2.4.2 The Process of Authentication Control and Access

This is the mechanism wherein we take into account a number of factors, such as control policy, which includes what is the administration, what is your description, and various policy decisions in order for the network system as a whole to function and infrastructure to be developed. The system as a whole will then go through authentication services while the network and security measurement parameters are being carried out.

Fortunately, the most recent authentication mechanism policy is significant for both management and access control for each individual, meaning that it will take into account both the enterprise network and the consumer network to ensure that the mapping between various machine-to-machine communication or the Internet of Everything device interaction process is done well [22].

2.4.3 The Network Enforce Scheme

This scheme will encounter various elements which play their role in the process of routing or transporting services to different endpoints so that traffic is manageable in all circumstances, made possible through control, management administration, or various data plans. In this way, all these elements will configure, by themselves, to proceed with two configurations and authentication control processes that have been defined in various network enforced schemes so that it's very easy to encounter the process of authorization and network control access. Like the other process for accessing control and authorization similar to enforcing policy, there are multiple locations of protocols that will take care of the security parameters well suited to the Internet of Everything in the machine-to-machine interaction environment.

2.5 The Top Five Security Protocols

Security is one of the most important considerations and demanded parameters for any type of network, therefore most of the security protocols are essential to bringing the network into a reliable state. The Internet of Everything has various possibilities but it may also consider a few of the risks relevant with the security and privacy parameters. Since it's an open environment network, control can take place by outside hackers to compromise the security of the Internet of Things devices. They can easily take over the network and take advantage of it. This feature is particularly important in the healthcare setting, where numerous heart-related devices and numerous other human organs have been connected directly through an Internet of Everything devices network to track and record the activity at different times [23].

2.5.1 The Message Queen Telemetry Transport Protocol

This is one of the most important protocols in the consideration of security for the Internet of Everything communication environment. It was introduced by Dr. Andy in 1999. The role of this protocol is to ensure secure communication between client and server by conducting it through messaging technology which protects the network from outside attackers. During communication this protocol works based on the TCP IP protocol, ensuring reliable connectivity assurance with security and bringing the network into a two-way association and framework for the lossless communication process.

- **Features of MQTT**

Several security features have been provided by this protocol, therefore it's very important in the Internet of Everything that all the features have been listed as follows:

1. It is the fastest data transmission protocol that adopts the easiest way to transmit data to achieve simplicity and it is also considered a lightweight protocol.
2. This protocol was created specifically for unreliable, low bandwidth channels, a network with significant latency, and numerous other Internet of Everything devices.
3. The protocol is very simple and less complex which consumes a lot of power to consume during the process of communication and will save the battery life of all the interconnected devices, making it perfect for wireless communication devices like tablets and mobile phones.
4. Since it works on transmitting data through messages, this protocol is extremely fast, working along with the assurance of reliability.
5. It offers an ease-of-service feature that will make it applicable for all Internet of Everything devices.

2.5.2 Constraint-based Application Protocols

This protocol is specially designed for managing the security to communicate with various IoE best communication devices like microcontrollers and the network devices which are considered low bandwidth and faulty. Therefore, it is also one of the most powerful and reliable protocols for managing security.

- **Features of CoAP**

The major features provided by CoAP Protocols are:

1. This is similar to the HTTP protocol because it brings services to its client based on URL-provided resources using various communication commands like put, post, get, etc.
2. The Internet of Everything application devices using microcontrollers to handle billions of nodes less expensively, given that the design protocol is extremely effective
3. One of the major benefits of this protocol is that it can work with fewer resources which makes it simpler, less complex, and easy to use for The Internet of Everything network infrastructure.

2.5.3 Datagram Security Protocol

The datagram security protocols contribute to the Internet of Everything which was designed to protect communication systems among all the datagram-oriented applications that work with transport layer security protocols and bring the highest level of security for a large number of applications. The major objective of this protocol is to manage the problems coming from sequencing, reordering, and data loss issues in various user datagram applications since the semantic structure of the underlying protocol is unchanged, and therefore it will not provide any delay during the communication.

The major features of the datagram security protocol are as follows:

1. To resolve the packet loss from the network it will use the retransmission policy which works based on the timer scheme, so if we find the timer finished before the acknowledgment of reception of a message from the server side then the retransmission of the data frame will have taken place.
2. The ordering problem was overcome with this protocol's reordering pause by assigning precise numbers to each message, which made it much easier to identify the following frame in the sequence.
3. The protocol is not guaranteed in the delivery of data packets to the accurate destination therefore it is not very reliable in this case.

2.5.4 IPv6LPAN

The full form of this protocol is Internet Protocol version 6 power personal area network communication program and this is also the security protocol applicable for enabling devices to communicate among different wireless sensor nodes to exchange information securely. This protocol is especially used in the design part of the network which is of low power capacity, for example the Internet of Everything and wireless sensor-based communication networks. The following are the major characteristics of this [24]:

1. The protocol has been used to manage data protection using Internet Protocol version 6 for various categorized networks.
2. It is enabled for direct communication with a variety of networks along with end-to-end data delivery services based on demand.
3. The protocol is especially used to protect communication between multiple users and sensor networks.

2.5.5 ZigBee

ZigBee is a cost-effective, less complex, open-source wireless communication network that enables machine-to-machine communication up to 10–100 meters long using low-powered communication devices therefore we can say ZigBee is the security protocol for Internet of Everything devices and various applications and it will support the following two security models.

2.5.5.1 The Centralized Network Models

This particular type of network system uses an intermediate device called a trust centre. This has been provided for various services and runs on the application trusted only in the range of the ZigBee network. The entire communication network has been working

TABLE 2.1
Comparison between various security protocols

Factor of Difference	MQTTP	CoAP	DSP	IPV6LPAN	ZigBee
Efficiency	High	Moderate	Moderate	High	High
Support to low bandwidth network	High	High	Moderate	High	High
Complexity	Less	Less	Moderate	Less	Less
Reliability	High	High	Moderate	High	Moderate
Easy of Service	High	High	Less	Moderate	Moderate
Cost	Moderate	Moderate	Moderate	Moderate	Less
Support Retransmission Policy	Less	High	Less	Less	Less
Scalability	Less	Less	High	High	High

on the concept of sharing the TC link key so that all the eligible network devices who want to join ZigBee network services can join through the TC link key [25].

2.5.5.2 The Distributed Network Models

In this category of the network, there is no centralized node required, therefore it is simpler and less complex and it also belongs to the low-security communication category compared to the others where every router can initiate the process and distribute the request as their own whenever a new node joins the ZigBee network, it will receive the network key holder.

The major features of the ZigBee network as per the security point of use are:

1. The Zigbee network brings standardization among all the layers, enabling compatibility among different vendors in manufacturers of communication devices.
2. The ZigBee network offers multipoint connectivity and technology has been involved so that every device in the network can connect to all the other devices in the same network at the same time.
3. This energy-efficient protocol supports the concept of "Green Power Management" which offers low energy wastage and consumption.
4. It requires less cost.
5. It supports a large number of devices, therefore it is highly scalable.

The comparison among all the discussed protocols is presented in Table 2.1, which defines the idea of different levels of services received from various protocols in the context of IoE.

2.6 THE MAJOR INTERNET OF EVERYTHING SECURITY SOLUTION

The Internet of Everything market is anticipated to reach $700 billion in use of IoT devices by the end of 2025 if you talk about the US nation alone. This is because the business domains and their users are more comfortable using IoT devices. According to the most recent report released in the US, there could be up to 6.5 billion enterprises and individual devices of the Internet of Everything device-enabled technology by that time. In order for everyone to benefit from the Internet of Everything in many different ways, we would like to see its goods in every element and need in life. A comfortable lifestyle always encourages people to employ various sensor-enabled intelligent devices around the clock.

However, there are multiple benefits and valuable IoE-enabled featured devices are vulnerable to various cyber-crimes and attacks since as usage increases, so do fraud and criminal activities per the latest published report of IoE in 2022. Various security research associations have proposed that up to 68 percent of IoE devices are at high risk as a part of security.

According to the latest incident of such loss of security discovered in Singapore, where attackers obtained unauthorised access to private home camera footage and then uploaded it to an online platform, this could be extremely dangerous for the users of IoE. From the perspective of customers, the laws of security and privacy are being countered [26].

2.7 THE REASON BEHIND VULNERABILITY AND REQUIREMENTS OF SAFETY

In this section, we try to find the reasons why IoE-enabled devices attract attackers and why we need to care about it. One of the biggest reasons behind it is the lack of security standards and technologies, and many times we can find that most IoE-enabled devices have low hardware configuration issues and low-capacity resources that will compromise various cybersecurity applications and services. Many types of human errors have also contributed to a major role in the process of allowing cyber-criminal activities.

In addition to this, many IoE-enabled devices deal with sensitive and confidential data relative to any corporate, government agency, or industry that could be observed by others therefore highly professional cybercriminals watch the activities and security considerations under the specification of these devices to use them and take advantage in many ways. This study advances the notion that "reasonable security features and its upgradation" has been considered to raise the security level of IoT devices to the desired level.

2.8 MAJOR SECURITY SOLUTIONS

Given the numerous vulnerabilities, we must take care of various security solution types that should work together with IoE-enabled device operations to maintain the combination of vulnerability detection, prevention, and isolating security solutions across various platforms [27]. Table 2.1 compares various security solutions.

- **Palo Alto Security Solution**
  The network offers various security solutions at network level operations, and the proposed Palo Alto solution includes hardware and virtual firewall-based

security solutions along with device detection and identification tools to bring protection for various IoT devices.

- **First Point Security Solution**
  This especially offers security services for preventing cellular-based IoE-enabled devices from various cellular network attacks originated throughout cellular network communication, for example cellular mobile networks need to prevent and secure themselves from cellular threats. This solution also considers the security tools and services for covering large areas with private 4G and 5G networks.
- **Trustwave Security Solution Framework**
  This offers security solution services for analysing networks to find the weakness of different connected devices and servers, integrated application interfaces, and various virtual-level cloud facilities. The solution offers facilities to troubleshoot and fixes the network resources against attacks.
- **Nano-Lock Security Solution**
  This offers a protective layer for the industry-oriented IoE-enabled devices. The objective of the solution is to protect the system from various trojan horses, malware, and many other cyber threats, and the suggested approach involves implementing security measures as a first line of defence to shield the system against hardware-based solutions by keeping items out of flash memory ends.
- **Armis Agentless Device Security Solutions**
  This is specially designed for considering the security state of many critical industries like healthcare and manufacturing units. In the process of securing, the technique uses a scanning approach across traffic patterns connected through various managed and non-managed devices. In this way the system carefully monitors the entire communication process of the system to find vulnerabilities and deploy the necessary action against them to rectify the system security policies [28].
- **Bastille Security Solutions**
  This category of security solution especially considers cellular networks as well as RF threats by the way of detection, analyses, and categorization of multiple devices of the network along with the suggested tools that have been provided by the system which offers the rectification services and multiple actions to be taken against those threats. This system is considered best suited for bringing security solutions within a specific network range.
- **Broadcom**
  Since frequent physical interaction between multiple IoE devices and other operators are the major source of malware spreading and other breaches, Broadcom is specially designed to address this through a scan procedure for finding the threats and preventing USB and other weaknesses of the interfaces, while at the same time making the network more protected and efficient [29].
- **Centri Protected Security Solution**
  This approach brings security solutions using a protective session in which the assurance of security comes through a pre-installed set libraries to achieve data security. It works for the devices throughout the entire communication network of the session, moreover, and it will also support and integrate various

IoE-enabled cloud servers to provide advanced and encrypted data sharing services for securely transmitting to multiple endpoints.

- **Trusted Object for Security**
  This approach follows the verification and validation-based approach for marking the identities of devices so that system creates multiple trusted objects where every object has its own created identities for IoT devices through which they have been connected to IoE clouds. This approach ensures protection both physically and from the outside world as attacks threaten.
- **Over Watch Security solution**
  This is a devices migration approach in which the Overwatch system is dedicated to migrating threats at a run time throughout all the IoE devices connected through the IoE cloud network. This approach is different in that it provides a warning indicator for upcoming malicious attacks. This process has been conducted through intermediate agents installed at every IoE device endpoint.
- **SecuriThings Security Solution**
  Specifically offering security for critical operation-based corporates like banks, airports, and official campuses run under the IoE network with their sensitive critical operation, this security framework brings a malware-free environment for the network nodes. In addition to this, it offers thread and malware detection along with isolation and maintenance of communication services in an agentless networking environment.
- **Secure Sensor Hound's Framework**
  This ensures automated observation, monitoring, and analysis tools for your IoE devices along with this parallely it offers the tracing of various software infections, intrusion detection, and forensic diagnostics across all the participative nodes in the network.
- **Tempered Airwall Security Solutions**
  This provides security services for Internet of Everything networks that enable servers and clients in the best possible service architecture. . It also offers high scalability in deployment frameworks which consider all the physical, virtual, remotely located, and virtual services networks with various accessing and controlling security policies to be deployed to manage guaranteed access with privileges to avoid unwanted access.
- **Vdoo Security Solution**
  This is specially designed for IoE integrated devices for transmitting security across multiple industries like healthcare, medical, and utilities. It uses default automated agent deployment services on IoE devices and in corporate security issues, making it more efficient and effective because it has the capacity to migrate the issues without making changes to the existing state of code. This process will automatically take that into account, reflecting the proactive nature of the operation that will shield the system from hacking activities while it is being carried out.

The main technological framework that provides numerous security solutions in the current IoE-enabled commercial market environment is as follows.

TABLE 2.2
Comparisons among various security solutions in the context of IoE

Security Solutions	Support Virtual network	Hardware-based security	Detection and identification of attacks & Malwares	Support 5G security	Securing Cellular network
Palo Alto Security Solution	Yes	Yes	Yes	No	No
First Point Security solution	No	No	Yes	Yes	Yes
Trustwave Security Solution Framework	Yes	Yes	Yes	No	Yes
Nano-Lock Security Solution	No	Yes	Yes	No	No
Armis Agentless Device Security Solutions	No	Yes	Yes	No	No
Bastille Security Solutions	Yes	Yes	Yes	No	Yes
Broadcom	Yes	No	Yes	No	No
Centri Protected Security Solution	Yes	Yes	Yes	No	Yes
Trusted Object for Security	Yes	Yes	Yes	No	Yes
Over Watch Security solution	No	No	Yes	No	Yes
SecuriThings Security Solution	Yes	No	Yes	No	Yes
Secure Sensor Hound's Framework	Yes	Yes	Yes	Yes	Yes
Tempered Airwall Security Solutions	Yes	No	Yes	Yes	Yes
Vdoo Security Solution	Yes	Yes	Yes	No	No

2.9 CONCLUSION AND FUTURE SCOPE

This chapter considers the fundamentals of the Internet of Everything (IoE). The first section describes IoE using different terminologies, along with an introduction to and description of its basic elements. In the second section we described the essential security requirements in IoE since IoE technology is growing, and integrating with many other applications, interfaces, and tools to fulfill human life needs more comfortably. IoE offers numerous benefits, but one also needs to keep safe and make it stronger and prevent attacks and threads with a reliable security wall. In this section we provided all the detail for security perception. In the third section we discussed the IoE security framework and its components since IoE is the extension of IoT. Therefore, we added a few recommended changes sufficient to deploy in IoE. The fourth section describes IoE security protocols that will be required and that play an important role in the process of communication in IoE; the suggested protocols greatly help in the connection and execution of the process to exchange the data, in the expansion of IoT devices, application and tools and will discourage outside hackers to attack the network and take the advantage of and misuse the information for their personnel benefits. This can only be detected and prevented through IoE security protocols. In the fifth section we discussed various available security solutions for IoT devices, networks, connections, and applications for various domains.

According to the latest report released in the USA, the advancement and expansion of facilities and services will increase to around 700 billion users by 2025. This huge number is enough to define the role of IoE in near future. IoE offers various contactless and interaction-oriented services for effortless facilities on our doorstep but there are a few barriers we need to keep in mind and that we need to work for, this being IoE security, and researchers and computer scientists need to work in this area because without security IoE is unable to gain the trust of its users. In this case, it would be good for many non-commercial application domains but not for all. This study reflects the security challenges and barriers in the uses of IoE tools, devices, networks, and applications but further work on this limitation to use the IoE at the universal level is needed.

REFERENCES

[1] H. Guo, J. Liu, and J. Zhang (2018) "Computation offloading for multi-access mobile edge computing in ultra-dense networks," IEEE Commun. Mag., vol. 56, no. 8, pp. 14–19, Aug. 2018.

[2] H. Guo, J. Liu, and H. Qin (2018) "Collaborative mobile edge computation offloading for IoT over fiber-wireless networks," IEEE Netw., vol. 32, no. 1, pp. 66–71, Jan. 2018.

[3] H. Guo, J. Zhang, and J. Liu (2019) "FiWi-enhanced vehicular edge computing networks: Collaborative task offloading," IEEE Veh. Technol. Mag., vol. 14, no. 1, pp. 45–53, Mar. 2019.

[4] Y. Mao, J. Zhang, Z. Chen, and K. B. Letaief (2016) "Dynamic computation offloading for mobile-edge computing with energy harvesting devices," IEEE J. Sel. Areas Commun., vol. 34, no. 12, pp. 3590–3605, Dec. 2016.

[5] J. Zhang (2018) et al., "Energy-latency tradeoff for energy-aware offloading in mobile edge computing networks," IEEE Internet Things J., vol. 5, no. 4, pp. 2633–2645, Aug. 2018.

[6] G. S. Vernam (2019) "Secret signaling system," U.S. Patent 1 310 719, Jul. 12, 2019.

[7] Y. Wu, A. Khisti, C. Xiao, G. Caire, K.-K. Wong, and X. Gao (2018) "A survey of physical layer security techniques for 5G wireless networks and challenges ahead," IEEE J. Sel. Areas Commun., vol. 36, no. 4, pp. 679–695, Apr. 2018.

[8] J. M. Hamamreh, H. M. Furqan, and H. Arslan (2019) "Classifications and applications of physical layer security techniques for confidentiality: A comprehensive survey," IEEE Commun. Surveys Tuts., vol. 21, no. 2, pp. 1773–1828, 2nd Quart., 2019.

[9] Advanced Encryption Standard, Federal Information Processing Standards Publication Standard FIPS PUB 197, 2001. Accessed: Apr. 21, 2020. [Online]. Available: http://csrc.nist.gov/publications/fips/ fips197/fips-197.pdf.

[10] B. Yang, L. Guo, F. Li, J. Ye, and W. Song (2020) "Vulnerability assessments of electric drive systems due to sensor data integrity attacks," IEEE Trans. Ind. Informat., vol. 16, no. 5, pp. 3301–3310, May 2020.

[11] M. S. Mahmoud, M. M. Hamdan, and U. A. Baroudi (2019) "Modeling and control of cyber-physical systems subject to cyber-attacks: A survey of recent advances and challenges," Neurocomputing, vol. 338, pp. 101–115, Apr. 2019.

[12] A. F. Moreno Jaramillo, D. M. Laverty, J. M. del Rincon, J. Hastings, and D. J. Morrow (2020) "Supervised non-intrusive load monitoring algorithm for electric

vehicle identification," Proc. IEEE Int. Instrum. Meas. Technol. Conf. (I2MTC), May 2020, pp. 1–6.

[13] M. Al-Saud, A. M. Eltamaly, M. A. Mohamed, and A. Kavousi-Fard (2020) "An intelligent data-driven model to secure intravesical communications based on machine learning," IEEE Trans. Ind. Electron., vol. 67, no. 6, pp. 5112–5119, Jun. 2020.

[14] H. S. Sánchez, D. Rotondo, T. Escobet, V. Puig, J. Saludes, and J. Quevedo (2019) "Detection of replay attacks in cyber-physical systems using a frequency-based signature," J. Franklin Inst., vol. 356, no. 5, pp. 2798–2824, Mar. 2019.

[15] H. Zhou, W. Xu, J. Chen, and W. Wang (2020) "Evolutionary V2X technology toward the Internet of Vehicles: Challenges and opportunities," Proc. IEEE, vol. 108, no. 2, pp. 308–323, Feb. 2020.

[16] D. Garcia-Roger, E. E. González, D. Martín-Sacristán, and J. F. Monserrat (2020) "V2X support in 3GPP specifications: From 4G to 5G and beyond," IEEE Access, vol. 8, pp. 190946–190963, 2020.

[17] M. Shukhej (2019) "Connecting vehicles today and in the 5G era with C-V2X," GSMA, London, U.K., Rep., 2019. [Online]. Available: www.gsma. com/iot/wp-content/uploads/2019/08/Connecting-Vehicles-Todayand-in-the-5G-Era-with-C-V2X.pdf

[18] W. Saad, M. Bennis, and M. Chen (2020) "A vision of 6G wireless systems: Applications, trends, technologies, and open research problems," IEEE Netw., vol. 34, no. 3, pp. 134–142, May/Jun. 2020.

[19] C. D. Alwis (2021) et al., "Survey on 6G frontiers: Trends, applications, requirements, technologies, and future research," IEEE Open J. Commun. Soc., vol. 2, pp. 836–886, 2021.

[20] P. Porambage, G. Gur, D. P. M. Osorio, M. Liyanage, A. Gurtov, and M. Ylianttila (2021) "The roadmap to 6G security and privacy," IEEE Open J. Commun. Soc., vol. 2, pp. 1094–1122, 2021.

[21] R. Verma and N. Jain (2021) "Cyber security and privacy fundamentals" in The Smart Cyber Ecosystem for Sustainable Development: Principles, Building Blocks, and Paradigms. Publisher Wiley-Scrivener; 1st edition (July 7, 2021).

[22] S. Gyawali, S. Xu, Y. Qian, and R. Q. Hu (2021) "Challenges and solutions for cellular-based V2X communications," IEEE Commun. Surveys Tuts., vol. 23, no. 1, pp. 222–255, 1st Quart., 2021.

[23] V. Sharma, I. You, and N. Guizani (2020) "Security of 5G-V2X: Technologies, standardization, and research directions," IEEE Netw., vol. 34, no. 5, pp. 306–314, Sep./Oct. 2020.

[24] Mayur R. Bhoyar and Ravi Verma (2018) "Review Paper on a Lightweight Secure Data Sharing System for Mobile Cloud Computing", Jour of Adv Research in Dynamical & Control Systems, Vol. 10, 02-Special Issue, pp. 2174–21279.

[25] M. Harounabadi, D. M. Soleymani, S. Bhadauria, M. Leyh, and E. Roth-Mandutz (2021) "V2X in 3GPP standardization: NR sidelink in release-16 and beyond," IEEE Commun. Stand. Mag., vol. 5, no. 1, pp. 12–21, Mar. 2021.

[26] M. B. Mollah (2021) et al., "Blockchain for the Internet of Vehicles towards intelligent transportation systems: A survey," IEEE Internet Things J., vol. 8, no. 6, pp. 4157–4185.

[27] R. Verma and B. Bhushan (2014) "QoS model for intranet area network based on MAC control protocol over TCP connection." IEEE International Conference on IT in Business, Industry, and Government (CSIBIG), pp. 1–5. IEEE.

[28] R. Verma (2021) et al., "IoT: Fundamentals and challenges" in Integration of WSNs into the Internet of Things: A Security Perspective Book series titled: Internet of Everything's (IoE): Security and Privacy Paradigm, Boca Raton: CRC Press, Taylor & Francis Group, USA.

[29] R. Verma (2020) et al., "Integrating Secured Crypto System with Cloud for Enhancing Cloud Based Encrypted Data Sharing Services" in Emerging Technologies in Data Mining and Information Security. Singapore: Springer. Series title Lecture Notes in Networks and Systems, May 2020.

3 IoE and Blockchain Convergence for Enhanced Security

Rahul Samanta, Arindam Biswas, Atul Bandyopadhyay, and Gurudas Mandal

CONTENTS

3.1 INTRODUCTION

Commercial sectors are on a verge of redirection toward a digital economy in the 21st century. Technologies like the *Internet of Everything* (IoE) and *Blockchain* have impacted both financial and commercial sectors to enhance security in the digital economy [1]. Blockchain is an integral part of the fintech technology that brings both security and advancement to this sector [2]. As per the opinion of Pass, Seeman, and Shelat [3], blockchain technology brings a protection mechanism through different cryptographic algorithms that overhauls economic diligence completely. Blockchain is now considered a potential disruptor of the core status in the commercial sector that converges to innovation and development, especially in sectors like logistics,

DOI: 10.1201/9781003366010-4

industries, energy, and financial transactions that drive economic change on a global scale [4]. Industry 4.0 revolution follows the path of Industry 1.0, Industry 2.0, and Industry 3.0. The 4th revolution in the industry has arrived after the revolution in water and the generation of steam power from it, a revolution in electricity generation, and, most important, revolution in electronic devices and internet service. Industry 4.0 is mainly characterized by an integration of cyber-physical environments thrust by a technological array [5]. It enabled the improvement of the automated and digital industry [6-7]. The technology effectively follows the egalitarian ideologies related to modern society where openness, mutual trust, equality, consensus, and direct dealing play a key role. According to the research of Sikorski, Haughton, and Kraft [8], blockchain technology with immense potential can effectively lead the engineering industry to ever-changing procurement and manufacturing technologies. During the 3rd industrial revolution, the construction industry developed the least in terms of technology adaptation [9]. For that, industrialists and researchers are in a conundrum of whether blockchain technology can help to advance security in real life [10].

The concept of the Internet of Energy can be described as a combination of "cloud computing," "information and communications technology" (ICT), "encryption," "cyber-security," and many more [10]. The convergence of smart sensors, implementation of IoT technology, and most importantly data analysis process altogether create the concept of IoE [11-13]. The key motto of IoE is to maximize the potential efficiency of transmission, utilization, and digitization of technology, especially in the financial and commercial sectors [14]. IoE creates a smart network using smart sensors that consists of various applications related to the smart grid [15]. These also include demand-side energy management, renewable energy integration, power monitoring, and secure OT/IT convergence. In the foreseeable future, "Distribution Network Operators" (DNOs) can be progressed toward high diligence, where IoE can be driven by "artificial intelligence," "communications of high bandwidth," and "secure mobile edge devices" with real-time data fletching and analysis [16-19]. These highly advanced technologies lead us toward high security [20-22], virtualization [23, 24], automation [25, 26], and intelligence [27]. In "operational technology" (OT) these developed capabilities enable suitable resources for management systems. Despite all the advancements, financial sectors are still facing issues like cybercrime, software, and integrating data subsystems that effectively require highly complex security algorithms and protocols [28].

In order to unlock communication architectures [29-31] and to enable security codes to get desired functionalities, these complex protocols are highly needed. The major challenge mainly lies in the virtualization, security framework, data storage, intelligent cloud-based platforms, and implementation of grid applications [32-33]. The main focus of the industries in the context of security is to take care of both company and customer data and it depends upon the requirement of level of redundancy and also the required number of devices in each substation [34]. In order to provide a cost-effective solution and continuous hardware integration, factors like virtualization, identification, and verification need to be deployed automatically to generate secured cyber access [35]. The study highlighted the key factors of blockchain and its various characteristics that help to provide more secure cyber access.

3.2 CURRENT STATUS OF IOE IN THE MARKET

The market of IoE can be divided by real-time data analytics of both software and hardware [36], management of data storage, remote monitoring, security frameworks, and many more [37]. From the mentioned types, the software regarding remote monitoring in the energy industry has been projected in spreading throughout the forecast at the annual rate of substantial compound growth. It can be because of the increasing demand for systems of remote monitoring with activated devices through the digital network system. Management of energy can effectively make it very much possible for several end-use industries like oil, gas industries, mining, and electricity industries in order to reduce shortages of energy to reduce running costs [38]. Driving the need for IoT applications and devices can be considered the main reason for increasing thoughts regarding management of energy.

In the year 2015, the international market effectively hit almost 6.8 billion USD, hitting 26.5 billion USD by the year 2023, with a CAGR of around 15.5% in the period 2016 to 2023. Components like enhancing globalization have been combined with projected urbanization through the forecast period to expand the market through major sales. The market of IoE sharing applies to national and interconnected IoT devices. The connected devices to the internet in this situation surpass the number of people in the universe. From the review by analyst firm Gartner, 8.4 billion devices are associated with the Internet in 2017. It can predict IoT devices will rise to almost 20.4 billion around the world.

Furthermore, the growth is not only at home, with the number of devices like smart refrigerators, smart TVs, smart lights, smart fans, and security systems extremely high at industry level. Improving business can be stated as connecting gadgets to the digital world and also turning those into smart assets which may help in driving huge efficiencies, enhancing competitiveness, making new models of business, and giving proper solutions to problems. New low-power IoT technologies, like low-power wide-area networks (LP-WAN), can help to fuel the next level of IoT adoption by 2023, with the assurance of enhanced network security and low-cost seamless integration. Additionally, IoT may assist organizations in incorporating new technologies like lightweight encryption, cloud services, machine learning, Artificial Intelligence, and blockchain.

The energy sector has become the highest customer of digital artificial intelligence-based devices [37], totaling 1.17 billion devices in the year 2019, and it increased by 17% in 2020 and reached 1.37 billion devices. Western Europe and China have household smart energy metering that has been employed in forecasting and intelligent metering in 2020, with 12% and 26% of all IoT devices, respectively, compared to 2019. Nevertheless, in spite of the breakthrough made on technological standards, that is, LPWAN, edge-computing, bandwidth allocations, cloud-integration, data analytics, and private mobile-connectivity, ecosystems of IoT still remain very complex mainly because of lightweight communication or security protocols, the lesser availability of unified standards, and, most importantly, unified firmware platform which can provide the coordinates among the multivendor equipments [39]. The existing dispersion of technological approaches and IoT connectivity has increased equipment prices and decreased interoperability. The latest survey of the market aimed at 12 main regions of growth in IoT and saw the complicated kind of ecosystem regarding IoT but huge enhancement in the connections. The amount of IoT linkages in the listed industries is expected to increase in the UK from roughly 13 million in 2016 to more than 150 million by the year 2024.

3.3 BLOCKCHAIN TECHNOLOGY

In the modern digital world, security concern has become a key issue in big data and fintech industry. Thus, the introduction of new technologies like blockchain can effectively provide a major relief to these sectors. In order to protect data from cyber criminals this technology can provide major security. Some of the major factors of blockchain technology are discussed below.

3.3.1 Several Kinds of Blockchain Networks

The performance and functionality of the blockchain are based on the deployed network architecture. On the basis of implementation, blockchain networks can be categorized broadly as Public networks [40] and Enterprise networks [41]. Figure 3.1 describes several kinds of blockchain networks that help to keep the data secured.

3.3.1.1 Permission-less Blockchain

A public network in blockchain can be described as an architecture that can send records after reading and is expected to be included in every sector in the world where cyber threat becomes a major concern for companies [42]. Crypto-economics

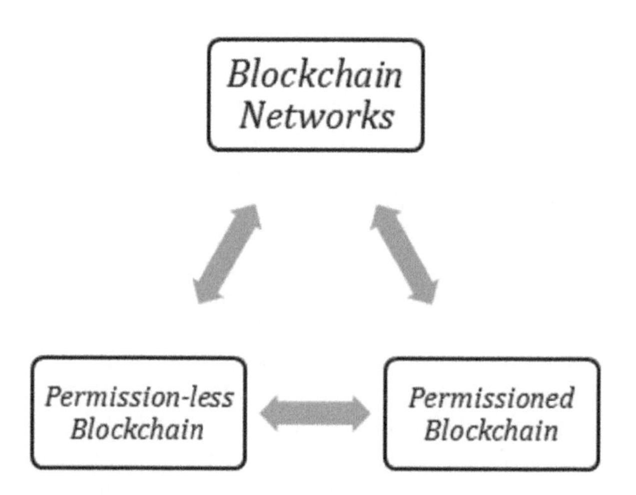

FIGURE 3.1 Several kinds of Blockchain networks.

is mainly concerned with the combination of monetary incentives with cryptographic authentication mechanisms like proof-of-stake (Ethereum) or proof-of-work (Bitcoin). These blockchains can be appraised to be "fully decentralized" [43]. Major disadvantages are detected as the number of resources (computing power) needed in order to endure a large-scale distributed ledger [44]. Distributed Blockchain networks mean the end-user shielding from some developers, through the establishment of stringent requirements.

3.3.1.2 Permissioned Blockchain

Enterprise networks can be described as private entities networks or consortiums. An enterprise can be a blockchain in which consensus is generally managed by several nodes that were pre-selected [45]. Decisions for the entire network are made by the consortium when it meets. Consequently, it is referred to as a consortium blockchain or federation blockchain. The fact that rights have been owned by a single entity is the one negative of a private blockchain. The Blockchain enterprise effectively decreases some major expenses on processing and removes obsolete processes, and data redundancies simplify record management and help to avoid frameworks of semi-manual enforcement [46]. As read permissions are generally restricted, enterprise blockchains provide a much higher degree of protection and privacy [47].

3.3.2 Consensus Algorithms (CA)

A CA can be a referred process that effectively allows machines or users for coordinating in a distributed network [48]. These algorithms effectively take care that all the nodes present in the network system can surely agree on the verification of a single source, even if some nodes may sometimes fail [49]. In order to enable resiliency over node failures inside an infrastructure, mirrored database systems were developed in response to the requirement for one single truth source. These types of database-systems significantly established that data is not lost in such cases where some nodes may fail [50].

3.3.2.1 PoW

According to the "Proof of Work" (PoW) algorithm, the miner's computing effort affects the likelihood of discovering a block (verified node). The requestor and verifier are two distinct parties (nodes) that make up a PoW method (provider). The Prover carries out a task related to resource-intensive computation in indentation for achieving a goal and presents it to an authenticated verifier or multiple sets of verifiers to validate which requires much less resources [51]. For such a block to be considered legitimate and validated in PoW, several requirements must be met. It could specify, for example, that only blocks whose hash starts with 00 are valid. The miner can only produce one which matches that combination by using brute-force inputs. They can experiment with the settings in the data pool to get a different result for each estimate until they find the proper hash. Blockchain is cumbersome for competing with other miners, as special hashing hardware (ASICs) and high-performance computing [52] are very much required for achieving the required validation.

3.3.2.2 PoS

In the "Proof of Stake" (PoS) algorithm, the probability of validation of a new block can be determined by the size of the stake of individual posses [53]. The main idea revolves around a concept that can be explained as the participating nodes present in the blockchain architecture [54]. Besides, the nodes must lock a major portion of their stakes into the escrow account [55, 56] to participate in some creative process related to the block. A commitment that it can effectively follow the protocol's principles is made by the stake.

3.3.3 Blockchain Standards

In the present scenario, various standard development projects under different societies are progressing on a global scale. "Global Standards Developing Organizations (SDOs)" in association with an array of business clusters are adopting these methods where partnerships require cooperation from the end users of specific platforms and the implementation sector. Most specifications are still in the process of development, whereas only a few models are released. Moreover, the majority of the versions are in the early blueprint stage. However, a plan is developed to liberate the work and other introductory information on "blockchain specifications" for the financial year 2020. Currently, the most significant blockchain/DLT standards under analysis are represented in Figure 3.2.

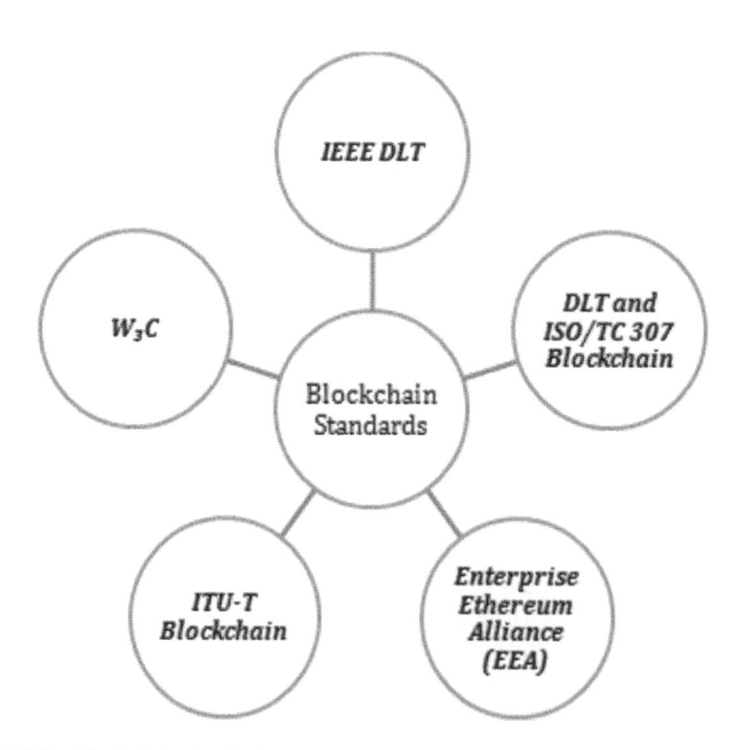

FIGURE 3.2 Blockchain standards.

- *Blockchain Standards/IEEE DLT:* The IEEE is responsible for developing the IEEE P2418 series while focusing on interoperability, generic framework, "vertical industry standards," architecture, and "building for enabling technology" [57, 58]. Furthermore, the system handles issues like security, privacy, and scalability in service and implementation [59]. It is observed that the standard features and facets of Blockchain involving intelligent contracts, tokens, and storing of data can be both permissions as well as permissionless.
- *DLT and ISO/TC 307 Blockchain:* "TC 307" is a significant global initiative that is driven by the "Australian Standards Body" along with the "International Organization for Standardization" (ISO). The TC 307 is the initial stage where taxonomy, architecture, privacy and protection, smart contracts, usage cases, governance, identification, and interoperability seen between blockchain apps are focused for ISO 307. It can be noted that the first credential architecture, privacy explanation, and security of publicly identified data are going to be available by the year 2021.
- *Enterprise Ethereum Alliance (EEA):* This collaboration is a member-driven partnership of standards that aims to structure unambiguous blockchain particulars facilitating harmonization along with interoperability of different companies globally and their customers. More than 500 participants are working on standardized architecture, transparency, and needs that can speed up the acceptance process of "Enterprise Ethereum." EEA is considered a major industry association that focuses on creating software requirements and enterprise credentials.
- *ITU-T Blockchain:* For examining the standardization parameters of DLT-based services and applications, ITU has organized an "extended participation ITU-T" focusing DLT group. This focus group emphasizes on identification and review of "DLT-based services and applications". Moreover, it develops best practices and provides recommendations to ease the global deployment of these services and applications.
- *W_3C:* Such an aspect has a "Blockchain Community Group" which works on the principles of "Network Ledger Protocol (WLP)" for the development of "ISO 20022-based" Blockchain "message format." This works in addition to the creation of guidelines for storage usage by including public and private Blockchain, torrent, and "side chain storage." This amalgamation is responsible for reviewing and examining emerging blockchain-based technologies by referring to cases like cryptocurrencies, interbank communications, and many more.

The requirement mentioned above can be categorized into four different types of DLT/Blockchain. The divisions are made by focusing on the degree of depth, demarcation and boundaries, and opinions of the specific industry. Each element of the device and its subsystem using this technology are expanded by including opinions from the industry. In the upcoming years, it is predicted that they will give important consideration to constructing "cross-chain interoperability" across various "public and enterprise grade" Blockchain systems. Different platforms of Blockchain are supposed to interact with each other in order to develop DApps to become easier and more pervasive [60]. It is predicted that "multiple side chains" and interoperability between networks need to be specified by using a common protocol. This common protocol is considered an edge gateway that is required for growth and DLT technology convergence in order to address decentralized services and networks in the future.

3.4 CYBER RESILIENCE OF PUFS

PUFs are effectively proposed to be building blocks used for the cryptographic protocols as well as security architecture for IoE and IoT systems [61]. It is observed that the security of PUFs depends on assumptions made on physical features for a subject of interest, unlike the classical "cryptographic primitives." SRAM PUFs are noted to "smooth out the IoE security" with applied and delicate "cryptographic primitives." This is done for authentication and providing certification without making substantial modifications in the structuring process. In order to prevent the occupancy of pitfalls in IoE networks, there need to be extremely cautious strategies followed for developing an IoE network that can be used for endangered protection. Cyber attackers enabling accurate responses to different challenges are considered potential threats to PUF devices. Taking the semiconductor properties of PUF edge devices into consideration, it is difficult to replicate such devices. However, with the help of computational resources, attackers can predict CRPs. In the context of the review, two significant attackers' models are discussed where the first attacker expropriates the communication framework and the second attacker can physically access the device.

3.4.1 Man-in-the-Middle (MITM)

The security of PUFs relies on the size of CRPs, therefore, it is important to make a strong PUR by involving a large number of CRPs that are vulnerable to machine learning and can assist in cyber-attacks. These kinds of attacks originate when hackers get a hold of a large set of CRP databases. Hackers use "the statistical model" to precisely forecast how the PUFs will respond to all of the flaws presented by the layers for the purposes of identification and confirmation. Furthermore, the attack can penetrate the "deployed network in IoE" in the OT field. It is observed that the MITM launch impersonates an attack on nearby edge tools and gets access to data under the "secret-key generation protocol." These attacks are done with comparatively little data since devices are seen to connect to the central gateways dynamically. Any extrinsic spammers can present a computing device of decreasing cost close to the attached device and set it to join the "same wireless network." Therefore, risks related to this kind of attack are higher and defense systems like Blockchain or lightweight cryptography need to be used for prevention.

3.4.2 Side Channel Attacks

Such ranges of ambush are seen when hackers get access to the device physically and are categorized as "invasive" in nature. Along with such, it can also be seen that such an approach also acts as "semi-invasive and noninvasive." In invasive attacks, attackers can physically access the device song with its interior alignment of circuits.

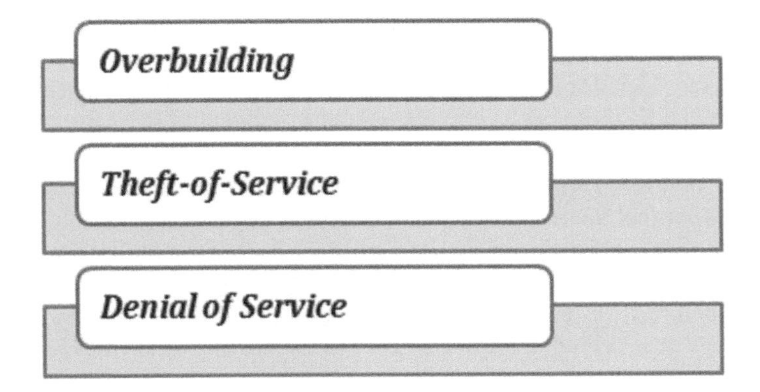

FIGURE 3.3 Side channel attacks.

Even though such a procedure is uncommon, such a kind of attack is where one can physically damage a PUF chip. This kind of attack requires proper knowledge of interfaces, "penetration testing skills," and communication protocols. However, in the modern system, these cyber-attacks are less common once a smart grid application is implemented. Substantial cloning involves synchronization, time, and error correction to make the right combination for getting the exact CRPs. There are different types of attacks seen including "timing attacks," "differential fault detection," "electromagnetic attacks," and "power monitoring attacks" which are some of the prominent attacks of this group [62]. Figure 3.3 describes some potential attacks types concerning PUF-incorporated devices.

A timing attack involves mathematical analysis of the time required for a CPU to carry out different cryptographic operations and find out the secret key. PUFs are considered more challenging for the calculation of delays in timing in a circuit system. Since the number of latches and zeroes is constant in PUFs, it is considered more stable against attacks. However, electromagnetic attacks are done by exploiting PUFs, which include the insertion of faults in protection hardware by displaying them in the external environmental conditions as done in fault analysis. This kind of attack can be mitigated by incorporating a "delay-based modification" while designing an environmentally sensitive PUF [63]. A few potential attacks types concerning PUF-incorporated devices are listed below:

- *Overbuilding:* Such kinds of cybercrimes arise due to the absence of "intellectual property (IP)" for the devices to store information related to chips. The additional gadgets are then sold to third parties or in the open market. Moreover, the PUF devices can be replicated easily by other parties.
- *Theft-of-Service:* Due to the expansion of digitization methods, edge instruments are seen to be virtualized with the help of different services and logarithms from a cloud base. However, in the case of the adversary on IoE nodes, services are mostly not authorized which can be used to make a clone of CRPs.
- *Denial of Service:* Periodic updates occurring in the edge tools are to be encrypted as well as audited in IoE. Moreover, for a malicious node, it is

important to get respective upgraded codes in order to launch a cyber-attack. For such cases, PUF developers need to design a firework update process securely such that all the information cannot be used without authentication.

The modules of PUF beds are to be modified in a way that can diminish the frequency of exterior cyber charges. Amongst the various kinds of tools implemented, it can be seen that the integration of "Controlled Physical Unclonable Functions (C-PUFs)" [64]. This scheme is accomplished by incorporating several layers into the PUF devices. For example, one hash function selected randomly can be modeled before PUF in order to avoid any kind of MITM attack. This is a critical strategy that does not allow attackers to extract any kind of PUF attributes through the use of a "model-based adversary attack." Another layer provided to the PUF edge devices includes "presenting an error correcting code" after the output obtained from PUF. This is done for diminishing any kind of chaotic measurements seen in the output. This resulted in the production of strong and resilient responses. This solution is provided by the Public Unclonable Functions (P-PUFs) in order to mitigate the risks of cyber-attacks [65]. Moreover, this method prevents attacks through side channels to moderate the array of feedback achieved through the XOR gates.

Designers who work on the IoE network have the requirement to include a system of end-to-end approach through the application of "a fused hardware-software model." The encryption of software has been seen to be a vital cost-effective method that is uncomplicated during up-gradation and implementation [66]. Furthermore, the inclusion of security of software is for further modifications seen concerning "malware penetrants" or infiltrating into electronic devices. However, hardware protection is seen as more powerful for IoE. Hardware protection makes it difficult for attackers to breach the OT networks as it is hard to protect ICs and adjust physical layers [67]. CRPs need to be encrypted and installed in a safe ROM that promotes "trustworthy functionality." It is used to reliable and validates the signatures of a software program [68].

3.5 PROOF-OF-PUF AUTHENTICATION IN THE INTERNET OF EVERYTHING

Data-security and security of data, devices and most importantly SCADA systems are key aspects of IoE architecture that effectively needs to be sorted. Integrating blockchain technology into an IoE environment effectively can help in solving the issues and achieving the security of data technology [69]. The key attributes that mark the algorithms leading in security of devices and concurrent data in the IoE network are mentioned here:

- Key vault and storage in mass-volume and high-security edge devices
- Security of cloud integration (IT Convergence/OT)
- Prevention of reverse engineering of software and IP
- Combating cloned devices and counterfeiting
- Securing ASICs and FPGAs

For an IoE-based environment, maximum instruments are seen to be low computational devices and power. Different characteristics of devices like privacy and security, onboard memory, power consumption, and hosting computational algorithms have become bottlenecks while integrating Blockchain technology in the IoE environment [70]. A Blockchain is seen to be computationally concentrated in an impenetrable network. Therefore, there are certain difficulties faced before integrating into the IoE industry. In order to overcome these limitations, PUFChain was proposed for "resource constraint devices" [71-73]. Combining hash and PUF decreases the complexity of the process while structuring it to be important for integration in various scenes. PUFs have an ultra-low power design which is drastically minimized to attain a "reduced-carbon network". "Proof-of-PUF authentication" is considered to be an algorithm of "lightweight consensus" which guarantees privacy and defense [74]. With the help of initial findings [71], it is seen that "PUF-based authentication" exhibits 1000 times more "processing times" than PoW. It is noted to be five times more than previous hardware results. The time of the transaction reduced concerning hardware-dependent solutions was 79.15%. This is a result of the hybrid model of PUFChain that uses both hash and PUF systems for hardware processing.

Hardware-Assisted Proof-of-PUF Authentication: IoE system is considered deployed in a section known as Operational Technology (OT) that is used to track a huge amount of crucial "energy infrastructure data". Both hash and signature PUF are applied to the systems present in OT phases of the PUF chain. This significantly lowers computing pressure on edge gateways and decreases bandwidth. Moreover, it is noted that PUFChain architecture heightens the "latency needs of IoE networks" [75]. Different transactions can be validated, initiated, and connected to Blockchain for identification and authentication purposes. Furthermore, two important steps are used in the procedure that highlight a device that needs to undergo an enrollment phase after the application of a new IoE system, in the network.

The nodes present in PUF-Chain architecture are categorized into two different types that include client nodes and trusted nodes. Client nodes are used for collecting information and broadcasting the data in the network. Moreover, trusted nodes are allocated to "authenticate end-nodes and broadcast back to the network." This is used to obtain response outputs from an encrypted database by using the ID of a device. At a trusted node, the response extracted, the ID of the system and sensor information from the block is broadcasted in a network. When the hash in the block and created one are correlated with each other, the rest devices connect with their local blockchains. However, when the hash does not match, the whole procedure is replicated with the ID for other keys that are present in "the encrypted database." It is seen that a lack is discarded when "none of the hashes" correlate with each other.

PUF-Chain is seen to guarantee a safe IT/OT integration in the IoE networks by reducing complexity. Moreover, it provides data protection for concurrent devices by promoting a safe integration strategy. By implementing "Proof-of-PUF authentication" blockchain is seen to stride data that is encrypted with identical "PUF obtained keys and attributes." Furthermore, it offers immutable confirmation about the data is not being corrupted. Additionally, it enables trackable and upfront auditing abilities

[76]. The protection provided by PUFChain relied on the uncertainty of PUF technology as highlighted in previous segments. In order to make the PUF responses more stake as well as immune to "overseas cyber criminals," the number of 1s and 0s needs to be equal: 128 output keys are required for the development of "64 1-bits" and "64 0-bits." It is considered important as it makes the responses from PUF technology more immune and stable to numerous cyber-attacks. Together with Blockchain, PUF is considered an "exciting cryptographic primitive." In contrast to this, there are commercial problems related to the extensive application of the action plans in the "IoE" network system. The issues offered by this technology include stability properties and randomness of the PUFs. Therefore, it is important to mitigate the risk factors as well as develop an effective strategy to reduce errors for PUFChain.

3.6 APPLICATIONS

Identification, attestation, authentication, anti-counterfeiting, key agreement protocols, and secure boot are some common applications of PUFs that are proposed in the relevant literature [77, 78]. Security of the device as well as data can be possible by combining PUFs with entropy sources, other security primitives and by this procedure in the overall architecture, a divergent unique response can be produced to combat the same challenges occurring at different times. As a cost-effective monitoring solution for IoE [79], PUFs and edge devices can be incorporated into the network and by this technique some more interesting following applications of the proposed PUFs can be achieved. Figure 3.4 illustrates a few of this technology's most important applications to demonstrate its ability to undermine the security of digital transactions.

- *Authentication:* Required authentication for binding hardware to software platforms can be done using PUFs on the related devices. Further, it provides valid authentication to solve the purpose of secure key storage and keyless secure communication to integrate IT/OT systems. Based on CRPs, the two most used authentication schemes in today's platforms are client authentication and server authentication can also be achieved appropriately.
- *Device Identification:* For solving the purpose of effective identification for devices, PUFs can become a perfect match as they can easily and effectively tune the related devices into authentication tokens. Further, it can reduce the need to store cryptographic keys and lookup tables within the devices, which may create cyber attacks, and also can extract cryptographic information from nonvolatile memory. PUF devices are also useful to generate transient keys "on-the-fly" based on unique fingerprints that can minimize surface attack to extract the key to a greater extent.
- *Random Number Generation:* Random number generation is the key factor in many cryptographic systems and to generate such random numbers with high entropy, PUFs become the most effective alternative. Due to the semiconductor manufacturing process, PUFs become an interesting readily available source of randomness. Therefore, the property of PUF can be exploited as the number-generator.

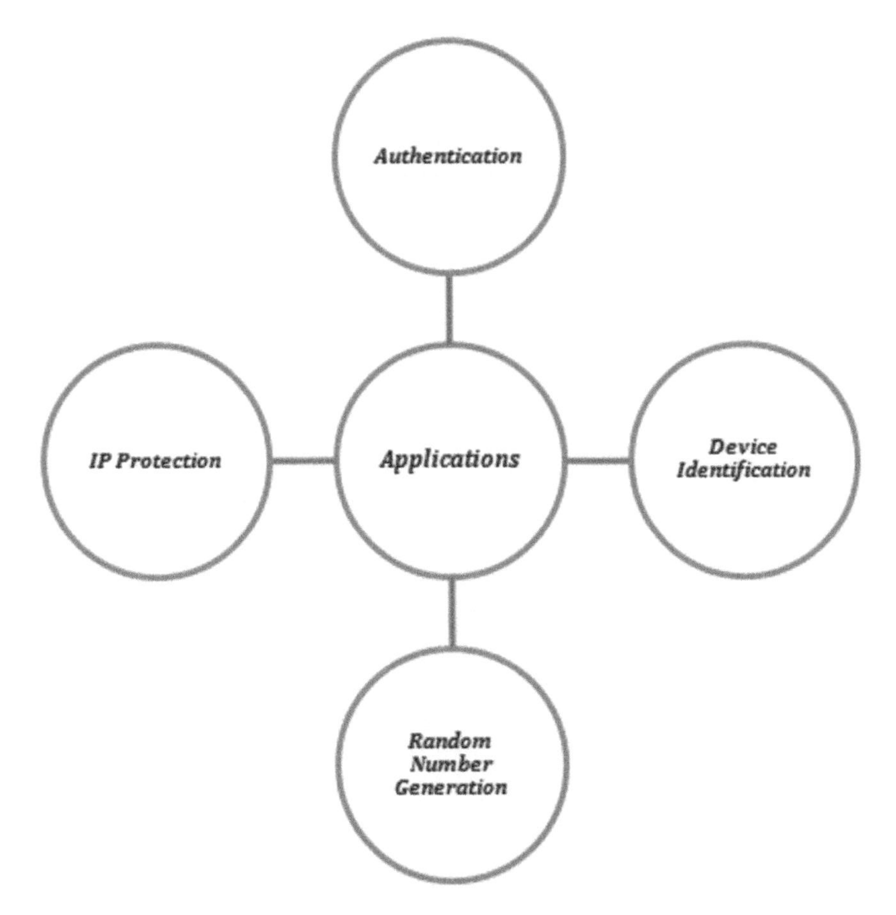

FIGURE 3.4 Some of the key applications of blockchain technology.

- *IP Protection:* PUFs are also helpful in IP protection which becomes a real concern for many IP providers. The integrated circuit (IC) design flow is globalized due to the increased cost of IC design, fabrication, testing, and verification. Besides providing cost benefits and reducing the time-to-market, globalization has also introduced several non-avoidable challenges such as piracy, counterfeiting, and malicious modifications. Using unique and unclonable, that is, PUF fingerprints in the IP of the devices can solve the purpose of ultimate protection.

3.7 EXISTING CHALLENGES

The commercial implementation of a hybrid approach like PUF and Blockchain has become the leading security application of smart grids. Despite some key advantages and its widespread use especially in IoE systems, some major challenges in the development still exist and those are shown in Figure 3.5.

FIGURE 3.5 Existing challenges of blockchain technology.

The circles for Design Types of blockchain, Scalability, and Standardization and Interoperability surround a central circle for Existing Challenges:

- *Design Types of blockchain:* It is very important to understand whether the system is public or private and according to that its design needs to be done. Privacy law and jurisdiction related to data should be considered with prior attention.
- *Scalability:* Integration of the network and technical stability are the major challenges of PUF and blockchain. In recent times multiple numbers of edge devices have started integrating into the energy network and for that its orchestration, verification, and authentication have become challenging.
- *Standardization and Interoperability:* Among all the challenges, another key challenge is the lack of reciprocity between a large number of existing networks in the same domain of smart grid.

3.8 FUTURE PROSPECTS

The disruptive potential of blockchain and IoE technologies can be used to address and resolve the current problems. Figure 3.6 shows its key future prospects that effectively help the financial and commercial sector to mitigate the cyber crime related issues.

Non-fungible tokens (NFTs): The year 2021 saw the rise of NFTs. For that, blockchain technology is the major factor in terms of security. Now, NFTs have started gaining prominence, especially in the field of gaming and sports.

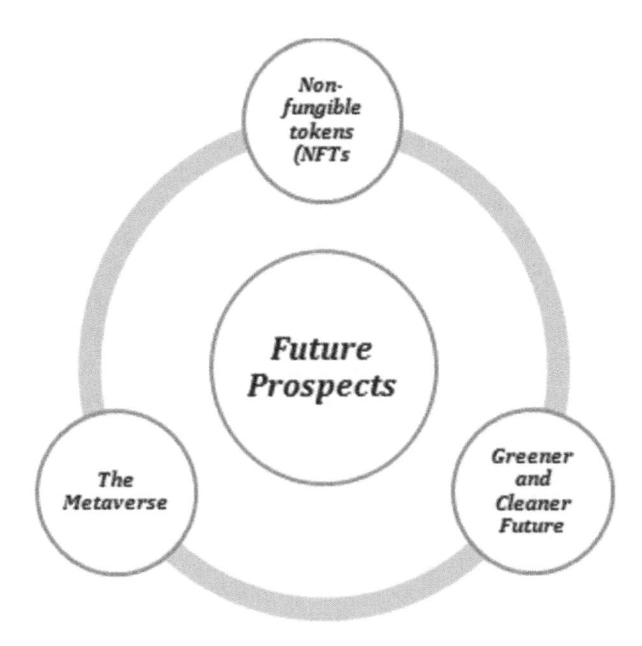

FIGURE 3.6 Future prospects of blockchain technology.

- *Greener and Cleaner Future:* One of the major arguments related to the use of blockchain technology is an increment in carbon footprint. As people become more aware of their carbon footprint and several pieces of research have been performed in order to reduce their carbon footprint to keep the environment safe, now blockchain technology can be used freely [80].
- *The Metaverse:* In 2022, the metaverse become a virtual reality and it can be considered an extended part of science fiction. Blockchain can ensure higher transparency and its applications can improve the data management process of industries as well.

3.9 ADVANTAGES OF THE STUDY

In order to increase security, traceability, transparency, and most importantly trust in digital transactions, blockchain technology is very effective. As it provides new efficiencies and cost saving ways of data security, commercial and financial sectors have opted for the technology more frequently. The PUF is integrated with hardware security in order to solve integration, latency, bandwidth, energy requirements, and scalability for IoE systems. It is a hybrid approach also known as the PUF-chain that enables data and device provenance which effectively records a history of data generation, data origins records, and its processing. As it is embedded with clone-proof device authentication and identification, for hackers it is now made impossible for them to hack. Nowadays, for ensuring privacy solutions and end-to-end security for a highly valued large amount of data, a device consisting of IoE technology has become much more popular. In this regard, seamless integration of IoE and blockchain in financial and commercial sectors not only helps to make a secured data storage framework but also ensures risk-free technological implementation in this digital era. The existing authentication mechanism cannot be measured as the safest any longer as it tends to be leaked or was stolen in the past. In this regard, IoE and blockchain provide a secure way of data storage where proof of authentication or a unique fingerprint has been used. The fact that PUF key generation is not sent across a communication channel in the consensus mechanism (Proof-of-PUF) is one of the PUF chain's fundamental advantages. It builds a resilient system to prevent communication attacks, ensuring latency and bandwidth requirements. A major advantage of this technology is it can be shared by multiple business partners and, thus, the charges have reduced by a significant margin. The study helps to understand several aspects of blockchain technology and its importance to make the data world more secure.

3.10 CONCLUSION

In order to comprehend a secure IoE development, this chapter provides the current position of blockchain technology. The convergence of IoE and blockchain can effectively enhance the security of the digital era. Some of the key aspects like end-to-end encryption blockchain help us to understand the data can be securely restored from cybercriminals. On the other hand, the PUF-chain helps to generate authentication and verification of data that helps to grow the digital industry globally. Now, it enables authentication proofs like a fingerprint and face identification which helps it to be trusted in both commercial and financial sectors. As it enables high security in a digital era, the demand for blockchain technology has started increasing significantly.

REFERENCES

[1] Asif, R., Ghanem, K., Irvine, J. (2020). Proof-of-PUF enabled blockchain: Concurrent data and device security for internet-of-energy. Sensors, 21(1), 28.
[2] Perera, S., Nanayakkara, S., Rodrigo, M. N. N., Senaratne, S., Weinand, R. (2020). Blockchain technology: Is it hype or real in the construction industry? Journal of Industrial Information Integration, 17, 100125.
[3] Pass, R., Seeman, L., Shelat, A. (2017, April). Analysis of the blockchain protocol in asynchronous networks. In Annual International Conference on the Theory and Applications of Cryptographic Techniques (pp. 643–673). Springer, Cham.
[4] Hamida, E. B., Brousmiche, K. L., Levard, H., Thea, E. (2017, July). Blockchain for enterprise: overview, opportunities, and challenges. In The Thirteenth International Conference on Wireless and Mobile Communications (ICWMC 2017).
[5] Brettel, M., Friederichsen, N., Keller, M., Rosenberg, M. (2014). How virtualization, decentralization and network building change the manufacturing landscape: An Industry 4.0 Perspective. International Journal of Information and Communication Engineering, 8(1), 37–44.
[6] Lee, J., Kao, H. A., Yang, S. (2014). Service innovation and smart analytics for industry 4.0 and big data environment. Procedia cirp, 16, 3–8.
[7] Xu, L. D., Xu, E. L., Li, L. (2018). Industry 4.0: State of the art and future trends. International Journal of Production Research, 56(8), 2941–2962.

[8] Sikorski, J. J., Haughton, J., Kraft, M. (2017). Blockchain technology in the chemical industry: Machine-to-machine electricity market. Applied Energy, 195, 234–246.

[9] Agarwal, R., Chandrasekaran, S., Sridhar, M. (2016). Imagining construction's digital future. McKinsey & Company, 24.

[10] Bui, N., Castellani, A. P., Casari, P., Zorzi, M. (2012). The internet of energy: A web-enabled smart grid system. IEEE Network, 26(4), 39–45.

[11] Moness, M., Moustafa, A. M. (2015). A survey of cyber-physical advances and challenges of wind energy conversion systems: Prospects for internet of energy. IEEE Internet of Things Journal, 3(2), 134–145.

[12] Shit, R. C., Sharma, S., Puthal, D., Zomaya, A. Y. (2018). Location of Things (LoT): A review and taxonomy of sensors localization in IoT infrastructure. IEEE Communications Surveys & Tutorials, 20(3), 2028–2061.

[13] Bedi, G., Venayagamoorthy, G. K., Singh, R., Brooks, R. R., Wang, K. C. (2018). Review of Internet of Things (IoT) in electric power and energy systems. IEEE Internet of Things Journal, 5(2), 847–870.

[14] Ma, M., Wang, P., Chu, C. H. (2013, August). Data management for internet of things: Challenges, approaches and opportunities. In 2013 IEEE International conference on green computing and communications and IEEE Internet of Things and IEEE cyber, physical and social computing (pp. 1144–1151). IEEE.

[15] Ma, Z., Xie, J., Li, H., Sun, Q., Si, Z., Zhang, J., Guo, J. (2017). The role of data analysis in the development of intelligent energy networks. IEEE Network, 31(5), 88–95.

[16] Vermesan, O., Blystad, L. C., Zafalon, R., Moscatelli, A., Kriegel, K., Mock, R., ... Perlo, P. (2011). Internet of energy: Connecting energy anywhere anytime. In Advanced microsystems for automotive applications 2011 (pp. 33–48). Springer, Berlin, Heidelberg.

[17] Fang, X., Misra, S., Xue, G., Yang, D. (2011). Smart grid: The new and improved power grid: A survey. IEEE communications surveys & tutorials, 14(4), 944–980.

[18] Farhangi, H. (2009). The path of the smart grid. IEEE power and energy magazine, 8(1), 18–28.

[19] Gungor, V. C., Sahin, D., Kocak, T., Ergut, S., Buccella, C., Cecati, C., Hancke, G. P. (2011). Smart grid technologies: Communication technologies and standards. IEEE transactions on Industrial informatics, 7(4), 529–539.

[20] Wang, K., Wang, Y., Hu, X., Sun, Y., Deng, D. J., Vinel, A., Zhang, Y. (2017). Wireless big data computing in smart grid. IEEE Wireless Communications, 24(2), 58–64.

[21] Combe, T., Martin, A., Di Pietro, R. (2016). To docker or not to docker: A security perspective. IEEE Cloud Computing, 3(5), 54–62.

[22] Blenk, A., Basta, A., Reisslein, M., Kellerer, W. (2015). Survey on network virtualization hypervisors for software defined networking. IEEE Communications Surveys & Tutorials, 18(1), 655–685.

[23] Ruland, K. C., Sassmannshausen, J., Waedt, K., Zivic, N. (2017). Smart grid security: An overview of standards and guidelines. e & i Elektrotechnik und Informationstechnik, 134(1), 19–25.

[24] Dorri, A., Kanhere, S. S., Jurdak, R., Gauravaram, P. (2017, March). Blockchain for IoT security and privacy: The case study of a smart home. In 2017 IEEE international conference on pervasive computing and communications workshops (PerCom workshops) (pp. 618–623). IEEE.

[25] Hellaoui, H., Koudil, M., Bouabdallah, A. (2017). Energy-efficient mechanisms in security of the internet of things: A survey. Computer Networks, 127, 173–189.

[26] Zhabelova, G., Vyatkin, V. (2011). Multiagent smart grid automation architecture based on IEC 61850/61499 intelligent logical nodes. IEEE Transactions on Industrial Electronics, 59(5), 2351–2362.
[27] Zhabelova, G., Vyatkin, V., Dubinin, V. N. (2014). Toward industrially usable agent technology for smart grid automation. IEEE Transactions on Industrial Electronics, 62(4), 2629–2641.
[28] Yan, Y., Qian, Y., Sharif, H., Tipper, D. (2012). A survey on smart grid communication infrastructures: Motivations, requirements and challenges. IEEE communications surveys & tutorials, 15(1), 5–20.
[29] Lundstrom, B., Chakraborty, S., Lauss, G., Bründlinger, R., Conklin, R. (2016, September). Evaluation of system-integrated smart grid devices using software-and hardware-in-the-loop. In 2016 IEEE Power & Energy Society Innovative Smart Grid Technologies Conference (ISGT) (pp. 1–5). IEEE.
[30] Metke, A. R., Ekl, R. L. (2010). Security technology for smart grid networks. IEEE Transactions on Smart Grid, 1(1), 99–107.
[31] Fan, Z., Kulkarni, P., Gormus, S. et al. (2012). Smart grid communications: Overview of research challenges, solutions, and standardization activities. IEEE Communications Surveys & Tutorials, 15(1), 21–38.
[32] Saputro, N., Akkaya, K., Uludag, S. (2012). A survey of routing protocols for smart grid communications. Computer Networks, 56(11), 2742–2771.
[33] Lytras, M. D., Chui, K. T. (2019). The recent development of artificial intelligence for smart and sustainable energy systems and applications. Energies, 12(16), 3108.
[34] Moslehi, K., Kumar, R. (2010). A reliability perspective of the smart grid. IEEE transactions on smart grid, 1(1), 57–64.
[35] Sharma, P. K., Chen, M. Y., Park, J. H. (2017). A software defined fog node based distributed blockchain cloud architecture for IoT. IEEE Access, 6, 115–124.
[36] Verma, S., Kawamoto, Y., Fadlullah, Z. M., Nishiyama, H., Kato, N. (2017). A survey on network methodologies for real-time analytics of massive IoT data and open research issues. IEEE Communications Surveys & Tutorials, 19(3), 1457–1477.
[37] Hossein Motlagh, N., Mohammadrezaei, M., Hunt, J., Zakeri, B. (2020). Internet of Things (IoT) and the energy sector. Energies, 13(2), 494.
[38] Liu, Y., Yang, C., Jiang, L., Xie, S., Zhang, Y. (2019). Intelligent edge computing for IoT-based energy management in smart cities. IEEE network, 33(2), 111–117.
[39] Patel, P., Ali, M. I., Sheth, A. (2017). On using the intelligent edge for IoT analytics. IEEE Intelligent Systems, 32(5), 64–69.
[40] Lei, K., Du, M., Huang, J., Jin, T. (2020). Groupchain: Towards a scalable public blockchain in fog computing of IoT services computing. IEEE Transactions on Services Computing, 13(2), 252–262.
[41] Fu, Y., Zhu, J. (2019). Big production enterprise supply chain endogenous risk management based on blockchain. IEEE access, 7, 15310–15319.
[42] Peck, M. E. (2017). Blockchain world: Do you need a blockchain? This chart will tell you if the technology can solve your problem. IEEE Spectrum, 54(10), 38–60.
[43] Zheng, Z., Xie, S., Dai, H., Chen, X., Wang, H. (2017). "An overview of Blockchain technology: Architecture, consensus, and future trends". In IEEE International Congress on Big Data (pp. 557–564).
[44] Marsalek, A., Zefferer, T. (2019, August). A correctable public blockchain. In 2019 18th IEEE International Conference on Trust, Security and Privacy In Computing and Communications/13th IEEE International Conference on Big Data Science and Engineering (TrustCom/BigDataSE) (pp. 554–561). IEEE.

[45] Huang, D., Ma, X., Zhang, S. (2019). Performance analysis of the raft consensus algorithm for private blockchains. IEEE Transactions on Systems, Man, and Cybernetics: Systems, 50(1), 172–181.
[46] Zhang, X., Chen, X. (2019). Data security sharing and storage based on a consortium blockchain in a vehicular ad-hoc network. IEEE Access, 7, 58241–58254.
[47] Fan, M., Zhang, X. (2019). Consortium blockchain based data aggregation and regulation mechanism for smart grid. IEEE Access, 7, 35929–35940.
[48] Bach, L. M., Mihaljevic, B., Zagar, M. (2018, May). Comparative analysis of blockchain consensus algorithms. In 2018 41st International Convention on Information and Communication Technology, Electronics and Microelectronics (MIPRO) (pp. 1545–1550). IEEE.
[49] Gramoli, V. (2020). From blockchain consensus back to Byzantine consensus. Future Generation Computer Systems, 107, 760–769.
[50] Sankar, L. S., Sindhu, M., Sethumadhavan, M. (2017, January). Survey of consensus protocols on blockchain applications. In 2017 4th international conference on advanced computing and communication systems (ICACCS) (pp. 1–5). IEEE.
[51] Kumar, G., Saha, R., Rai, M. K., Thomas, R., Kim, T. H. (2019). Proof-of-work consensus approach in blockchain technology for cloud and fog computing using maximization-factorization statistics. IEEE Internet of Things Journal, 6(4), 6835–6842.
[52] Cho, H. (2018). ASIC-resistance of multi-hash proof-of-work mechanisms for blockchain consensus protocols. IEEE Access, 6, 66210–66222.
[53] Niya, S. R., Schiller, E., Cepilov, I., Maddaloni, F., Aydinli, K., Surbeck, T., Stiller, B., et al. (2019). Adaptation of proof-of-stake-based blockchains for IoT data streams. In 2019 IEEE international conference on blockchain and cryptocurrency (ICBC) (pp. 15–16). IEEE.
[54] Bentov, I., Lee, C., Mizrahi, A., Rosenfeld, M. (2014). Proof of activity: Extending bitcoin's proof of work via proof of stake [extended abstract]. ACM SIGMETRICS Performance Evaluation Review, 42(3), 34–37.
[55] Denning, D. E., Branstad, D. K. (1996). A taxonomy for key escrow encryption systems. Communications of the ACM, 39(3), 34–40.
[56] O'Neil, P. E. (1986). The escrow transactional method. ACM Transactions on Database Systems (TODS), 11(4), 405–430.
[57] Lima, C. (2018). Developing open and interoperable dlt\/blockchain standards. Computer, 51(11), 106–111.
[58] Di Silvestre, M. L., Gallo, P., Ippolito, M. G., Sanseverino, E. R., Sciumè, G., Zizzo, G. (2018, June). An energy blockchain, a use case on tendermint. In 2018 IEEE International Conference on Environment and Electrical Engineering and 2018 IEEE Industrial and Commercial Power Systems Europe (EEEIC/I&CPS Europe) (pp. 1–5). IEEE.
[59] Anjum, A., Sporny, M., Sill, A. (2017). Blockchain standards for compliance and trust. IEEE Cloud Computing, 4(4), 84–90.
[60] Gordon, W. J., Catalini, C. (2018). Blockchain technology for healthcare: Facilitating the transition to patient-driven interoperability. Computational and structural biotechnology journal, 16, 224–230.
[61] Chatterjee, B., Das, D., Maity, S., Sen, S. (2018). RF-PUF: Enhancing IoT security through authentication of wireless nodes using in-situ machine learning. IEEE Internet of Things Journal, 6(1), 388–398.
[62] Merli, D., Schuster, D., Stumpf, F., Sigl, G. (2011, October). Semi-invasive EM attack on FPGA RO PUFs and countermeasures. In Proceedings of the Workshop on Embedded Systems Security (pp. 1–9).

[63] Zhang, J., Wu, Q., Lyu, Y., Zhou, Q., Cai, Y., Lin, Y., Qu, G. (2013, November). Design and implementation of a delay-based PUF for FPGA IP protection. In 2013 International Conference on Computer-Aided Design and Computer Graphics (pp. 107–114). IEEE.
[64] Shamsoshoara, A., Korenda, A., Afghah, F., Zeadally, S. (2020). A survey on physical unclonable function (PUF)-based security solutions for Internet of Things. Computer Networks, 183, 107593.
[65] Potkonjak, M., Goudar, V. (2014). Public physical unclonable functions. Proceedings of the IEEE, 102(8), 1142–1156.
[66] Mohammed, N. M., Niazi, M., Alshayeb, M., Mahmood, S. (2017). Exploring software security approaches in software development lifecycle: A systematic mapping study. Computer Standards & Interfaces, 50, 107–115.
[67] Potlapally, N. (2011). Hardware security in practice: Challenges and opportunities. In 2011 IEEE International Symposium on Hardware-Oriented Security and Trust (pp. 93–98). IEEE.
[68] Jin, Y. (2015). Introduction to hardware security. Electronics, 4(4), 763–784.
[69] Islam, M. N., Kundu, S. (2019). Enabling IC traceability via blockchain pegged to embedded puf. ACM Transactions on Design Automation of Electronic Systems (TODAES), 24(3), 1–23.
[70] Buchanan, W. J., Li, S., Asif, R. (2017). Lightweight cryptography methods. Journal of Cyber Security Technology, 1(3-4), 187–201.
[71] Mohanty, S. P., Yanambaka, V. P., Kougianos, E., Puthal, D. (2020). PUFchain: A hardware-assisted blockchain for sustainable simultaneous device and data security in the internet of everything (IoE). IEEE Consumer Electronics Magazine, 9(2), 8–16.
[72] Patil, A. S., Hamza, R., Yan, H., Hassan, A., Li, J. (2019, December). Blockchain-PUF-based secure authentication protocol for Internet of Things. In International Conference on Algorithms and Architectures for Parallel Processing (pp. 331–338). Springer, Cham.
[73] Yanambaka, V. P., Mohanty, S. P., Kougianos, E. (2018). Making use of manufacturing process variations: A dopingless transistor based-PUF for hardware-assisted security. IEEE Transactions on Semiconductor Manufacturing, 31(2), 285–294.
[74] Zhu, F., Li, P., Xu, H., Wang, R. (2019). A lightweight RFID mutual authentication protocol with PUF. Sensors, 19(13), 2957.
[75] Zhang, J., Hu, X., Ning, Z., Ngai, E. C. H., Zhou, L., Wei, J., Hu, B., et al. (2017). Energy-latency tradeoff for energy-aware offloading in mobile edge computing networks. IEEE Internet of Things Journal, 5(4), 2633–2645.
[76] Braeken, A. (2018). PUF based authentication protocol for IoT. Symmetry, 10(8), 352.
[77] Herder, C., Yu, M. D., Koushanfar, F., Devadas, S. (2014). Physical unclonable functions and applications: A tutorial. Proceedings of the IEEE, 102(8), 1126–1141.
[78] McGrath, T., Bagci, I. E., Wang, Z. M., Roedig, U., Young, R. J. (2019). A puf taxonomy. Applied Physics Reviews, 6(1), 011303.
[79] Yilmaz, Y., Gunn, S. R., Halak, B. (2018). Lightweight PUF-based authentication protocol for IoT devices. In 2018 IEEE 3rd international verification and security workshop (IVSW) (pp. 38–43). IEEE.
[80] Extentia.com, (2022), *An Overview of Blockchain Technology*. Accessed September 2022 from www.extentia.com.

Part II

Blockchain-Based Security Mechanisms for IoE

Part II

4 Security Model and Access Control Mechanisms for Attack Mitigation in IoE

Neha Mathur and Shweta Sinha

CONTENTS

DOI: 10.1201/9781003366010-6

4.1 INTRODUCTION

4.1.1 Background of IoE

The Internet of Everything (IoE) is the contemporary concept for data transaction in the telecommunication field. The concept of IoE comprises all the fundamental domains, such as the data transfer network of humans, IoT, and Industrial IoT [1]. The Internet of Everything (IoE) is characterized as the advanced extension of IoT, which emphasizesdevice-to-devicee (D2D) or machine-to-machine interchanges and characterizes a more complicated framework that consists of procedures and individuals. Both IoE and IoT consist of different gadgets, which includes PCs and well-equipped processing gadgets as well as resource-limited gadgets. [2]. IoT concentrates on the physical elements while IoE concentrates on entire elements around them such as data, people, things, and process. The basic component of IoE is illustrated in the Figure 4.1. IoE intellectually attempts to gather and process the information from IoT as well as from different innovations and even treats individuals as a node [3].

IoE expands on the hypothesis of IoT which concentrates on consolidating the network gadgets outfitted with particular sensors through the Internet. The sensors can distinguish and react with the alterations in their current circumstance, which includes temperature, light, vibration, and sound. IoE decisively extends the capacity of IoT by enumerating the components that can intensively provide more extravagant encounters to organizations, people, and nations [4][5]. The predominant abilities of IoE, while contrasted with IoT, include: the recognition with the unique circumstance, an expansion in power processing, autonomous energy supply, and expanding of the enrollment and utilization of new kinds of data that are associated with the environment. IoE is found to be heterogeneous, which accommodates both the objects

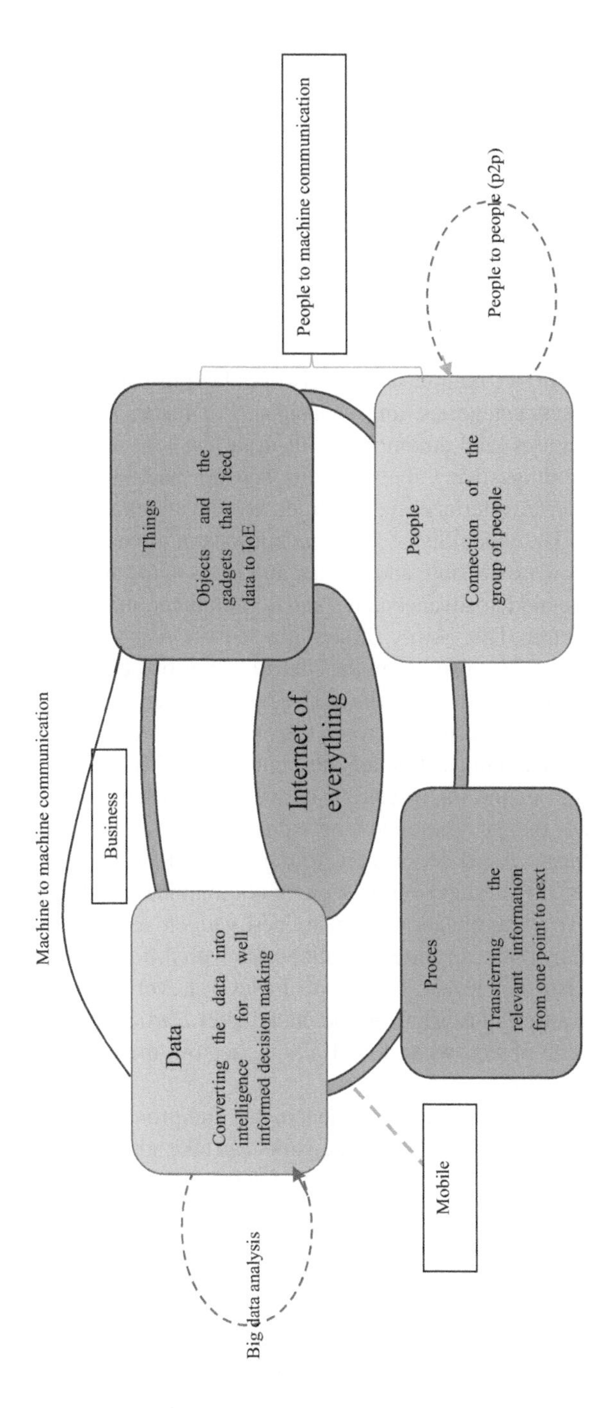

FIGURE 4.1 Basic components of Internet of Everything.

and administrations for wired and remote objects, indoor and open air pools, and is occupied by different products that categorize the scope of uncomplicated gadgets to complex processing gadgets [6]. Thus, we are abruptly entering the IoE cycle. Unlike IoT, IoE interfaces billions of people, things, and processes in important and critical organizations [7]. Additionally, IoE mainly concentrates on smart organization association and innovation in view of IoT framework. In this way, IoE has turned into a promising organization worldwide, and it offers support in different applications, such as transportation [8][9], industry [10][11], education [12], and trade [13][14]. IoE is a data specialized term that amalgamates commutating, monitoring, calculation, interaction abilities and data extraction, and communication functionalities together in a gadget. IoE permits non-identical electronic gadgets with various abilities to detect the surrounding area and to impart data for information trade. IoE is the extensive category of remote sensor networks [15] in which the IoE hubs might have various classes, categories, and capacities. For instance, cell phones, tablets, PCs, home apparatuses, and automobiles illustrate the IoE nodes. These nodes can detect the surroundings using their various sensors and interaction information, retrieve valuable data, interface via the Internet, and manage their functions with modifications [1]. The capability of IoE class consists of three aspects like communication or interaction, processing, and storage and leads to the storage and the computation power of the hidden advancements and communication.

a) Communication: This ability directs the sensors' capacity to locally interact and exchange data. This capacity might change at various degrees of coordination between IoE sensors and frameworks and be categorized as disconnected, which states that there is no connection between the enablers, specialized (essential organization network), grammatical (fundamental interoperability and information trade), semantic (figuring out the interpretation of the information), realistic or effectual (appropriateness of the data), applied (shared perspective on the unavoidable world) [16], or hierarchical (coordination and arrangement of business processes across authoritative limits) [17]. Moreover, in view of communication abilities, IoT gadgets are ordered into two classifications: constrained gadgets and the gateway gadgets. Also, as indicated by their capacities to collaborate with different objects, IoT articles can be grouped into four levels: Level 0 to Level 3. Level 0 items just receive the data, and Level 1 articles just transmit the data. Level 2 articles can perform the two activities within one object, while Level 3 stretches out the cooperation to another article [18] [19].

Different systems administration advancements and protocols provide organizing interoperability in IoT [20][16]. IoT frameworks can take advantage of a few kinds of organizations with various qualities in terms of size, convergence, data transmission, delay necessities, limit, and upheld reachability [18]. The central systems administration and correspondence innovations are local area network (LAN), wireless local area network (WLAN), metropolitan area network (MAN), wide area network, body area network, wireless metropolitan network, mobile communication network, the satellite communication (e.g., GPS) [21], IPv6 over low-power individual region organizations (6LowPAN), Neul, low-range remote region organizations, narrowband-IoT, cell Sigfox, and string or cross-section advancements, such as SDNs and Zigbee [22]. There are three categories of communication protocol that empower IoT to communicate and interrelates:

(a) D2D interaction, which is established to correspondence between mobile accessories and futuristic versatile organizations;
(b) device-server, for the detected information is dispatched to servers, and the distance from or near to the gadgets is utilized to the cloud processor;
(c) server-to-server, for the servers to send the data between one another —chiefly utilized for portable organizations [23] [19].

b) Processing: The gadgets and sensors utilized in data assortment also differ do to processing abilities. The research conducted by Mon et al. [24] characterizes sensors as top of the line or low-end gadgets, based on their assets and computational capacities. Low-end gadgets are asset contrived concerning energy, communication limit, and processing power. The processing ability refers to the sensors' ability to locally deal with collected information. For IoE frameworks, information is naturally processed to obtain the knowledge and create significant experiences. Generally, information processing strategies are either authentic or proactive. Historical information processing is connected with information disclosure, in which the proactive information processing provides the prescient and significant bits of knowledge. A general classification of utilizations takes part in the consistent generation of steam and evaluation of heterogeneous data [25]. The evaluation strategies allude to the methodical computational evaluation of changing various information from various sources into data [26] to assemble on smart choices at the accompanying circulation levels:

(1) The gadget level, where gadgets are liable for storing and registering process;
(2) The network level, which requests distant correspondence to haze figuring hubs (center points, base stations, doors, switches, and servers); and
(3) Cloud level, which requires remote level interaction within the gathering of associated servers [27].

Fog, cloud, and edge computing are basic attributes of the incorporated IoE environment, taking into account that gadgets that have confined process and memory limit need to assign these capabilities [22]. It effectively restrains the processing issues connected with enormous edge data. Since in the edge processing worldview, the information is at the edge of the organization [28]. Different categories of distributed computing and edge processing standards are versatile fog computing, distributed computing, and portable edge computing [29] [19].

c) Storage: This ability alludes to an IoE framework's storage capability, with respect to the prototype where its storage capability depends on fog, cloud, or edge [27]. A storage medium public, private or virtual, provides the adaptability and versatility that an IoE application needs, from advancement to sending [30]. Storage alludes to saving the data inside the network, and it alters arduously, starting with one object then onto the next. Storage connections between IoE enablers are recognized significantly based on the storage abilities of the object. A few objects can have confined capacities and store the least data. Most portable phones at the edge of the organization are asset constrained with respect to battery duration, storage, and computation ability [28]. Even though every object can store inserted codes to work internally, they diverge in storing collected and processed information. The storage capacity of the object directly depends on the sensitivity of the stored information [31][32].The evaluation process needs the real-time data stream processing to support the data arrival time, data storage, and data control [6].

The aforementioned abilities make the IoE the transpiring innovation, which helps them to attain remarkable progress in smart devices, smart city concepts, and enhanced demands of voice-depended service. Further, this innovation requires processed data in the field of machine learning, artificial intelligence, deep learning techniques, and data analytics [3].

It is predicted that the IoE will attain the achievement of 30 billion of IoE units or user by the end of 2022 [3]. For most of the people in the world, IoE is generally observed as basic Web-related devices, smart homes, smart refrigerators, cars, indoor controllers, toasters, and ovens [10]. Hence, in any case the ceaseless communication with individual to machine and machine to machine communication is established IoE [29]. This makes IoE an ideal for any country that is building a structure without any prior planned infrastructure [33]. For example, the report states that more than fifty percent of the current IoE advancement is engaged in data gathering. Different organizations utilize different schemes with better procedures to deal with mine data and improve work processes while growing functional viability, which makes the particular organization attain a foremost position in the world market. More developed nations observed the utilization of the IoE as a critical driver of running viable metropolitan networks through splendid cross-sections and data organization. The IoE utilities are the systems that can screen water systems, actually look at air quality, and direct waste or sewage with IoE-related devices. The ITU report refers to the way that response times to humanitarian fiascoes can be snappier due to the IoE, with related devices prepared to record and impact the data that can be helpful in dealing the outcome [34][35]. Robots can be utilized as domains of cultivating, distant sensors can follow crop improvement, and automated vehicles can reduce the medical stage work [36].

4.1.2 Advantage of IoE

Numerous people accept IoE as strategies for organizations and producers to improve functional efficiencies, for example, through better asset utilization [36]. Different routes and disciplines are examined for IoE along with related progressions, for instance, big data and cloud can engage new business frameworks with the existing ones in which isolated clients (organizations with advanced ideas) can be the main target, to provide better and efficient solutions to upgrade the feature. For example, GE can assist transporters with upgrading their organization's quality by extending the availability of its plane engines through better perceptive help, by, for example, removing determinations considering sensors that recognize, express, assortments in engine turn speed or oil weight [37]. IoE applications are not bound to associations, clearly; buyers can benefit by a broad assortment of related things running from activity trackers to quick indoor controllers to related vehicles [36].

- The Cisco/ITU report states that the IoE provides significant advancement in the health care, which requires special consideration in the modernization of the world. For healthcare the factors such as clean water, balanced food, power, social insurance, clean environment for the individual are precautionary steps to resist the pandemic outbreak.
- The advancements in IoE in the medication and the healthcare system are indispensable. IoE provides the on-time diagnosis of the disease through proper consultation with the physician [36].

4.1.3 Challenges in IoE

Though the IoE is utilized in various domains, such as healthcare and commercial organizations, it still has to experience lot of challenges related to the privacy of the data. Figure 4.2 depicts the challenges in IoE environment.

4.1.3.1 Authenticity

In the End hub level, IoE nodes and the sensors are connected to gather information from the real-world and transacts the data with the connected node device. Yet, in Edge Cloud-level, malignant cloud nodes can process the data deliberately to interrupt the activity of the framework. Consequently, the presence of malignant nodes in Edge Cloud-level can undermine the framework more than those in the End Node-level. Thus, the authenticity of the nodes needs to be verified before processing the data.

4.1.3.2 Reliability

In most of the IoE models reliability is considered the prime challenge, which is accompanied by security threats. The multi-level reputation system should be

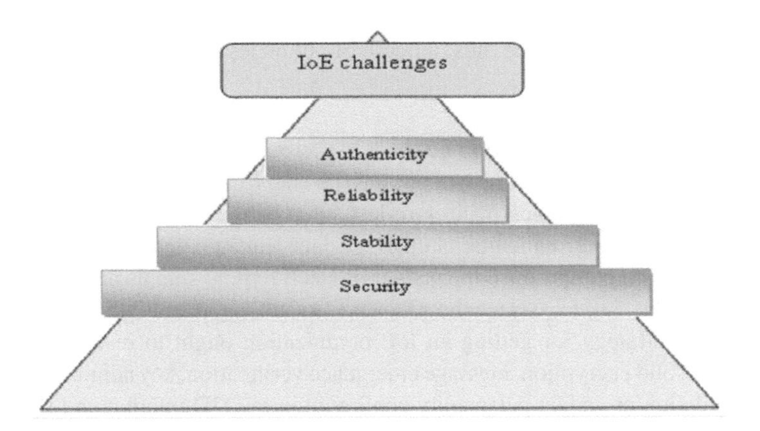

FIGURE 4.2 Challenges in IoE.

included in the system, in which the scores (rating) are provided to the different nodes (by other nodes in the network) to gain trust, where the malicious node finds it difficult to obtain the high standing score. The difficulties of the malicious nodes that are affected by collision and the SPOF attack to obtain high score are due to the following reasons:

- the reputation penalty was provided by the edge node that detects the malicious nodes providing the fake data.
- the subjective evaluation is carried along with the objective evaluation in order to prevent the damage caused by the malicious nodes.

4.1.3.3 Stability

In the IoE platform, the main challenges are the storage of huge volumes of information that consists of reputation data and also traffic data. The decentralized document framework, known as IPFS, is utilized for data storing. While evaluated with hypertext transfer protocol (HTTP), IPFS has a prevalence in handling SPOF issue. Further, the storage program that includes the hash address of record stores the similar content of information once included in the organization to limit the data over repetitiveness.

The emergence of malware, for example, known as "Mirai," creates a huge organization of botnet intruders to mount a surge of computerized intruders on designated destinations. Specifically, this implies that the development of smart gadgets integrated with internet (i.e., IoTs) currently permits a digital attacker to make a large network of bot-nets to attack a specific site. The remarkable expansion in the quantity of IoT gadgets that can be utilized to mount an assault prompts an area of expanded vulnerability for any associated site. In the principal occasions, this sort of Distributed Denial of Service (DDOS) assault against designated sites is adequately enormous to crash a site however humble were the endeavors. The initial attacks that generate the approaching surge of traffic were at a level of a few gigabits/second. Today with the Mirai malware an enormous scope bot-net of exceptional size can be gathered and such assaults are now expanded to tremendous dataflows. These data streams are by generated through the convention of an enormous net of bots all getting to a unified site. These data streams are adequately enormous to attack the biggest business destinations or even crack the most modern of public legislative sites [38].

4.1.3.4 Security Challenges

The end gadgets and organization nodes in numerous IoE applications are conveyed in open environment or hostile without any physical or actual assurance. The data transactions between gadgets with wire-less interchanges are exceptionally vulnerable to the assaults of malicious tampering, data leakage, and threat. An assailant might endeavor to compromise the gadget or catch remote signals and execute more chipper text-like attacks, select the plaintext assault, or potentially replay assault. A productive strategy for getting an IoE organization ought to restrict the threats, make use of solid encryption, message emergence verification, key administration, and client validation or gadget personality confirmation in D2D interaction [2]. Because of the description of IoE applications, it is advisable that the security techniques are versatile, that is to say, the security resistance and effectiveness can be altered by the security prerequisites and gadgets' capacities. The key length decides the security stability as well as the productivity measure as speed and asset utilization [2].

4.1.4 Requirement in IoE

This section provides a detailed description of the basic IoE requirements, such as security and privacy of data.

4.1.4.1 Security in IoE

From the security perception assaults, systems and administrations are discussed [38], [36]. Any activity that compromises the confidentiality or the security of data addresses a security assault. This activity can be detached, which means an arrival of message components or traffic evaluation or dynamic, such as Replay, Masquerade, Denial of Service – DoS, Modification of message contents, Distributed Denial of Service – DDoS. Security schemes were intended to identify, forestall, or recover the content from a security assault. Encipherment, advanced signature, data morality, access control, authenticated transaction, routing control, traffic padding, and authentication are the segments of the principal security components that provide sufficient security. Security administration upgrades the security of information processing frameworks and data transmission. A security administration utilizes at least one security scheme. The fundamental security administrations are the data confidentiality that includes association, connectionless, traffic stream and selective field, authentication that includes data origin and peer substance, data morality, non-contradiction with evidence of conveyance or origin, access control, and accessibility [6]. Table 4.1 provides the application of IoE and security concerns.

One of the central components in obtaining the security of IoE establishment depends entirely on gadget character and the components to validate it. The present validation and encryption procedures are supported on cryptographic suites, for instance, Advanced Encryption Suite (AES) for ordered Rivest-Shamir-Adleman (RSA) for electronic imprints, and key vehicle and data transport and Diffie-Hellman (DH) for key exchanges and organization. While the transitions are private, they require a high register stage, a resource that may not exist in all IoE-associated gadgets. Consequently, verification and authorization are required to legitimate re-planning to suit the new IoT associated world [13].

1) *Components for the secure architecture:* The fundamental components and predominant methodology of IT security is to devise secure information rather having extra layers of the current design [39]. With regards to devise a safe IT foundation for cross-culture communication, the following terms need to be considered:
   a. *Arrangement of Business Domains and Security Requirements:* A conventional IT framework is developed in association with the business activities and spaces. Specifically, while considering the retail organizations, their domains are always inserted on the integrated value chain from store administration to repository network organization [40]. To the contrary,

TABLE 4.1
Security concerns in IoE application

IoE application	Misuse	Technical standard	Administrative actions
Geo-fence tracking of IoT-devices for the physical security access control	Destruction of the entire framework	Requires prominent Encryption	The geo-defense invasion is treated as the global felony. Further, a penalty is imposed if the theft involves tampering with the IoT device
Security camera system for governmental, industrial and home organization	The surveillance provides the detail about when to attack the system	Encryption	Such abuses are defined as the crime
Autonomous car	It debases the vehicle or make it to crash	Encryption	These abuses are categorized as felony and possible actions should be carried out if there is loss of life or murder
Electric power grid monitor	Debase the power system	Encryption	These are also categorized as the felony if there is a loss of life
Gas and water supply or distribution system	Disable or any other form of assaults	Encryption	Considered as felony if there is any loss of life
Transport, security system, service or transport	Any form of attack	Code word access or encryption	Treated as crime in case of destruction of life or property

the IT framework configuration needs to take a look at both the points of view of chance openings to existing resources and business processes in every space. The security component ought to be installed and made a basic piece of the engineering instead of making it more complicated in the wake of adding more and more security layers [38].

b. *Gathering according to the capability*: The ICT framework is made sensible and secure based on direct reviews from the clients. The privileges are allotted to specific gatherings of business and safety domains. The menace is estimated (based on resources and procedures of the associations) through the capability level and task for which the data is required. Hence, for more reliable and sufficient protections, the ICT tasks are allotted to the different gatherings so that each gathering can focus only on the allotted task [22]. The homogenous degree of assurance is acquired in the wake of adding abilities to security areas.
c. *Measured quality*: The seclusion part manages changing the security level of spaces without influencing different areas. The business includes

different domains with various security levels and particular designs as this serves to quantify the menace and simultaneously also offers the protection. The framework security could be expanded by sending the essential focuses at different nodes to screen the innovation. Formulating a solid point of interaction just between a corporate organization and public web is inadequate. The dangers of hijacking the organization in the wake of associating and entering the infrastructure is increasing day by day. These dangers are not protected by the external organization guards and require a few inside geographies to be contrived to keep it secure through triggers. When a few users get associated with the IoE, an additional security layer ought to be initiated which distinguishes assaults. The framework ought to be planned in an intellectual way, which reliably notices within exercises, identifies client conduct change, and alarms the foundation. When the organization is partitioned into security spaces, it carries different advantages to recognizing dangers. Data is an important source and most present-day organizations depend on powerful utilization of data for their market reach, consumer loyalty, competitive advancement, and process. This demand for the important data overburdens security and information connected with individual preference, despising, and conduct. The data framework has gained the huge accomplishment of organizations completing their objectives. The data framework assembles process, disseminates, uses and connects with data. The progress of data frameworks is subject to diverting correspondences between various parts of such frameworks, including individuals. The data security is a laid-out discipline with obvious methods and measures with this impact [33].

4.1.4.2 Privacy in IoE

The term privacy conveys various thoughts like security of possessions, exercises, decisional security, and so on. The type of privacy alluded to in this domain is data security. More specifically, data security concerns an institutional control over the procurement, utilization, and disclosure of individual data [41]. The potential of a person to access and alter their own data is contemplated as a key factor and the basic privacy challenge of the information era [42]:

a. *Individual data*: A proper description of individual data is obtained as per the following elements. Any data connecting with an individual, who can be recognized as the person, in an indirect or the direct way, using that data and specifically by reference to an ID number or at least one variable well defined for their physical, physiological, mental, financial, social, locational, or social character. Privacy of individual data includes the option to control when, where, how, to whom, and to what extent the specific individual transfers their own data, as well as the option to access individual data provided to other individuals, to address it, and to guarantee it is defended and discarded properly.

b. *Individual privacy*: The option to control the stability of one's own data content, this covers such things as actual prerequisites, medical conditions, and required clinical gadgets.
c. *Social protection*: The rights of people to pursue their own decisions about what they want to share and what they want to hold private [42].
d. *Individual transmission privacy*: The option to interact without unjustifiable reconnaissance, control or monitoring [42].

Privacy or security has been a great concern of every individual from the beginnings of the Internet. IoE is now intensifying the issue of privacy due to the fact that numerous applications produce detectable signatures of the area and conduct of the people. Security issues are especially significant in the smart home that is categorized under the domain of IoE. In this circumstance, it is fundamental to determine the data source while determining the gadget used by the owner. Shadowing [13] is a system which is developed to accomplish security aspects. Basically, computerized shadows empower the client's objectives to follow up for his/her sake, storing only a virtual personality that contains data about his/her credits [43].

4.1.5 Main Concepts and Terms in IoE

This section introduces the main concepts and terms involved in IoE environments, which ensure enhanced quality of service to the users.

4.1.5.1 Cyber Security

According to the International Telecommunication Union (ITU), the network security is characterized as "the assortment of apparatuses, security ideas, policies, security safe-guards, rules, risk management strategies, activities, best practices, affirmation, training and advances that can be utilized to safeguard the digital environment association and the client's resources associated for specific "interlinked computing gadgets", like Internet of Everything (IoE)." Eventually, the motivation behind network protection is to guarantee that the security properties of authoritative and client resources are accomplished and kept up with against applicable security that takes a chance in the digital environment. The general security targets incorporate those of a) accessibility, b) integrity, and c) confidentiality, otherwise termed as the CIA group of three in data security applications. Confidentiality includes keeping data from being inappropriately revealed to unauthorized people, gadgets, or processes. Confidentiality implies that data is safeguarded from unapproved alteration or annihilation. Accessibility alludes to quick and dependable admittance to information and data for legitimate users [44].

4.1.5.2 Cyber Threats

A few researchers in datamining have distinguished a few attempts at characterizations whose fundamental goal is to recognize and comprehend the qualities and the threat sources so as to preserve the framework resources. Yet, the threat is determined as the capacity of an attacker to tamper with a framework or what an aggressor or attacker attempts to accomplish with a framework. It is further characterized as a strategy that assailants use to take advantage of vulnerabilities in framework components or the effect of dangers on a resource. Two primary classes of dangers have been referred to in the conventional methods. The categorization with respect to the assault procedures is known as the three symmetrical layered models wherein the threat domain is partitioned into three symmetrical aspects called inspiration, agent and location. The researchers also refer to the threat cube grouping model, which considers three fundamental standards: resource, frequency, and activity. In the pyramid model there are three variables: basic region, assailants, and losses. In the categorization with respect to the effect of dangers, the users refer to the STRIDE model, made by Microsoft, in light of spoofing personality, tampering with information, information disclosure, repudiation, DOS and elevation of perquisite. There are five significant effects and administrations of safety dangers recorded by the ISO standard (ISO 7498-2) which are as per the following: destruction of data, defilement or change of data, data stealing, evacuation or loss of data, revelation of data and interference of administrations. Table 4.2 provides the various types of attacks and their impact.

The researcher adds one more sort of arrangement which is the direct danger classification that recognizes dangers by specialists, apparatuses, and types [44].

TABLE 4.2
Various attacks and their impacts

Description	Vulnerabilities	Impacts on the resource
Spoofing	IP spoofing, ARP spoofing, session hijacking, replay attacks, DNS spoofing and sensor data spoofing	Change the characteristics of the temperature device, spoofed to transmit commands from the trusted device
Tampering	The apps that are utilized to alter the devise properties, altering the IoT device files. Installing the unnecessary or backdoor program, malicious software attack.	The ransomware attacks encrypt the data file devices into the bot's private data.
Repudiation	The fake data is transmitted in the identity of the genuine users	It orders the new products and services
Information- disclosure	Encrypts the network congestion, loss of private details, clear text account accreditations, observing the profiling behavior.	It discloses the private data such as personal preference, and private details such as videos, conversations, and attitudes along with confidential status.
DoS attack	Jamming, due to weakness in the protocols	Interrupts the home decision function.
Elevation privillage	Buffer overflow, default account setting, weak password, outdated version of sources.	It can provide the path for DDoS attack.

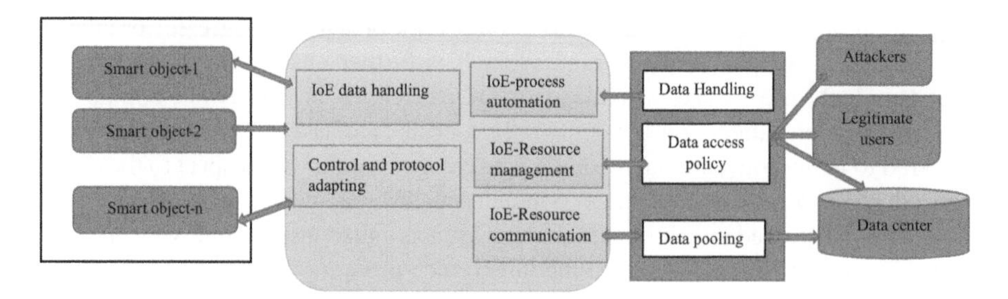

FIGURE 4.3 Framework of secure IoE.

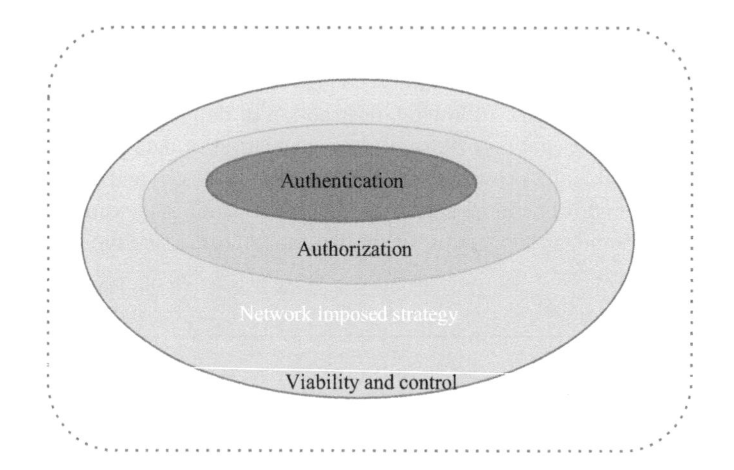

FIGURE 4.4 Secure framework for cross culture interaction.

4.1.6 Formulating Secure IOE Frameworks

The capacities of secure organizational frameworks are recognized through an initial security evaluation and categorized through the security framework. The secure IoE framework is depicted in the Figure 4.3 and the secure framework for cross-culture interaction is shown in Figure 4.4. The most basic business resources, such as endorsing data and exchanging algorithm, are evaluated to recognize security drawbacks. The compromise of the security drawbacks could prompt reputational harm and also material loss. The resources and cycles of high-esteem and high-risk nature are isolated based on threat depended categorization, yet crossculture interaction actually benefits through virtual conditions and shared frameworks [21]. Different applications and servers could be utilized to run the hierarchical site through a different authorization motor to handle the high-esteem monetary exchanges inside the cross-culture interaction. The attitudes to help process and information ventures for online cash transactions or other deals are categorized under discrete abilities [14]. The satisfactory degree of chance and assurance are not entirely settled through the examination of safety zone engineering inside the cross-culture correspondence. The gamble effect of penetrating can be assessed through the administrative, critical, monetary, reputational, and functional process of the association [45]. The security risk can always be assessed through the process downtime as this misuse of clients' individual data could prompt administrative fines [7].

4.1.6.1 Authentication

The authentication or validation layer is the focal point of the structure which is utilized to distinguish and validate the IoE element data. When the IoE gadgets begin to establish the association with one another, they need to be associated with the IoE framework [1]. The personality of the gadget ought to decide the trust relationship. Different IoE gadgets might have significantly various approaches for storing, overseeing, and introducing the data. It is noted that qualified clients in associations control the network for both cross-cultural and local via the human credentials of secret phrase and username [7]. As far as IoE, the endpoints need to be constructed through finger impressionsso do not need human collaboration. The inserted sensors inside the IoE gadgets need to be artificially intelligent to produce output and afterward perceive client identity on the basis of a specific gadget storage system [33]. The X.509 credentials are likewise be utilized to lay out major areas of strength for a framework to lay out this character. The new modalities and the shape factors increase the challenges of formulating more modest credentials on the basis of less comprehensive standards of cryptographics at the confirmation layers [46].

4.1.6.2 Authorization

Authorization is the subsequent layer that follows the authentication layer. The authorization layer controls all gadget access through the framework environment [2]. The center authentication layer is additionally implanted by coordinating the substance's personality data. The trading of proper data begins when a trust relationship is laid out among approval and validation parts [36]. A similar vehicle salesman can foster a trust coalition between the vehicles, so one intelligent vehicle can distribute specific protection abilities-related data to another vehicle. This determines the collusion connection among vehicles and their sellers might permit sending and trading extra data, for example, their odometer readings and the last maintenance records [46]. The process of user accessibility and administration in the business accounts is very much approved in the ongoing strategy design of IoE gadgets. Building a design taking care of correspondence of numerous of IoE gadgets with differing trust connections would be really difficult for cross-culture interchanges [43]. These difficulties are extended to the particular point for end-to-end interaction with traffic strategies and suitable controls to section and synchronize information traffic. The central point to be cared for in this engineering would be the reduction of data exploitation.

4.1.6.3 Network Imposed Strategy

The network-imposed strategy layer includes the traffic of everything that transports and routes on the framework safely. It also includes controlling and the executives of the information trade over IoE gadgets. Different systems and conventions are, as of now, settled in regards to controlling the upheld strategy to get the foundation of an organization when IoE gadgets [44].

4.1.6.4 Secure Evaluation (Visibility and Control)

This is the most common way of controlling the IoE environment determined to acquire visibility, and the service is characterized through the secure analytic layer through which network frameworks, server farms, and all endpoints participate in ensuring telemetry [6]. Monstrous parallel data set (MPD) platforms can be utilized as they would deal with huge volumes of data productively [44]. The inconsistencies of the acquired data can be selected and real-time qualitative evaluation could be performed while coordinating analytics with this innovation [7]. The IoE gadgets generate huge information that is just significant assuming the right security process and analytical algorithms are utilized to recognize and determine the threats [47]. The security algorithms are utilized on different layers of this model and the data gathered from those sources could create a better analytical outcome for managing the security threats. Consistently new innovation is advancing and network textures are turning out to be more complicated in nature. The topological information is moving to private and public clouds and this move requires defense abilities along with the resolution and threat intelligence recognition simultaneously on clouds. The derivation of precise knowledge requires control, setting, and perceptibility [33] [12].
Research questions:

1) What are the factors that generate concerns about the security issues in transmitting the data through the IoE system?
2) What are the threat scenarios caused by the insider attack, which negatively influence the organizational growth?
3) What are the attacks or threats to be identified and mitigated to increase the quality of service of the D2D communication?
4) How does the intrusion detection model preserve the security and privacy of the transmitted data and what are the current techniques involved in IDS?
5) What are the recent optimization algorithms employed to enhance the accuracy of the classifier in intrusion detection model?

4.2 INTRUDER DETECTION IN IOE

Most of the security attacks mentioned in Chapter 1 takes place when the intruders obtain access to IoE networks and steal the sensitive information of the user. This chapter provides the deep insights about the intruders and the way they affect the organization and the intruder detection system.

4.2.1 The Ways for an Intruder to Make an Attack in the Organization

There are various ways for intruders or attackers to get sufficiently close to the organization.

1) Wireless Signals: The assignments are now easier and more effective due to the availability of Wi-Fi. Yet, large numbers of home APs have frail security settings, which incorporate the reception of no or old encryption and the

utilization of feeble or default passwords. Bluetooth and Wi-Fi are prevalent in almost every home environment. People are starting to procure a rising quantity of Bluetooth gadgets for individual purposes, such as headsets, Fitbits, scales, and so on. The Bluetooth gadgets can undoubtedly become passages to a Wi-Fi or wired network if not secured. There are likewise different kinds of remote signals like ZigBee, yet they are substantially less far reaching than Wi-Fi and Bluetooth.

2) Wired Connections: A simple connection can be established as a switch associated with a modem consisting of a public IP which is consequently accessible through Internet. Therefore, an Internet-router acts as significant equipment in the home establishment and is vulnerable to different interruption attempts [5].

The intruders make attempts to compromise the gadget or block remote signals and execute ciphertext-only attacks, selected plaintext assault, or potentially replay attacks [2]. Digital assaults cover all potential threats that target at least one of the IoE elements, yet in addition which can also target the process that emerges due to their interception [47]. An expansion in the attack of 76% for 100 Gbps to the 400 Gbps has been observed since 2018–2019, with the complete number of DDoS attack increasing from 7.9 million in 2018 to 15.4 million in the year 2023 [48]. To limit the expenses incurred due to different digital attacks, a security procedure needs to be designed to reduce a wide range of vulnerabilities that can influence an IoE environment [14]. Such a procedure depends on the security efforts that can help the IoE clients (leaders) in preventing the security of their industries, organizations, data, frameworks, and application against attacks [44].

4.2.2 Threat Scenarios

IoE threat scenarios are described in the following section.

Scenario 1: Compromised cyber-attacks are categorized as the critical threat to IoE devices. IoE devices utilize the huge variety of system software with their variants and applications. The subsequent heterogeneity builds the intricacy, particularly for home users, and in this manner confounds the assurance. Provoking this present circumstance is the broad interconnectedness of IoE gadgets. To remain interoperable with any remaining IoE frameworks, it is most of the time essential to utilize unsecured interaction channels, which further increases the probability of attack, which further enhance the assault surface. When an IoE gadget is compromised, the opponents involve it in more ways than one for directing digital crimes. First and foremost, IoE gadgets under an aggressor's influence can be unified to a botnet, for example, Mirai botnet. Besides, compromised IoE gadgets can disregard a client's protection by uncovering data like discussions, video accounts, or conduct use. Third, utilizing ransomware, assailants can control or deactivate a gadget's capacity to coerce clients. Weaknesses include unreliable record settings, obsolete working frameworks or weak applications.

Scenario 2: Eavesdropping and Information Leakage: home IoE gadgets tends to screen and control virtually every feature of our future life. Associated webcams track client conduct, smart gadgets control the home environment (for example warming, cooling, lighting, and so forth) or associated security frameworks (for example entryway locks, alert frameworks, and so on) intended to safeguard the protection of the family continuously collects data and these may lead to data leakage. As discussed earlier, data leakage or monitoring results in the serious protection breaks. Video frameworks, which are executed to screen the security of a family, could release private pictures or recordings. Further, game control centers, smart TVs or other voice-controlled gadgets can be utilized to listen in on discussions. The rich source details of data (e.g., the TV channels they watched, shopping patterns utilizing Alexa or comparative gadgets, and so forth) about the home and behavior of clients could be an advantage to any custom-fitted marketing exercises or the social designing attacks. To further develop interoperability, IoE gadgets frequently utilize powerless or no encryption. Further, weaknesses in applications could cause data spillage.

Scenario 3: Jamming, Interference and DoS Attacks against IoE: more and more gadgets will influence our regular routines in future. The vision of IoE is to interface a multitude of existing family gadgets and assist developments with arising. A denial of basic IoE administrations could prompt extreme harm contingent upon the sort of gadgets. Jamming gadgets, which are just associated by remote means (for example remote associated observation and disturbing gadgets) cause a disavowal of administrations of these brilliant homes administrations. Missing security against actual assaults, like sticking of remote signs, is a huge danger to future IoE frameworks. Further, the enormous increment of remote gadgets sending different sorts of signs could prompt more obstruction among IoE gadgets [5]. Fundamentally there are two insiders like malicious insiders and un-deliberate insiders.

An insider threat is a noxious danger to the network that comes from individuals inside the organization, for example, workers, previous representatives, project workers, or business partners, who have inside data concerning the organizations security practices, data and PC frameworks [43]. The interests of insider threats are from representative harm, notoriety or association harm, information spillage, and information exfiltration and information access for abuse purposes. Insider objectives of treachery, undercover work, and getting IP locations can be related with such regions [1]. An insider will disclose the organization's delicate data or organization setup to an outsider for monetary benefit and these means help in recognizing the moves made in going after any resources [19]. The term insider-assault vector is constructed which incorporates a blend of these malicious advances that could determine the scheme of assault. When the category of attacks is distinguished, solutions are accomplished to counter the assault, finally preserving the enterprise organization, framework, delicate data, and information [44]. A valuable context concerning attacks is given from the initial three regions which assist the enterprises by demonstrating, coordinating, and figuring out the key threats, and in addition distinguish resources that could be impacted [16]. Significantly more alarming is the way that the organizations are frequently unaware of the nature and quantity of the IoE gadgets around us, also the potential chance of a security risk which they represent. The current security events emerging from the IoE security vulnerabilities authenticates this perception. Specifically, one of them is a Distributed Denial of Service (DDoS) attack against Dyn [49] observed in October 2016. This event included a botnet known as Mirai, comprising of roughly 100,000 IoE hosts, including routers and computerized cameras. The Mirai botnet sent off DDoS assaults against Dyn and cut down its Domain Name Servers (DNS). This process in turn resulted in the abuse of significant commercial sites, such as Netflix and CNN [5].

These DDOS assaults are generally categorized under any one of the subsequent classes:

- Synchronize flood (or synchronize flood attacks). The goal is to create a stream of ACK or the affirmation messages. With a SYN flood attack the aggressor can directly utilize irregular, fake, non-implicating source addresses for the packets that come from the large amount of bot-net.
- GET/Post Flood assault: This occurs at the point when an HTTP client or otherwise known as the web programmer converses with an HTTP server or the Web server. For this situation the information flood transmits the request which is distinguished in several categories, the two principals being GET and POST. A GET request is utilized for "typical connections" that includes the image. These GET requests are intended to recover a static piece of information, and the URL accentuates that source of information. At the point when you enter a URL in the URL bar, a GET is likewise finished. POST requests are utilized with structures. A POST demand incorporates the access parameters within the websites, which are normally taken from the information fields in total agreement. The intruders need to lower the objective server under a sequence of request, to elevate the server registering assets. Flooding provides better service when the server distributes a great deal of assets in response to a solitary request. Since POST demands incorporate the need to access the variables inside the website, they generally trigger somewhat complex processing on the server, for example most database assets. This includes more processing behavior for the server than processing a lot less difficult GET. In this manner, POST-based flooding in general is more compelling than GET-based flooding. It short, it takes less demands to suffocate the server in the event that the solicitations are POST. Then again, GET demands are substantially more typical. Subsequently, it is a simple for the attacker to join impulsively in the flooding exertion [38].

Because of these threats, it is essential to be concerned about potential IoE security threats between the end users through precise risk evaluation and powerful perceptions. Home users are particularly defenseless against the threat due to the fact that they are progressively encircled by IoE machines, such as child monitors, without a handheld surveillance camera and so on, yet the IoE model lacks the abilities and resources to recognize their own IoE-related security threats, remediate them, and reduce the potential security risks [5].

A preeminent scheme for fortifying IoE organization needs to address these attacks or threats, utilizing solid encryption, key administration, authentication of the data source, and data owner/purchaser verification or gadget identity validation in D2D correspondences [2].

4.2.3 Background of Intruder Detection System

The intruder detection system (IDS) emerged in 1980 and was mentioned by James Anderson in the article "Computer security threat monitoring and surveillance". The IDS innovation was gone through two different generations. The first was the intruder detection system, which is developed to distinguish between the intruders and the misusers. This system was known as the detective model, as it only alerts the incidence. The next generation was known as the preventive system as it prevents the intruders as well as detecting. [50].

4.2.4 Categorization of IDS

The intruder detection model is generally categorized into two types a) based on place of detection and b) based on approach of detection. The category of IDS is illustrated in Figure 4.5.

4.2.4.1 Based on the Place of Detection

The IDS based on the place of detection is generally categorized into two different types: Network-based IDS (NIDS) and host-based IDS.

The NIDS model is utilized to evaluate network traffic by implanting the Network Interference Card in the promiscuous node. The CiscoIDS, SNORT, and Symantec Netwprowler are some of the examples for NIDS. The sensors in the NIDS are distributed at tactical locations to determine the maximum ventures. The main objective of the NIDS is to observe the traffic congestion inside the network. While attaining the traffic the NIDS evaluates it and compares it with the derivable dataset related to the attack so as to identify the attack. The alert can be triggered while inscribing the abnormal behavior in the network. The NIDS are further categorized into two types: online NIDS and offline NIDS. The online NIDS is otherwise known as inline mode and it is responsible for real-time monitoring. The processor performance is the predominant factor that influences the system performance. The pros and the cons of the NIDS are illustrated as follows:

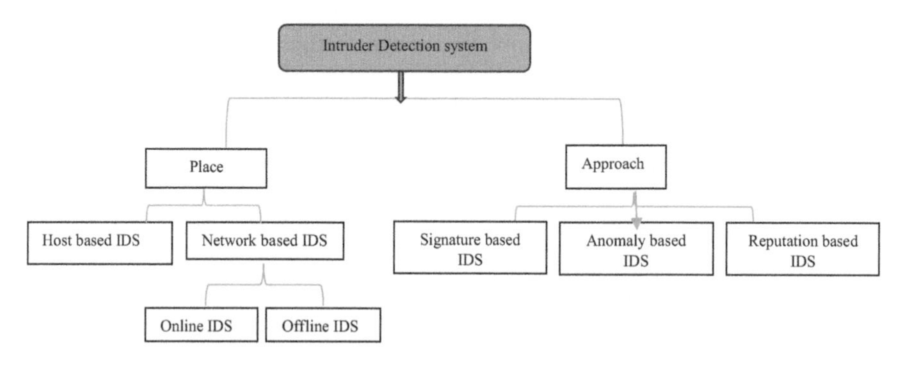

FIGURE 4.5 Category of IDS.

- The NIDS support numerous sensors to enhance monitoring abilities.
- The NIDS consists of blind spots, for example if the intruders are present on the host cells it fails to identify the categories of attack.
- The NIDS find it hard to decrypt the encrypted data.
- The NIDS fail to determine the categories of the attacks and their effectiveness.

The HIDS are utilized to track the elements present in both outbound and inbound data on the specific system. The HIDS tracks the events and figures out the controversial events, as well as validates the state of the intrusion detection system. The HIDS are employed to track and scan the attacker movements and to alarm the administrator. The HIDS is effective in tracking the local characteristics of the network while compared with NIDS. Yet, the degree of sensitivity to the illicit tampering is high while compared to the NIDS.

4.2.4.2 Based on Detection Performance

The IDS is categorized into three different types while considering the detection performance of the system:

a) Signature-based IDS model
b) Anomaly-based IDS model (ANIDS)
c) Reputation based IDS.

In the signature-based model the intruders are detected by comparing the real-time attack with the subsisting patterns, known as signature. The signature-based IDS is devised by the inspiration obtained from the antivirus and specifically the byte sequences which are called signatures. The major drawback of the system is that it fails to detect the new attacks and it requires timely modification. The limitation of the signature-based IDS is restrained through the ANIDS.

As mentioned above, the anomaly-based IDS are effective in recognizing the unknown attacks through the generated model that describes the authentic activities and these activities are utilized to compare the inbound behavior of the system. The main advantage of the system relies on the quality of training, which assists them to attain better detection performance in case of an unusual attack. However, the ANIDS exhibit some limitations such as increase in time consumption, increase in false positives and performance degradation.

The ANIDS are subcategorized into three types, while considering the techniques utilized to determine the variation in the characteristics. The statistical ANIDS is the first category that utilizes the general statistical techniques to recognize the variations in behavior. The next category is called knowledge-based ANIDS and it makes use of the previous knowledge about the behavior of intruders. The finite state machine or the expert system are a couple of examples of the knowledge-based ANIDS. The third category utilizes the machine learning technique to determine the variations in their behavior. These models utilize fuzzy logic or the Bayesian network, markovian models, or the neural network; genetic algorithm or the clustering are also utilized. These models are flexible and can be scaled as per the requirement. The ANIDS provide the robust solution while contrasted with the SIDS. However, the main limitation relies on the high level of dependence on the presumed model. The ANIDS may be either variate or multivariate [51].

4.3 USER ACCESS SECURITY BASED ON BLOCK CHAIN TECHNOLOGY

Blockchain technologies appear to be the best solution for the security and privacy issues of IoT. These models are utilized to direct the IoT gadgets and effectively solve issues of intervention by intruder into the network [50].

4.3.1 Block Chain Advancements

The current IoE always depends on centralized servers for their organizational interaction. The centralized cloud servers enable organizational interaction through validating the IoE gadgets. Hence, the current solutions for IoE in smart cities use distributed computing and organization assets, which results in high maintenance and infrastructural costs. In the smart cities that consist of a versatile environment, the gadgets are increased or reduced routinely. It suggests that smart sensors are utilized more regularly in a smart city framework as an adhoc network. The current framework fails to support the huge IoE gadgets because of the versatility issue. As the number of assets increased, the interaction among gadgets and servers are also increased. The cloud servers likewise experience an issue of a single point of failure. In a smart city, a centralized framework is a failure, and in this manner, the distributed model can be more successful. Hence, in these scenarios, the Blockchain is suitably applied since it is a decentralized and secured framework. The Blockchain thus connects the millions and billions of gadgets utilized in the organization. Further, it minimizes the expense of server management and installation process. It further saves the smart city IoT gadgets from man-in-middle attack since there are multiple channels of correspondence. The smart contracts and agreements are executed on the basis of organizational function. The data obtained from the smart sensors are preserved in Blockchain [52]. The application of Blockchain technology is demonstrated in Figure 4.6.

4.3.2 Blockchain

Satoshi Nakamoto originally formulated the idea of Blockchain in 2008 [53]. Basically, blockchain is a computerized record distributed on an organization without repository or centralized authority. Blockchain comprises of the coordination of the block chain, in which the principal block is termed as the genesis block and other as block header. The block header addresses the hash value of the preceding block, while block information records the data transaction over the limited time period. Every node in the organization can access the recorded data from the block chain. In the current era, the blockchain frameworks like Ethereum and Hyperledger are widely utilized in various pragmatic applications like transportation, training, web-based business [15], and IoT [16][17]. The IPFS is a globally distributed document framework, shared, content addressable, open source that can be utilized for sharing

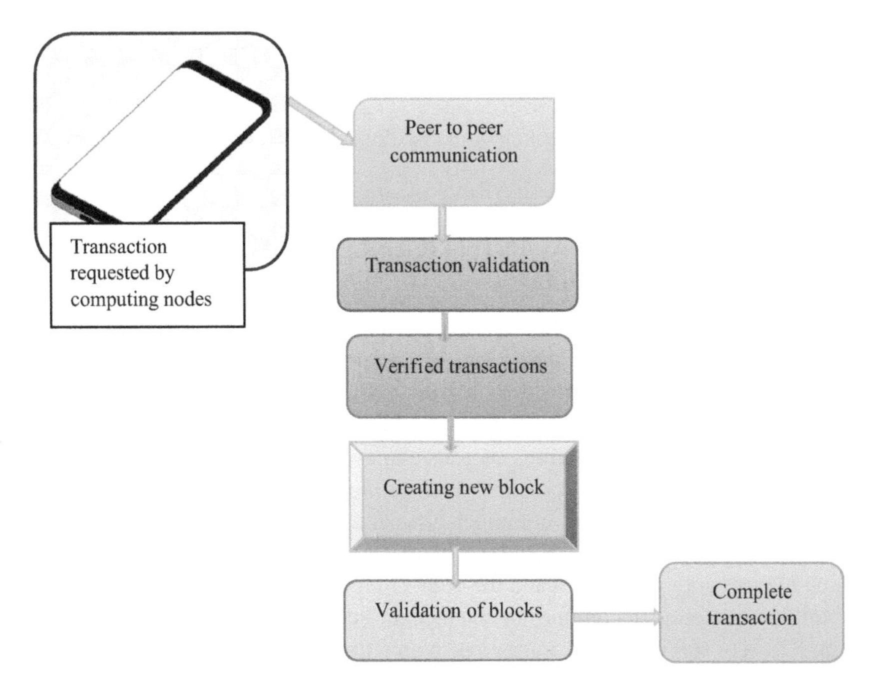

FIGURE 4.6 Processing flow of blockchain technology.

and storing a huge volume of records to high throughput [43]. The IPFS is a globally distributed document framework, shared, content addressable, open source that can be utilized for sharing and storing a huge volume of records to high throughput [43]. Rather than depending on a focal server, IPFS is independent of the centralized servers, and it disseminates the information to various nodes of the framework. In IPFS, each record will be relegated a distinct hash esteem and coordinated through Merkle DAG structure. When the data providers transfer a document to IPFS, the provider can acquire an exceptional location returned by IPFS. By utilizing the record address, the data providers download the document from IPFS.

4.3.3 The Smart Contract

The term smart contract was first devised by Szabo in the year 1994 [18], which is characterized as "a modernized exchange protocol which leads the conditions of an agreement" [18]. Eventually, Szabo suggested changing the details of an agreement to a line of code and implant it into a proper environment, in order to run the code consequently. Hence, contrasted with the conventional agreements, the smart agreement or contract can upgrade the exchange effectiveness between the members essentially and minimize the event of unintentional or malicious exemptions generally. Normally, the smart contracts are predefined and sent on the blockchain, and every node of the organization can utilize the smart contracts by sending an exchange to it. By getting the exchange information, the smart contracts lead consequently onto the blockchain [13]. It is prominent that all nodes executing an equivalent smart contract obtains a similar outcome from the execution, and these outcomes are recorded on the blockchain [14].

Hence the integration of blockchain technology in IoT enhances the security and transmission of the data.

4.4 ROLE OF ARTIFICIAL INTELLIGENCE IN ATTACK DETECTION MODEL

Most of the security frameworks devised for IoE need the authentication or packet encryption models to handle the access of a malicious user. Further, various other security-based solutions like key transmission, secure routing, and authentication preserve the data from various potential attacks. Yet, these security methods fail to provide protection against the various security threats and attacks against the IoE [54]. The IDS is described as the dynamic security model utilized to recognize, prevent and effectively react to those computer attacks and is determined as the effective component element in the security framework. It tracks the activities of the selected source like network traffic data and the audits in the computer or the network system and utilizes numerous techniques to enhance the security of data. The main intention of the IDS is to sort out all the intrusions in an effective manner. The implementation of IDS enables the network administrators to identify the violations in the security. The violations in the security threats vary from the external attackers' attempts to gain illegitimate access to network security frameworks or to make the resources inaccessible for insiders who exploit the system. IDS are observed to be submissive in nature, which means that they have ability to detect intrusion. However, they fail to mitigate or forestall the attacks [54]. The exploration communities now concentrate to devise advanced techniques for the detection of intrusion in IoE. Artificial intelligence is considered to be the significant and low-cost method to devise IDSs, while considering energy consumption [55][56].

People succeed in several tasks and learn in the course of time. Artificial intelligence takes into consideration the algorithms that perform critical thinking as done by humans and acquire the capacity to perform tasks by preparing model. The expert framework handles issues utilizing a model of skilled human reasoning. However, every expert framework requires continuous upgrading for better performance [51]. The artificial intelligent system is categorized into three classes such as artificial neural network (ANN), artificial immune system (AIS), and genetic algorithm.

4.4.1 Artificial Immune-based IDSs

Artificial immune systems ensure abnormality-based identification of security threats or attacks against IoE. The human immune framework is a complex security framework which defends the human body against numerous imperceptible creatures. It is extremely perplexing and comprises of T cells, B cells, and dendritic cells (D cells). D cells progress in blood and gather data about the dead cells and antigens. These cells fundamentally actuate the response framework. Immune system microorganisms are produced in bone marrow and are utilized to obliterate contagious cells existing in blood. B cells are white platelets which are accountable for the creation of antibodies. These days, artificial immune frameworks have numerous applications in computing like frameworks advancement, data characterization, and intrusion identification. There are numerous works that have been established in intrusion detection and utilized in the artificial immune system. The researchers guarantee that the immune dependent framework can be insusceptibly utilized in the wireless communication domain. It depends on dendritic cell algorithm (DCA). Moreover another cyber-attack known as interest cache poisoning (ICP) attack also exist. ICP is an organization layer attack which is equipped for disturbing routing packets. Both the DCA and coordinated dispersion are carried out on each sensor node to execute two assignments: recognition of malicious nodes and identification of antigens. The immediate dispersion system is comprised of the following components: (i) keep two tables, or at least, interest store and data reserve; (ii) handle two categories of packets, that is, data packets and interest packets.

4.4.2 Artificial Neural Network Based IDSs

Fake brain organization (ANN) emerges from the inspiration obtained from the human nervous system, which is associated through neurons. NN have the ability to comprehend and advance via learning and can be utilized to distinguish complex patterns. There are two sorts of ANN designs, like feedback ANN and feedforward ANN. In feedforward ANN, the signs move in just a single direction that is from input to output, whereas in feedback ANN, the signs proceed in two directions. ANN schemes are useful in numerous areas like intrusion detection and pattern recognition. ANN-based intrusion identification is utilized to get rid of the limitations in rule-based IDSs. Yet, ANN-based IDSs are found to be more productive if appropriately prepared with both typical and unusual informational indexes.

4.4.3 Genetic Algorithm-based IDSs

Genetic algorithms are broadly utilized in numerous computing areas to handle the perplexing issue in computing. This ensures strong, versatile, and ideal solutions for the majority processing related issues. Genetic algorithm computing is motivated from the natural process like genetic inheritance, theory of mutation, evolution, and natural selection.

4.5 UPCOMING TRENDS IN AI USING OPTIMIZATION ALGORITHMS

Intrusion detection in the view of ANN is developed by utilizing the assembled features about a few categories of assaults. Typically, building information in view of assembled data requires adequate measures of information and are extensive in nature. Unfortunately, in the use of intrusion recognition, it is impractical to generate adequate knowledge for learning or possibly adjust learning between the various classes. Consequently, learning algorithms must be optimized based on the characteristics of the dataset. This leads us to research about how to recognize the optimization parameters of the NN algorithm [57]. More accurately, the issue is the manner by which to track down the parameters of the hidden layer neurons in ANN to keep up with the most elevated exactness of testing.

The Feature Selection is the complex process in ANN, considering that the IDS have to manage the huge quantity of data. FS is the most common way of choosing the important features. Only the selected features create the subset of the entire features involved in classification. The computational overhead is the significant issue with an IDS [58]. Accuracy is also determined as an essential concern [59], as the IDS can distinguish an enormous category of interruptions progressively. The premier issue in devising an IDS is ranking and choosing the subcategory of extremely discerning features. In recent years, FS strategies and optimization procedures provide significance for choosing the main features [60].

Optimization is defined as the method involved in tracking down an ideal set from the arrangements of all potential solutions for any given issue. An algorithm regularly evolves so as to track down solutions. Due to their specific design, such algorithms are expected to contrast two solutions obtained at some stage in order to conclude with the best one. The objective functions that are frequently known as wellness or cost functions are utilized to assess the value of every solution

Provided with the reality with that the volume of organization traffic information is expanding quickly, the requirement for more productive FS techniques has developed. The three previously mentioned FS methods utilize complex computations, which makes them somewhat ineffective for a lot of information. The algorithms that emerge are said to be Bio-inspired algorithms. They provide solutions for the complicated issues involving direct techniques that exist in nature. Bio-inspired algorithms are characterized into three categories: evolutionary, swarm-based, and environment based [61] [60].

4.5.1 Evolutionary Algorithm

Evolutionary algorithms are constructed on the basis of Darwinian standards of nature's capacity to advance living creatures very much adjusted to their environment [21]. Transformative algorithms can be GA (Genetic Algorithms), GP (Genetic Programming), ES (Evolutionary Strategies), or developmental programming. They use methods inspired by the biological advancement, like mutation, selection, reproduction, and recombination. The set of solutions assumes the role of people in a populace, and the fitness function assesses the nature of the solutions. The advancement of the populace occurs after the frequent utilization of cross over and mutation operators. Evolutionary algorithm frequently generates the appropriate solutions for a wide range of issues, and thus , they are effectively applied to different fields, for example, biology, engineering, marketing, arts, financial matters, physical science, operative research, science, sociologies, and hereditary qualities.

4.5.1.1 Genetic Programming (GP)

The GP suggested by Koza in the year 1992 is an organized, domain-specific method for PCs to consequently take care of issues. GP is an augmented form of GA and the main difference is that in GP the individual solutions in the populace are PC programs.

4.5.1.2 Genetic Calculation

The GA presented by Holland in 1973 is a transformative algorithm and follows a global search. The GA is the most productive variant of the evolutionary algorithm, inspired by the transformative characteristics of regular hereditary qualities. These algorithms create predominant individuals from a populace; every individual illustrates the solution for the issue. GA takes care of issues in a simpler way since it is liberated from numerical determination.

4.5.1.3 Evolution Procedures

ES are categorized as a global optimization algorithm inspired by the hypotheses of evolution and transformation. They were suggested by Bienert in Technical University, Berlin in 1964. Generally, this method is stimulated by species-level or macro- or micro-level cycles of advancement (aggregate, genetic, variety).

4.5.2 Swarm-based Algorithm

In an SBA (Swarm-Based Algorithm), the performance of the framework is made out of numerous singular swarms collaborating with the surrounding conditions and, furthermore, locally with each other. Decentralized administration and Self-association are utilized by swarms to accomplish their objectives. Swarm-based frameworks are devised to accomplish complex issues. Swarms are simple living organisms with barely engaged intellectual capacities and restricted methods of interaction. Swarms exhibit the perceptive attitudes and provide the relevant solutions for complex issues like predator avoidance, tracking down the shortest way and the "rucksack issue."

4.5.2.1 Ant Colony Optimization (ACO)

ACO, presented by Dorigo in the year 1990, is categorized as the nature-inspired algorithm that attains the best solutions for immutable combinatorial streamlining issues with a sensible computation time. The ants track down the shortest courses because of their capacity to store pheromones as they move; different subterranean ants follow, moving toward the path where the compound substance is most extravagant. Since the pheromone rots over the long haul, the less feasible the paths and we finally end up with less pheromone concentration. The crossing rate by subterranean ants is more in the feasible path [62].

4.5.2.2 Particle Swarm Optimization

Kennedy and Eberhart presented the bioinspired algorithm termed particle swarm optimization (PSO) in 1995. In PSO, a fitness value is utilized that estimates the quality of the current solution. It utilizes the schemes of the gathering conduct of birds to restrain the optimization issues. In PSO, more elements or particles are stochastically generated in the search space. The particles ripple through the issue space by succeeding the ongoing ideal solutions. All solutions consist of the fitness value that are assessed by the fitness function to be enhanced and have speeds that direct the searching of the particles. A swarm comprises of particles that are flying around in the forage region. Each molecule swarm has a particular sort of topology that describes the correlation among the particles. The significant features of PSO are inherent abilities and speed in tracking down the global best solution.

4.5.2.3 Bee Colony Algorithm

The BC (bee colony) algorithm depends on the ways of honey bees in nature, such as mating and foraging characteristics. BC indicates the intellectual searching characteristics of a honey bee. A BC consists of three sorts of bees, such as working drones, spectators, and scouts. Spectator bees stand by in the dance locale to come to a conclusion about choosing a food source. Working drones go in search of food from every food source. Scout honey bees engage themselves in random searches to decide new food sources. The place of a food source relies on the feasible solution of the streamlining issue, and the amount of nectar in a food source matches the class of the related solution. A group of virtual honey bees is generated and starts to move randomly in a two-layered search space. Honey bees act together when they find a new objective, the task (feasible solution) is accomplished by the force of honey bee (all three categroies discussed above) collaborations as the feasible solution is achieved by combining their work [61].

4.5.2.4 Fish Swarm Algorithm

The fish swarm algorithm is the swarm intelligent algorithm obtained by the natural attitudes of fish called FSA (Fish Swarm Algorithm). FSA has a strong capacity to avoid nearby essentials to accomplish global solutions. FSA imitates three natural behaviors of the fish such as swarming, preying, and following. Preying is an arbitrary movement of fish in search of food, with a tendency toward food fixation. Swarming attempts to fulfill food consumption requirement, connect with swarm individuals, and draw in new swarm individuals. To succeed to the next move, adjacent individuals follow the fish and obtain the food. The components involved with FSA are: visual distance (visual), greatest step length (step), and a group factor. The step and visual parameters generally impact the effectiveness of FSA [63].

4.5.2.5 Firefly Algorithm (FA)

Firefly Algorithm is the unique swarm-based algorithm, which is inspired by the flickering character of fireflies [64]. A populace-based iterative calculation with various specialists (saw as fireflies), FA simultaneously takes care of the enhancement issue. It effectively solves various optimization issues as it is a population dependent optimization algorithm. Fireflies interact with one another through bioluminescent shining that assists them in determining the cost function space more efficiently. This procedure depends on the hypothesis that solutions of an optimal issue can be viewed as the search specialists that sparkle relatively to their effectiveness in settling the issues. Subsequently, the more splendid fireflies draw the other search agents accompanied with them, which simplifies the searching process.

4.5.3 Ecology-Based Calculations

The software engineering and the engineering issues are easily restrained by the concepts of difficult designing and software engineering issues can be restrained by the natural process of regular biological systems. EBA (ecology-based algorithms) are established by the populaces of people wherein every populace becomes as indicated by a streamlining procedure. Thus, people of every populace are altered by the systems of diversification and intensification, provided that the initial components are well defined for every optimization methodology.

4.6 KEY FINDINGS

From the study it is observed that privacy and security is one of the main concerns of the users due the existence of various security attacks and intruders. The intruders remain as the major challenge in preserving the privacy of the data as it discloses delicate information of the organizational frame work. Hence, there is a need for an intruder detection system that recognizes and mitigates malicious users. Yet, the conventional intrusion detection methods are found to be expensive and require full-time monitoring. These issues are mitigated through applying artificial intelligence, which consists of the human reasoning ability. Yet, artificial intelligence requires proper tuning of hyper parameters for increasing detection ability, which is established by the advanced metahueritc algorithm. By analyzing various optimization algorithms, it has been observed that most of the algorithms face issues such as pre-mature convergence, trapped in local optima, and low convergence rate. Hence, there is a need for an advanced optimization algorithm, which can effectively restrain the aforementioned issues of the traditional algorithms.

4.7 CONCLUSION

The IoE systems are devised with numerous sensors and interconnected smart gadgets, in which the expanded number of connected devices generates more challenges on the security features of IoE. A large number of solutions concerned with layered architecture of IoE are therefore presented by the various researchers to restrain the security related issues. The perception layer is found to be in extreme danger as it undergoes confidentiality and the hardware insecurity issues. Further, the DoS and trust management negatively influence the security of the IoE. The solutions such as detection of intrusion, mitigation of intrusion and privacy preservation (for example encryption, authentication, and access control) are considered the significant tools to ensure security of the system. The block chain is considered as the futuristic approach that remains the ideal solution for security issues in the IoE domain. Further, this research illustrates the intrusion detection model and advancement in the IDS using artificial intelligence. Moreover, this research exhibits the detailed representation of the origin and fundamentals of various optimization algorithms. This chapter finally presents some of the most productive optimization algorithms. However, it leaves a future scope to explore more advanced techniques like ensemble learning, transfer learning, federated learning, and encryption techniques to avoid the security issues in IoE environment.

REFERENCES

[1] Sergey, M., Nikolay, S., and Sergey, E., "Cyber security concept for Internet of Everything (IoE)," in proceedings of Signal Synchronization, Generating and Processing in Telecommunications (SINKHROINFO), pp. 1–4, 2017.

[2] Wu, X.W. and Jolfaei, A., "Securing the Internet of Everything through adaptive and resource-efficient mechanisms," IEEE Internet of Things Magazine, vol.3, no.4, pp.16–19, 2020.

[3] Khekare, G., "Internet of everything (IoE): intelligence, cognition, catenate," edicon Engineering Themes, vol.1, no.2, pp.31–32, 2021.

[4] Masoud, M., Jaradat, Y., Manasrah, A., and Jannoud, I., "Sensors of smart devices in the internet of everything (IoE) era: Big opportunities and massive doubts," Journal of Sensors, vol.2019, no.6514520, 2019.

[5] Ryoo, J., Kim, S., Cho, J., Kim, H., Tjoa, S., and DeRobertis, C., "IoE security threats and you," in proceedings of 2017 International Conference on Software Security and Assurance (ICSSA), pp.13–19, 2017.

[6] Petac, E. and Duma, P., "Some aspects of intrusion detection in IoE," Ovidius University Annals, Economic Sciences Series, vol.15, no.1, pp.595–599, 2015.

[7] Singh, P., Nayyar, A., Kaur, A., and Ghosh, U., "Blockchain and fog based architecture for internet of everything in smart cities," Future Internet, vol.12, no.4, p.61, 2020.

[8] Liu, Y., Dai, H.N., Wang, Q., Shukla, M.K., and Imran, M., "Unmanned aerial vehicle for internet of everything: Opportunities and challenges," Computer communications, vol.155, pp.66–83, 2020.

[9] Queralta, J.P., Gia, T.N., Tenhunen, H., and Westerlund, T., "Collaborative mapping with IoE-based heterogeneous vehicles for enhanced situational awareness," in proceedings of 2019 IEEE Sensors Applications Symposium (SAS), pp.1–6, 2019.

[10] Hiriyannaiah, S., Matt, S.G., Srinivasa, K.G., and Patnaik, L.M., "A Multi-layered framework for Internet of Everything (IoE) via wireless communication and distributed computing in industry 4.0," Recent Patents on Engineering, vol.14, no.4, pp.521–529, 2020.

[11] Sanchez, M., Exposito, E., and Aguilar, J., "Industry 4.0: Survey from a system integration perspective," International Journal of Computer Integrated Manufacturing, vol.33, no.10-11, pp.1017–1041, 2020.

[12] Mohammadian, H.D. and Rezaie, F., "The role of IoE-education in the 5th wave theory readiness & its effect on SME 4.0 HR competencies," in proceedings of 2020 IEEE Global Engineering Education Conference (EDUCON), pp.1604–1613, 2020.

[13] Mohanty, S.P., Yanambaka, V.P., Kougianos, E., and Puthal, D., "PUFchain: A hardware-assisted blockchain for sustainable simultaneous device and data security in the internet of everything (IoE)," IEEE Consumer Electronics Magazine, vol.9, no.2, pp.8–16, 2020.

[14] Wang, M., Zhou, Z., and Ding, C., "Blockchain-based decentralized reputation management system for Internet of Everything in 6G-enabled cybertwin architecture," Journal of New Media, vol.3, no.4, p.137, 2021.

[15] Mainetti, L., Patrono, L., and Vilei, A., "Evolution of wireless sensor networks towards the internet of things: A survey," in proceedings of SoftCOM 2011, 19th international conference on software, telecommunications and computer networks, pp.1–6, 2011.

[16] Noura, M., Atiquzzaman, M., and Gaedke, M., "Interoperability in internet of things: Taxonomies and open challenges," Mobile networks and applications, vol.24, no.3, pp.796–809, 2019.

[17] Pliatsios, A., Goumopoulos, C., and Kotis, K., "A review on iot frameworks supporting multi-level interoperability: The semantic social network of things framework," International Journal on Advances in Internet Technology, vol.13, no.1, pp.46–64, 2020.

[18] Asghari, P., Rahmani, A.M., and Javadi, H.H.S., "Service composition approaches in IoT: A systematic review," Journal of Network and Computer Applications, vol.120, pp.61–77, 2018.

[19] Farias da Costa, V.C., Oliveira, L. and de Souza, J., "Internet of everything (IoE) taxonomies: A survey and a novel knowledge-based taxonomy," Sensors, vol.21, no.2, pp.568, 2021.

[20] Siow, E., Tiropanis, T., and Hall, W., "Analytics for the internet of things: A survey. ACM computing surveys (CSUR)," vol.51, no.4, pp.1–36, 2018.

[21] Gao, J., Lei, L., and Yu, S., "Big data sensing and service: A tutorial," in proceedings of IEEE First International Conference on Big Data Computing Service and Applications (pp. 79–88). IEEE.

[22] Yaqoob, I., Ahmed, E., Hashem, I.A.T., Ahmed, A.I.A., Gani, A., Imran, M., and Guizani, M., "Internet of things architecture: Recent advances, taxonomy, requirements, and open challenges," IEEE wireless communications, vol.24, no.3, pp.10–16, 2017.

[23] Mahdavinejad, M.S., Rezvan, M., Barekatain, M., Adibi, P., Barnaghi, P., and Sheth, A.P., "Machine learning for Internet of Things data analysis: A survey," Digital Communications and Networks, vol.4, no.3, pp.161–175, 2018.

[24] Mon, A., Del Giorgio, H.R., De María, E., Querel, M., and Figuerola, C., "Evaluation of technological development for the definition of Industries 4.0," in proceedings of 2018 Congreso Argentino de Ciencias de la Informática y Desarrollos de Investigación (CACIDI), pp.1–6, 2018.

[25] Mehmood, E. and Anees, T., "Challenges and solutions for processing real-time big data stream: A systematic literature review," IEEE Access, vol.8, pp.119123–119143, 2020.

[26] Rehman, M.H., Liew, C.S., Wah, T.Y., and Khan, M.K., "Towards next-generation heterogeneous mobile data stream mining applications: Opportunities, challenges, and future research directions," Journal of Network and Computer Applications, vol.79, pp.1–24, 2017.

[27] Yaqoob, I., Hashem, I.A.T., Gani, A., Mokhtar, S., Ahmed, E., Anuar, N.B., and Vasilakos, A.V., "Big data: From beginning to future," International Journal of Information Management, vol.36, no.6, pp.1231–1247, 2016.

[28] Zhang, J., Chen, B., Zhao, Y., Cheng, X., and Hu, F., "Data security and privacy-preserving in edge computing paradigm: Survey and open issues," IEEE access, vol.6, pp.18209–18237, 2018.

[29] Nezami, Z. and Zamanifar, K., "Internet of Things/Internet of Everything: Structure and ingredients," IEEE Potentials, vol.38, no.2, pp.12–17, 2019.

[30] Smutný, P., "Different perspectives on classification of the Internet of Things," in proceedings of 17th International Carpathian Control Conference (ICCC), pp.692–696, 2016.

[31] Alsamani, B. and Lahza, H., "A taxonomy of IoT: Security and privacy threats," in proceedings of International Conference on Information and Computer Technologies (ICICT), pp.72–77, 2018.

[32] Mohamed, A., Najafabadi, M.K., Wah, Y.B., Zaman, E.A.K., and Maskat, R., "The state of the art and taxonomy of big data analytics: View from new big data framework," Artificial Intelligence Review, vol.53, no.2, pp.989–1037, 2020.

[33] Majeed, A., Bhana, R., Haq, A.U., and Williams, M.L., "Devising a secure architecture of internet of everything (IoE) to avoid the data exploitation in cross culture communications," International Journal of Advanced Computer Science and Applications, vol.7, no.4, 2016.
[34] Keoh, S.L., Kumar, S.S., and Tschofenig, H., "Securing the internet of things: A standardization perspective," IEEE Internet of things Journal, vol.1, no.3, pp.265–275, 2014.
[35] Miraz, M.H., Ali, M., Excell, P.S., and Picking, R., "A review on Internet of Things (IoT), Internet of everything (IoE) and Internet of nano things (IoNT)," in proceedings of Internet Technologies and Applications (ITA), pp.219–224.
[36] Majeed, A., "Developing countries and internet-of-everything (IoE)," in proceedings of 2017 IEEE 7th Annual Computing and Communication Workshop and Conference (CCWC), pp.1–4, 2017.
[37] Park, J., Kwon, H., and Kang, N., "IoT–Cloud collaboration to establish a secure connection for lightweight devices," Wireless Networks, vol.23, no.3A, pp.681–692, 2017.
[38] Pelton, J.N. and Singh, I.B., "Challenges and opportunities in the evolution of the Internet of Everything," Smart Cities of Today and Tomorrow, pp.159–169, 2019.
[39] Su, X., Riekki, J., Nurminen, J.K., Nieminen, J., and Koskimies, M., "Adding semantics to internet of things," Concurrency and Computation: Practice and Experience, vol.27, no.8, pp.1844–1860, 2015.
[40] Karimi, K. and Atkinson, G., "What the Internet of Things (IoT) needs to become a reality," White Paper, FreeScale and ARM, pp.1–16, 2013.
[41] Wang, H., Sun, L., and Bertino, E., "Building access control policy model for privacy preserving and testing policy conflicting problems," Journal of Computer and System Sciences, vol.80, no.8, pp.1493–1503, 2014.
[42] Desai, S., Alhadad, R., Chilamkurti, N., and Mahmood, A., "A survey of privacy preserving schemes in IoE enabled smart grid advanced metering infrastructure," Cluster Computing, vol.22, no.1, pp.43–69, 2019.
[43] Nadargi, A. and Thirugnanam, M., "Addressing identity and location privacy of things for indoor: Case study on Internet of Everything's (IoE)," in Proceedings of the 3rd International Symposium on Big Data and Cloud Computing Challenges (ISBCC–16'), pp.377–386, 2016.
[44] Bokhari, S., Hamrioui, S., and Aider, M., "Cybersecurity strategy under uncertainties for an IoE environment," Journal of Network and Computer Applications, vol.205, p.103426, 2022.
[45] Miao, Y., Liu, X., Choo, K.K.R., Deng, R.H., Wu, H., and Li, H., "Fair and dynamic data sharing framework in cloud-assisted internet of everything," IEEE Internet of Things Journal, vol.6, no.4, pp.7201–7212, 2019.
[46] Majeed, A., Haq, A.U., Jamal, A., Bhana, R., Banigo, F., and Baadel, S., "Internet of everything (IoE) exploiting organisational inside threats: Global network of smart devices (GNSD)," in proceedings of IEEE International Symposium on Systems Engineering (ISSE), pp. 1–7, 2016.
[47] Padhi, P.K. and Charrua-Santos, F., "6G enabled industrial internet of everything: Towards a theoretical framework," Applied System Innovation, vol.4, no.1, p.11, 2021.
[48] Garzia, F., Lombardi, M., and Ramalingam, S., "An integrated internet of everything—Genetic algorithms controller—Artificial neural networks framework for security/safety systems management and support," in proceedings of 2017 International Carnahan Conference on Security Technology (ICCST), pp. 1–6, 2017.

[49] Woolf, N., "DDoS attack that disrupted internet was largest of its kind in history, experts say," The Guardian, vol.26, 2016.
[50] Pongle, P. and Chavan, G., "Real time intrusion and wormhole attack detection in Internet of Things". International Journal of Computer Applications, vol.121, no.9, 2015.
[51] Peng, S.L., Pal, S., and Huang, L. eds., "Principles of internet of things (IoT) ecosystem: Insight paradigm," Springer International Publishing, pp.263–276, 2020.
[52] Singh, P., Nayyar, A., Kaur, A., and Ghosh, U., "Blockchain and fog based architecture for internet of everything in smart cities," Future Internet, vol.12, no.4, p.61, 2020.
[53] Fadhil, M., Owenson, G., and Adda, M., "Locality based approach to improve propagation delay on the bitcoin peer-to-peer network," in proceedings of 2017 IFIP/IEEE Symposium on Integrated Network and Service Management (IM), pp.556–559, 2017.
[54] Alrajeh, N.A. and Lloret, J., "Intrusion detection systems based on artificial intelligence techniques in wireless sensor networks," International Journal of Distributed Sensor Networks, vol.9, no.10, pp.351047, 2013.
[55] Alrajeh, N. A., Khan, S., Lloret, J., and Loo J., "Artificial neural network based detection of energy exhaustion attacks in wireless sensor networks capable of energy harvesting," Ad Hoc & Sensor Wireless Networks, vol.2013, pp. 1–25, 2013.
[56] Frank, J., "Artificial intelligence and intrusion detection: Current and future directions," in proceedings of the 17th National Computer Security Conference, vol.10, pp.1–12, 1994.
[57] Ali, M.H., Al Mohammed, B.A.D., Ismail, A., and Zolkipli, M.F., "A new intrusion detection system based on fast learning network and particle swarm optimization," IEEE Access, vol.6, pp.20255–20261, 2018.
[58] Debar, H., Dacier, M., and Wespi, A., "A revised taxonomy for intrusion-detection systems," in Annales des télécommunications, vol.55, no.7, pp.361–378, 2000.
[59] Garcia-Teodoro, P., Diaz-Verdejo, J., Maciá-Fernández, G., and Vázquez, E., "Anomaly-based network intrusion detection: Techniques, systems and challenges," Computers & Security, vol.28, no.1–2, pp.18–28, 2009.
[60] Balasaraswathi, V.R., Sugumaran, M., and Hamid, Y., "Feature selection techniques for intrusion detection using non-bio-inspired and bio-inspired optimization algorithms," Journal of Communications and Information Networks, vol.2, no.4, pp.107–119, 2017.
[61] Karaboga, D. and Akay, B., "A comparative study of artificial bee colony algorithm," Applied Mathematics and Computation, vol.214, no.1, pp.108–132, 2009.
[62] Dorigo, M. and Blum, C., "Ant colony optimization theory: A survey," Theoretical Computer Science, vol.344, no.2–3, pp.243–278, 2005.
[63] Li, X.L., "An optimizing method based on autonomous animats: Fish-swarm algorithm," Systems Engineering-Theory & Practice, vol.22, no.11, pp.32–38, 2002.
[64] Yang, X.S., "Firefly algorithms for multimodal optimization," in proceedings of International Symposium on Stochastic Algorithms, pp.169–178, 2009.

5 Enhancement of Security of Messages in Blockchain-Based IoE with Modified Proof-of-Authentication (MPoAh)

Narendra Kumar Dewangan and Preeti Chandrakar

CONTENTS

5.1 INTRODUCTION

Blockchain is a technology that is immutable, transparent, quick, distributed, and decentralized. Blockchain ensures the non-repudiation and authenticity of the network's peers. The peers can transfer data from one peer to another directly using peer-to-peer technology. The blockchain maintains an immutable database for each transaction in the blockchain's form and chained hashes, which are special data structures. Blockchain is a mixed concept of database, data security, digital signature, storage, and networking. Blockchain is a decentralized, distributed, transparent, and end-to-end data storage system. It provides a facility of immutability, transparency, verifiability, authentication, and integrity. Blockchain technology enables distributed ledgers, a decentralized control structure, and peer-to-peer data

DOI: 10.1201/9781003366010-7

transactions. Using a cryptographic hash function, digital signatures, and encryption methods ensures privacy and anonymity in blockchain technology. Each transaction is recorded in the blockchain technology database, and transparency is maintained by displaying all transactions in the explorer. Every node has a copy of every block and transaction. Miners play an important role in the blockchain because they approve transactions and blocks. Miners are powerful nodes in the blockchain, and all nodes are powerful [1]. In this chapter, we are going to discuss the basic properties of the blockchain technology: the basics of the consensus algorithms in detail. Blockchain has the flexibility to solve various real-time problems, such as transparency and immutability. The Internet of Things is defined as the devices connected to the internet services to enhance the quality of services to humankind. The Internet of Everything is defined as the people, processes, and devices connected to the internet in synchronization with each other. The dependable communication devices share the messages between each other and send them to the cloud servers for processing and saving in the distributed database in the blockchain. These smart homes can help with any kind of accidental prevention or patient monitoring in the blockchain network. In [2], a patient feedback-based physical selection is proposed, which can be improved by using the IoE with IoMT devices in the smart home environment.

Inter-message communications between the IoE devices and verification of the transactions carried out by these devices need to be fast and authenticated. A device's action can depend on the other device's message. For example, an open window message for a smart home can be input data for the temperature sensor in the room. If the temperature outside is lower than the inside and a normal temperature must be maintained inside the room, then the window sensor has to send a message to close the window. In the second condition, if the room temperature is lower than normal, the automatic room heater can be switched on to maintain the room temperature at a normal level. IoE devices are very low-powered devices, and a processor system is not available in all IoE devices. So, these types of devices are required to process third-party data in the cloud environment of higher storage and processing units. This chapter includes the authentication between different dependable devices in the IoE environment and their connection to the processing unit. The security and privacy of the data transmitted through these devices are enhanced using the Ed25519 curve. The goals of the proposed scheme are as follows:

- Enhancing the security of the devices connected to each other in an IoE environment.
- Design of the secure algorithm for intercommunication between IoE devices.
- Design of the Modified Proof-of-Authentication (MPoAh) as blockchain consensus to approve transactions and blocks for IoE devices.
- Analyzing the security risks involved in this communication process.
- Validating the security and privacy of the messages.

The rest of the chapter is organized as the related works in Section 5.2. The proposed scheme of the MPoAh and security authentication protocol is discussed in Section 5.3. Section 5.4 is the implementation of the proposed scheme with security input and verification. Section 5.5 describes the results and comparison of the proposed scheme with the previously developed systems. Section 5.6 concludes the chapter with future possibilities and limitations and the references follow.

5.2 RELATED WORKS

Vivekanandan et al. [3] designed an authentication protocol for device-to-device communication in the IoT environment. Their system is specially designed for the Smart City Applications using 5G Technology. They also provide a random oracle model and security analysis of the proposed system using the proverif tool. Unial et al. [4] presented a practical approach to integrate federated learning (FL) with blockchain. They used fuzzy hashes to detect anomalies and variations in FL-trained models against poisoning attacks. Yang and Wang [5] designed system managing energy with privacy preservation in the smart homes and IoT devices. They also applied blockchain technology and smart contract to support holistic TEM system. They used modified PBFT (Practical Byzantine Fault Tolerance) consensus. Prakasam et al. [6] provide a brief introduction of 6G technology with the IoE device and their application in a real-time scenario. Siriweera and Naruse [7] presented a model for the internet of cross-chains driven by privacy, security, interoperability, and scalability (Chain as a service). Dhiman et al. [8] presented a survey of 6G technology with the challenges of implementation of IoE in 6G. They also discuss the architectural requirements and industrial challenges. A systematic review of recent research studies in BaaS models was presented by Li et al. [9]. The primary goal of this review is to categorize the applied scenarios, trends, evaluated Quality of Service (QoS) factors, new challenges, and open directions in IoT management. Yang et al. [10] proposed a synchronization engine with a formal workflow that is used at mobile and fixed-edge gateways to generate, validate, and synchronize data blocks with the cloud. A real-life case study on vaccine logistics is presented to validate our proposed approach, with results. Chai et al. [11] presented a blockchain named CyberChain to reduce cost and storage in the vehicular network with the privacy maintenance. They used DPBFT consensus algorithm. They also provide a lightweight cyber consensus algorithm with privacy-preserving smart contract. In order to investigate the use of blockchain as middleware between different participants in intelligent transportation systems, a blockchain-based solution for establishing secure payment and communication (PSEV) was proposed by Jabbar et al. [12]. Their results of computational tests revealed that the developed solution is faster and more scalable than the existing solutions. Manoharan et al. [13] proposed a novel idea for combining a privacy-preserving blockchain framework with a 6G communication network. They also proposed a blockchain radio access network integrated system as a reliable model for 6G networking based on blockchain technology with improved efficiency and security. Hossein et al. [14] proposed architecture for allowing data owners to define desired access policies for their sensitive healthcare data. BCHealth is made up of two distinct chains that store access policies and data transactions. They use clusters to address real-world problems. They also provide a node failure algorithm. Rodrigues et al. [15] present analyses of the three basic architecture of the blockchain with IoT applications. Analytic queuing models, simulations, and theoretical discussion are used to evaluate performance. The results show that when the blockchain system is properly configured, efficiency and security requirements can be met, primarily by observing a trade-off between data integrity and response time. Wang et al. [16] suggested a multi-server edge computing architecture-based ITS system for authentication using blockchain technology. This handover authentication scheme enables the authenticated server to assist users in subsequently authenticating with other servers, allowing users to interact with the server at any time and from any location with minimal overhead. Mehdinejad et al. [17] design a p2p blockchain for energy trading in a small-scale demand response system. All participants in the proposed market can conduct bilateral energy token transactions (P2P) at agreed-upon prices with each other and with the retail market by using smart contracts. Lee at al. [18] provided triple layer encryption to the blockchain transaction for the patient data or the EMR. For encryption, AES and RSA are used. A high-level architecture for a blockchain-based Internet of Things network that encourages shared mobility by combining car-sharing and leasing was proposed by Auer et al. [19]. This work also highlights how, in car-sharing platforms, the design of such an integrated platform hinges on striking the correct balance between the fundamental design principles of security and privacy, authenticity, traceability and reliability, scalability, and interoperability. Using the Hyperledger fabric to implement. Ye at al. [20] proposed a trust-centric privacy-preserving blockchain for DSA in IoT networks. To be specific, we propose a trust evaluation mechanism to evaluate the trustworthiness of sensing nodes and design a Proof-of-Trust (PoT) consensus mechanism to build a scalable blockchain with high transaction per second (TPS). Yang et al. [21] proposed a certificateless cryptosystem to encrypt keywords, which solves the certificate management problem and key escrow problem.

5.3 PROPOSED SCHEME

The proposed scheme is divided into three parts. The first section deals with device registration on the blockchain and device identity generation. The second part is about device intercommunication and dependable communication. The third part discusses the modified PoAh algorithm for the transaction verification approval and the block creation process.

5.3.1 Device Identity and Devices Registration Process

The term smart included a variety of devices with trustworthy characteristics. Let the types of devices in the smart home be n and the total number of devices be m, where $m>n$. The total number of dependable devices is x, where $x<m$. The total number of identity generations required for blockchain device registration is m. So, for each device, we take the basic device identifier, where k is the device and defined as $k=m(n)$. To generate the identity of the device, the ***Ed25519*** curve is used, and the private key of the device is generated as.

$$PrivK_{Dev_k = Ed25519(hash256(R_{num}).hash256(Dev_k).Tm)} \quad (1)$$

To generate the public key of the device, we applied the Ed25519 multiplication property, and the generated public key is

$$PubK_{Dev_k = Ed25519(PrivK_{Dev_k}.G_p)} \quad (2)$$

This is for the one device where the registration keys are stored in the local data storage of the device, which is not to be shared with any other devices. Now, for the registration of the devices with the cloud node of the blockchain, a home identifier is required, which is a combined identity of the devices available in the home. It is required only once for the identification of the home. The home identification is generated using the following formula for the smart home:

$$ID_{H_j} = \sum_{i=1}^{n} PubK(Dev_i) \quad (3)$$

To register a device with the cloud node, the registration request send messages such as $M_1 = ID_{H_i}, PubK_{Dev_k}, T_s >$ (here T_s is timestamp of current message). This message is received by the cloud node (CN_y), then it is signed by the private key of the device and shared with CN_y. It is verified using the Ed25519 verification algorithm and the new timestamp T_s' is calculated for authentication of the message M_1'. After the device authentication CN_y sends a message $M_2 = RegID_{Dev_k}, PubK_{Dev_k}, T_s >$. This message contains the registration ID of the device. Like this all the other available devices are registered with the CN_y. The registration ID of all devices are stored in the distributed database. Each node in the network has the copy of all the registered nodes with the CN_y. The overall proposed architecture is illustrated in Figure 5.1. The registration phase of the devices is illustrated in Figure 5.2. Notations used in this chapter are described in Table 5.1.

5.3.2 Device Intercommunication and Dependable Communications

Many devices depend on each other. When a communication is passing by the device to the CN_y and it requires the message from the other device to complete the communication, it sends the message $M_3 = Data_{Request}, PubK_{Dev_k}, T_s >$ to the device. In this message, data and public ID of the device are included. The device communicates with the $PubK_{Dev_k}$ and send the query's response and the device's current condition to the CN_y. The message send by the device is formulated as

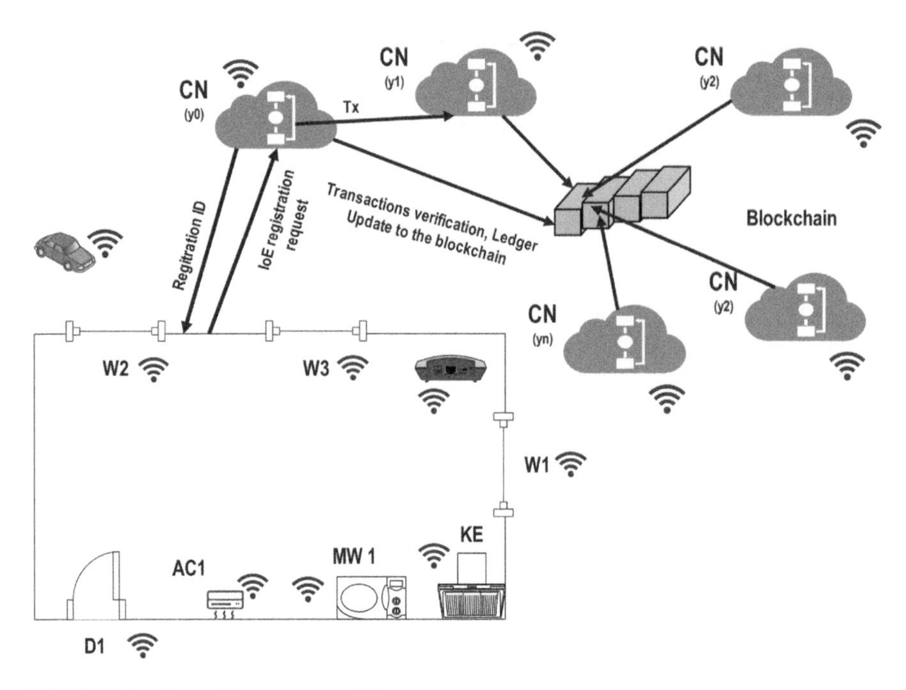

FIGURE 5.1 Overall proposed architecture.

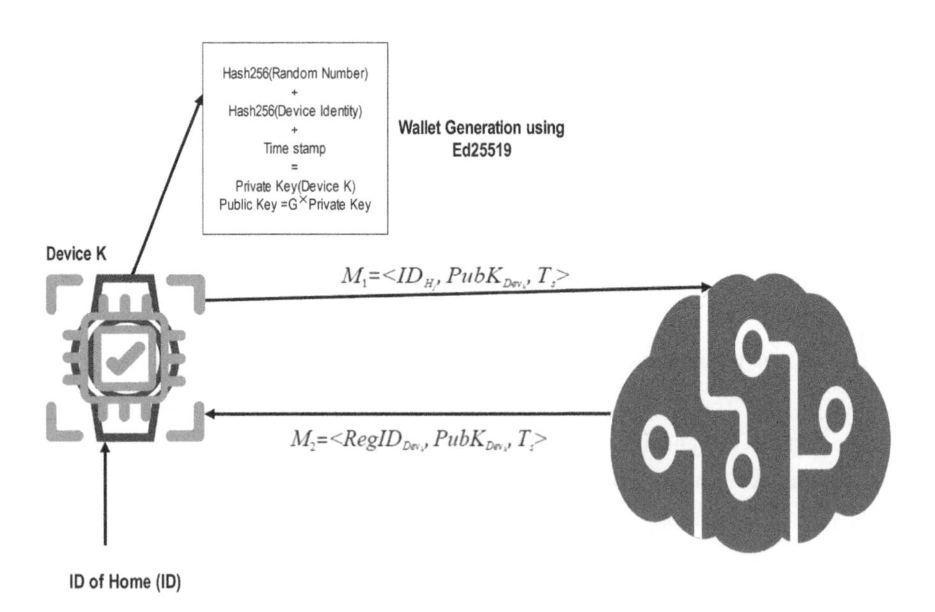

FIGURE 5.2 Registration of devices with Cloud Node (CN).

TABLE 5.1
Symbols used in this chapter

S.N.	Symbol	Description
1	$PrivK_x$	Private key of x
2	$PubK_x$	Public key of x
3	$Ed25519(x)$	Ed25519 curve algorithm for x
4	ID_x	Identity of x
5	$hash256(x)$	Hashing of x with SHA-256
6	$Data_{Request}$	Data request
7	$Data_{Reply}$	Data reply
8	T_x	Time stamp x
9	M_x	Message number x
10	CN_y	Cloud node y
11	$RegID_{Dev_k}$	Registration ID of device k
12	ΔT	Transaction life time

$M_4 = Data_{Reply}, PubK_{Dev_k}, T_{new}, CN_y >$. This message is verified by the CN_y. After verification of the message. CN_y processes data according to the reply and send the data to the other cloud node in the form of transaction. This transaction is verified at by the miner cloud nodes. Verification of transaction and MPoAh is discussed in subsection C. Now there is device dependency. If more than two devices are dependent in each other, then the cloud sends the requested data to all the dependent devices together and the reply is combined in the following formulae:

$$Data_{Reply} = \sum_{n=1}^{i} Data_{ReplyDev_i} \tag{4}$$

The device dependency can be seen in Table 5.3 and the request-reply system is illustrated in Figure 5.3.

5.3.3 Modified Proof-of-Authentication (MPoAh) Algorithm and Transaction Verification

Algorithm 5.1: Modified Proof-of-Authentication (MPoAh) and Transaction Verification
Input: Data from Devices to Cloud Node and Cloud Node to another Cloud Node.
Output: Verified and Approved Transaction to be added into blockchain.

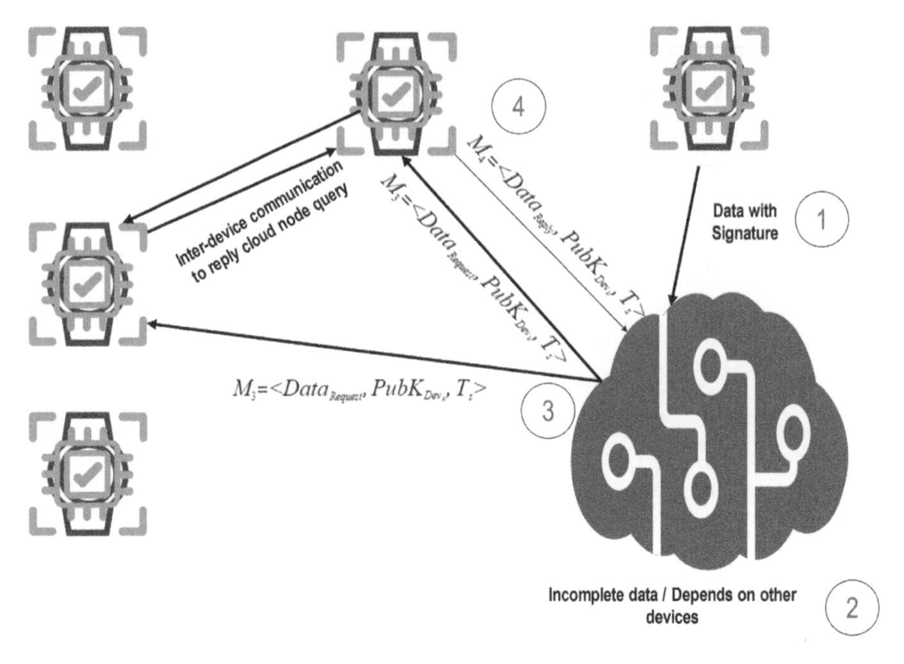

FIGURE 5.3 Inter-device communication, request and reply by the dependent devices.

Step-1: Check message sender
Step-2 If (Sender == Device and Receiver == Cloud Node) Then
    //If message sender are devices
    CN{Verify=Ed25519$(\text{Message})_{\text{PrivK}}$(Device)}
    // Signature on message is verified by the receiver cloud node and added in the verified transaction pool of current block
  Else If (Sender= CN and Receiver=CN) Then
    Verify(Ed25519$(\text{Message})_{\text{PrivK}}(CN_y)$)
    Calculate

$$\Delta T = \frac{Tx_{maximum} - Tx_{minimum}}{Tx_{minimum}}$$

{
  If(Number of Verification == $\frac{z-1}{3}$ in ΔT) Then
    Transaction is verified
}

```
            End IF
        End IF
    Step-3: If (Transaction == Verified) Then
            Block= Block+1;
        End If
```

The transaction is in the form of the messages and data in the proposed scheme. The cloud nodes send messages to the devices, these messages are also recorded by the blockchain. The messages $< M_1, M_2, M_3, M_4 >$ are considered as the transaction on the blockchain and verified by the receiving node on the cloud. If the messages are sent by one cloud node, then is verified by the other peer cloud nodes. Total number of the verification required for a verified transaction is (z-1)/3 where the z is the total number of active cloud node in transaction lifetime ΔT. The transaction lifetime is calculated as the difference between minimum transaction approval time and maximum transaction approval time.

$$\Delta T = \frac{Tx_{maximum} - Tx_{minimum}}{Tx_{minimum}} \tag{5}$$

After verification of authentication in time-based miner verification, the transaction is added in the current unmined block. The block is auto verified in each ΔT. The algorithm is shown in Algorithm 5.1.

5.4 IMPLEMENTATION OF THE PROPOSED SCHEME

As in the proposed scheme, the implementation part is also divided into three parts. The first part is node creation, device registration, and channel creation for the IoE devices. The second part is about the device intercommunication and data-sending process. The third part is transaction sending and approval with ledger updating. The system requirements for the implementation are discussed in Table 5.2.

TABLE 5.2
System requirement for the proposed system

S.N.	Hardware/Software	Details
1	System Configuration	AMD Athlon Processor, 12 GB RAM, 500 GB HDD
2	Operating System	Ubuntu 22.04
3	Front End	Python
4	Data Storage	MongoDB
5	Cryptographic Tools	SHA-256, EdDSA (ed25519)
6	IoE Simulation	Bevywise IoT Simulator, MQTT tool
7	Protocol Verifier Tool	Scyther

TABLE 5.3
IoE devices used for the simulation with dependent devices

S.N.	Device	ID	Dependent On	Description
1	Door Sensor	D1, D2, D#	NA	Door open or close
2	Window Sensor	W1, W2, W#	Temperature Sensors	Windows open or close
3	Air Conditioner	AC1, AC2, AC#	Smoke Sensors, Smart Electric Meters	Controlling temperature inside house
4	Room Heaters	RH1, RH2, RH#	Smoke Sensors, Smart Electric Meters	Controlling temperature inside house
5	Entertainment Sets	ET1, ET2, ET#	Smoke Sensors, Smart Electric Meters	For entertainment like Alexa
6	Kitchen Machines	KM1, KM2, KM#	Smoke Sensors, Temperature Sensors	Active and Non-active status of Machines
7	Cars	CR1, CR2, CR#	Weather Sensors	Car locations tracking
8	Internet of Medical Things	IoMT1, IoMT2, IoMT#	Smoke Sensors, Temperature Sensors, Weather Sensors	Recording Medical data of the human body

A. For IoE devices, we use the Bevywise IoT simulator. The Bevywise IoT Simulator is a GUI-based data simulation tool which is used to load and test the MQTT/IoT application to know the performance level of that application. In the IoT simulator, you can create a simulated IoT network and device, publish events with simple or complex JSON payloads, set up subscriptions, and more. You may learn how to mimic IoT networks and IoT devices from the IoT simulator help page. This creates 15 sensors of eight types of devices. Which are sending data to the MQQT server as the cloud server for these devices. This cloud server works as node and sends data to the other cloud server as transactions. The total devices used in the simulation are described in Table 5.3.

For registration of the devices, the key generated by the Ed25519 chilkat library is used. Example of the key generation process is shown below:

Device ID = AC1 (74d3d0716d5a45fdd79209d4f0496a0e1f6e9b8a0462a93aa75498270b87ee33)
Private Key = (74d3d0716d5a45fdd79209d4f0496a0e1f6e9b8a0462a93aa75498270b87ee33.
5a45fdd79209d4f0496a0e1f6e9b8a0462a6576980757654765aedfbc. 675347624167838213)
=510e17df8e6611b37fd7ad367e3ecdc8c88f70391fccf28a2b96ab53aef66106
Public Key = 374d3d0716d5a45fdd79209d4f0496a0e1f6e9b8a0462a93aa75498270b899e8

B. Device inter-device communication is implemented using the Python-based decision-making process. In the message setting of the IoT device, it is already mentioned that when a device needs input from another device, the cloud node awaits the dependent device's response.
C. The transactions are as hashes from one cloud server to the other, and the blockchain ledger is maintained by each cloud node. The MPoAh algorithm is working between the devices and the cloud server to maintain the proper transaction and authentication process. The transaction formats and approved transaction formats are shown below:

"id": "1",

"Transaction": "bef77f243602982a4ed260bff45\\f84a91d0eef119674d5f0923 0eb8e8bcefaed",

"Status": "Unapproved",

"To": "11bd2180c5cc001d3dde10434bd678b756cce83bc1ec508a01b18f7d95 24b31e",

"From": "99c77062235edb0353b9ccca83b676f88adaba94172eecfe8174a655a 82a2fe7",

"Data": "Active AC 25, Door open",

"Signature":"c213ddc5d8543b264b35f6487fca080acd254814cc14233b63ee6 ef1e34985c7f9fc7c2b2639b74d5499753dfa10a339f0c1fef851f0f326f706587d dadd4509",

"Time Stamp": "2022-07-15 10:01:25",

"Mining Status": "Unmined"

5.5 SECURITY ANALYSIS, RESULTS, AND COMPARISONS

This part discusses the security analysis of the proposed system regarding different attacks on the proposed system. Results and comparison sections discuss the experimental results of the proposed system as well as a comparison with previously developed systems.

5.5.1 Analysis of Security

Security analysis of the proposed system discusses the false device, delay transaction, man-in-the-middle attacks, and so on.

Security Analysis: The proposed scheme has been implemented in the cloud-based public blockchain. Devices can send data to the cloud node and the node can process this as a transaction to the blockchain. These processes need security in an open channel for data transmission. The system is vulnerable to the following attacks, and the following measures can be taken to prevent them:

- **False devices**: If any devices that are not member of the IoE network for the cloud node and want to send false data to the cloud node, the authentication and registration data is required for accepting and verifying the data sent by the device. In dependable communication type devices, fake devices failed to verify inter-devices' communication authentication. So attacks from fake or non-IoE devices fail in the proposed scheme.
- **Delay Transactions Approval:** If any cloud node causes a delay in transaction approval during the transaction's lifespan. This condition causes a delay in the processing and recording of the next transaction on the blockchain. To handle this attack, a consensus algorithm is designed in a way that the transaction can be approved by the other available cloud nodes in the network. Since it is an automatic process, then there is no need to worry about the delay transaction approval in the proposed scheme.
- **Man–in-the–middle attack:** The attacker in this attack has the ability to listen the cloud node's transaction and reveal it. Let adversary **A** is attacker in the network and has the ability to sniff out the data send by node CN_y at timestamp T_m. In this scenario, every value is composed with hash in the form of 256-bit hash. So the device and cloud node identity are safe using hash functions. Like this, **A** wants to attack on the data send by the device to the cloud node, sniffs out the data and want to use for any other purposes. It is not possible because of the hash form of the transaction and authentication of the data sent by one device to the other. In some conditions, the device data is incomplete and useless for the attacker until it has received data from other dependent devices.
- **Authentication:** For the authentication of device and cloud node the registration and device registration are done with the message exchanges and key exchange algorithms. Whenever data is sent from the device to the cloud node and cloud node to another cloud node, its signature is verified and if verification is successful, then transaction and blocks are approved to be added to the chain.
- **Integrity:** In the data sending and message communication process, nodes are verification points. So the transaction including the data and messages has a registration ID generated for the cloud node. The transaction is hashed using the SHA256 algorithm, and the approver point hash is included in the Markle tree. This tree is included in the block in the last stage. Therefore, these are the methods to determine a candidate's vote, and with each approval, a hash is formed and its integrity is checked.
- **DDoS Attack:** Allow adversary A to launch a DDoS attack against the cloud node in order to disrupt the network from remote data transactions. Because of this attacker, the cloud nodes were swamped with meaningless requests. At this point, the blockchain network closes the cloud node system as this is distributed network and decentralized. Then the cloud node moved to the new system with the same identity. DDoS attacks will be restricted from the cloud node in this

```
1   const exp: Function; const hash: Function; hashfunction h; const XOR:Function;
2   const h1:Function; const plus:Function;const mod:Function;
3   protocol devicelogin(Dev,CN)
4   {
5   role Dev {
6   const RegIDi, Pi, Bi, SIDk,k,s,b,g,p;
7   fresh N1: Nonce;
8   var N2: Nonce;
9   macro b = h(b);
10  macro k = mod(exp(g,b),p); macro Pij = h(XOR(XOR(h(RegIDi), h(h(RegIDi),
11  h(s))), plus(h(N2),1)),h(RegIDi,k)); macro Lij = XOR(h(N2) , h(RegIDi,k));
12  send_1(Dev,CN, XOR(h(SIDk,N1), h(RegIDi)),N1);//C1
13  recv_2(CN, Dev, XOR(h(N2), h(N1,1)));//C2
14  send_3(Dev,CN,h(XOR(XOR(h(RegIDi),
15  h(h(RegIDi),
16  h(s))),
17  plus(h(N2),1)),h(RegIDi,k)));//Pij
18  send_4(Dev,CN,Lij);
19  claim_i1(Dev, Secret, XOR(h(SIDk,N1), h(RegIDi)));//C1
20  claim_i2(Dev,Secret,XOR(h(N2), h(N1,1)));//C2
21  claim_i3(Dev,Secret,h(XOR(XOR(h(RegIDi),
22  h(h(RegIDi),
23  h(s))),
24  plus(h(N2),1)),h(RegIDi,k)));//Pij
25  claim_i4(Dev, Secret, h(s)); claim_i10(Dev,Secret,k); claim_i11(Dev,Secret,h(RegIDi));
26  claim_i12(Dev,
27  Secret,N1);
28  claim_i5(Dev,
29  Secret,
30  h(N2));
31  claim_i13(Dev,Secret,Lij);
32  claim_i6(Dev,
33  Niagree);
34  claim_i7(Dev,Nisynch);claim_i8(Dev,
35  Alive);claim_i9(Dev,Weakagree);
36  claim_i10(Dev, Commit, CN,N1,N2);
37  }
38  role CN{
39  const RegIDi,Pi,N2,Bi, SIDk,k,s,b,g,p;
40  var N1:Nonce; fresh N2: Nonce;
41  recv_1(Dev,CN, XOR(h(SIDk,N1), h(RegIDi)),N1);//C1
42  send_2(CN, Dev, XOR(h(N2), h(N1,1)));//C2
43  recv_3(Dev,CN,
44  h(XOR(XOR(h(RegIDi),
45  h(h(RegIDi),
46  h(s))),
47  plus(h(N2),1)),h(RegIDi,k)));//Pij
48  recv_4(Dev,CN,Lij);
49  claim_r13(CN, Secret,Lij); //Lij
50  claim_r1(CN,  Secret, XOR(h(SIDk,N1), h(RegIDi)));//C1
51  claim_r2(CN,  Secret, XOR(h(N2), h(N1,1)));//C2
52  claim_r2(CN,
53  Secret,
54  h(XOR(XOR(h(RegIDi),
55  h(h(RegIDi),
56  h(s))),
57  plus(h(N2),1)),h(RegIDi,k)));//Pij
58  claim_r3(CN,
59  Secret,h(s));
60  claim_r10(CN, Secret,k);
61  claim_r4(CN, Secret,h(RegIDi));
62  claim_r5(CN,  Secret, h(N2)); claim_r6(CN , Alive); claim_r7(CN, Niagree);
63  claim_i10(CN,  Running, Dev,N1,N2);claim_r8(CN, Nisynch);
64  claim_r9(CN,  Weakagree);
65  }}
66
```

FIGURE 5.4 Scyther input for login verification of device registration with cloud node.

manner. If a DDoS attack is launched in an insecure internet network, only approved devices can send data to cloud nodes, and the system is restarted automatically after a one-time restart if DDoS halting is detected. Security verification and verification results are shown in Figure 5.4 and 5.5. This verification is done by Scyther.

Claim				Status	Comments
devicelogin	Dev	devicelogin,i1	Secret XOR(h(SIDk,N1),h(RegIDi))	Ok	No attacks within bounds.
		devicelogin,i2	Secret XOR(h(N2),h(N1,1))	Ok	No attacks within bounds.
		devicelogin,i3	Secret h(XOR(XOR(h(RegIDi),h(h(RegIDi),h(s))),plus...	Ok	No attacks within bounds.
		devicelogin,i4	Secret h(s)	Ok	No attacks within bounds.
		devicelogin,i10	Secret mod(exp(g,h(b)),p)	Ok	No attacks within bounds.
		devicelogin,i11	Secret h(RegIDi)	Ok	No attacks within bounds.
		devicelogin,i12	Secret N1	Ok	No attacks within bounds.
		devicelogin,i5	Secret h(N2)	Ok	No attacks within bounds.
		devicelogin,i13	Secret XOR(h(N2),h(RegIDi,mod(exp(g,h(b)),p)))	Ok	No attacks within bounds.
		devicelogin,i6	Niagree	Ok	No attacks within bounds.
		devicelogin,i7	Nisynch	Ok	No attacks within bounds.
		devicelogin,i8	Alive	Ok	No attacks within bounds.
		devicelogin,i9	Weakagree	Ok	No attacks within bounds.
		devicelogin,Dev1	Commit CN,N1,N2	Ok	No attacks within bounds.
	CN	devicelogin,r13	Secret XOR(h(N2),h(RegIDi,mod(exp(g,h(b)),p)))	Ok	No attacks within bounds.
		devicelogin,r1	Secret XOR(h(SIDk,N1),h(RegIDi))	Ok	No attacks within bounds.
		devicelogin,r2	Secret XOR(h(N2),h(N1,1))	Ok	No attacks within bounds.
		devicelogin,CN1	Secret h(XOR(XOR(h(RegIDi),h(h(RegIDi),h(s))),plus...	Ok	No attacks within bounds.
		devicelogin,r3	Secret h(s)	Ok	No attacks within bounds.
		devicelogin,r10	Secret mod(exp(g,h(b)),p)	Ok	No attacks within bounds.
		devicelogin,r4	Secret h(RegIDi)	Ok	No attacks within bounds.
		devicelogin,r5	Secret h(N2)	Ok	No attacks within bounds.
		devicelogin,r6	Alive	Ok	No attacks within bounds.
		devicelogin,r7	Niagree	Ok	No attacks within bounds.
		devicelogin,r8	Nisynch	Ok	No attacks within bounds.
		devicelogin,r9	Weakagree	Ok	No attacks within bounds.

FIGURE 5.5 Result of device registration verification using Scyther.

5.5.2 Comparisons and Results

To prove the strength of our proposed scheme, we compared it with the previously developed systems. The comparison in Table 5.4 shows the potential of our proposed system over the previously developed system. In the resulting point of view, we discuss the following points as the strength of our proposed system:

A. **Storage**: In terms of storage, the suggested system storage utilized blockchain for key storage and required storage of 74 bits for device data and 74 bits for home id and cloud node keys: 75 bits of storage are required when data is transferred to the cloud node. Figure 5.6 depicts a comparison of previously

TABLE 5.4
Comparison between previously developed systems and proposed system

Property	[3]	[5]	[10]	[11]	Proposed system
Blockchain Used	Yes	Yes	Yes	Yes	Yes
Platform	NA	Quorum	NA	NA	Python-based with MongoDB
Security	Yes	No	No	Yes	Yes
Privacy	Yes	No	No	No	Yes
Device Verification	Yes	No	No	Yes	Yes
IoE	Yes	No	Yes	No	Yes
Public /Private	NA	Private	NA	Private	Public
Consensus algorithm	NA	PBFT	NA	Diffused Practical Byzantine Fault Tolerance (DPBFT)	Modified Proof-of-Authentication

Note: Yes – feature supported, No – feature not supported by system, NA -Feature not available

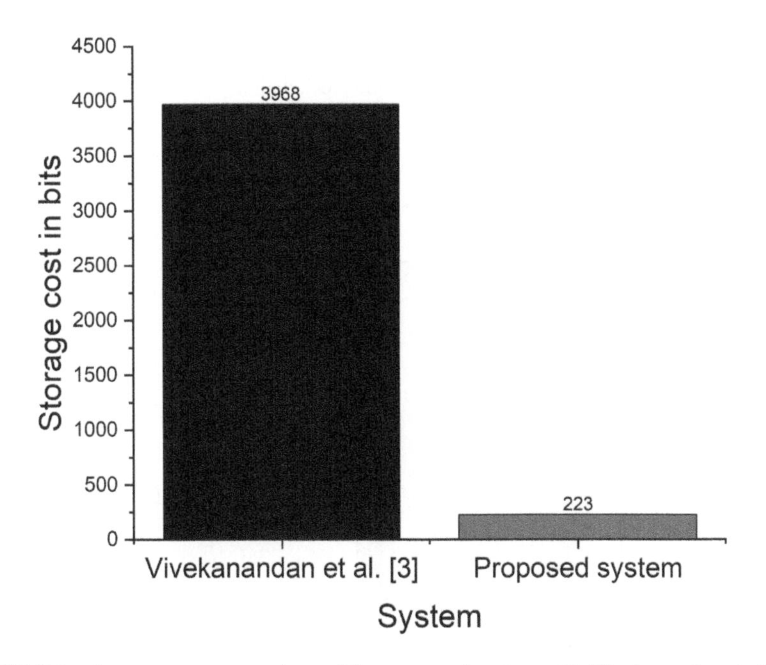

FIGURE 5.6 Storage cost comparison of the proposed system with Vivekanandan et al. [3].

produced system storage and the proposed system storage. The total amount of data storage required is 223 bits.

B. **Cost:** The proposed system is implemented using the customized Python-based blockchain with the MPoAh. A one-time installation cost is required for this blockchain and no transaction cost is needed for the data sending to the

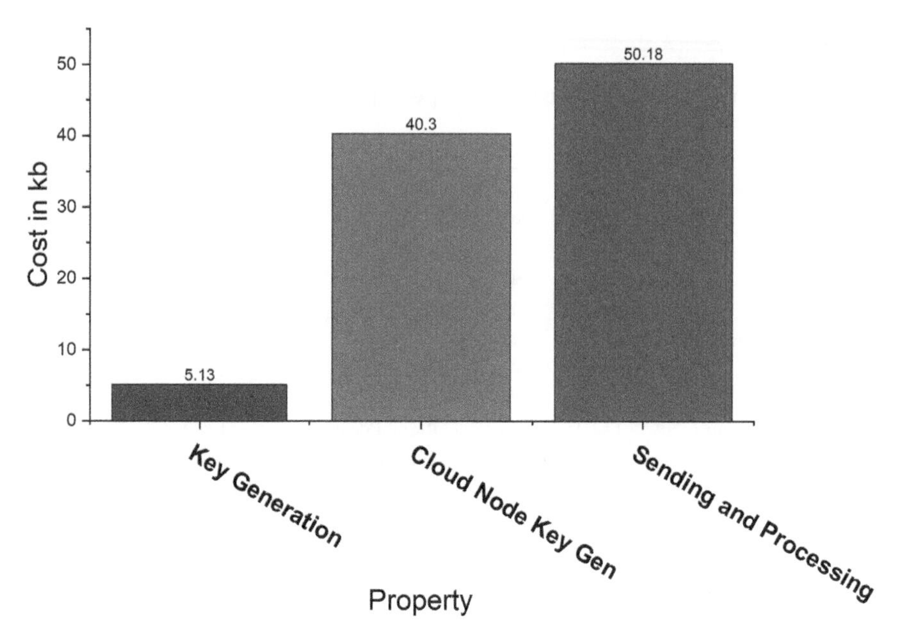

FIGURE 5.7 Cost of different operations in the blockchain.

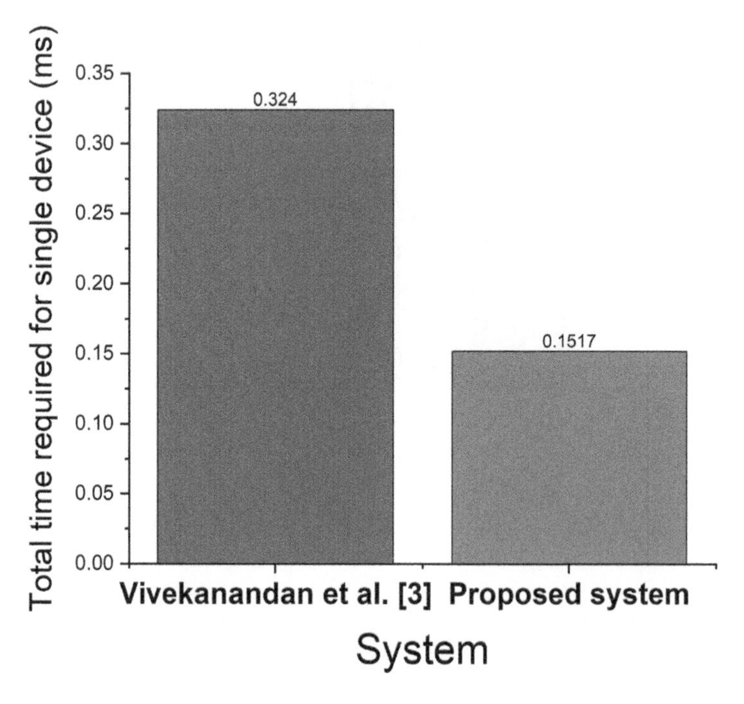

FIGURE 5.8 Total time required for a single device setup compared with Vivekanandan et al. [3].

cloud node and transacted to the other cloud node. The cost of the developed system is shown in Figure 5.7. For setting up a system, the running script requires 5.13 kb for key generation of device registration, 40.3 kb for key generation of cloud node, and 50.18 kb for data sending and processing. The total cost of script installation is 95.61 kb.

C. **Time for transactions**: The time required for the key generation per voter is 0.09245 ms. The time required for registration of a device is 0.0142 ms. The time required for a transaction from the cloud node is 0.0451 ms. The total time required for a single device through the complete process is 0.15175 ms. Since we are using fifteen devices, the total run time for the simulation is 2.27625 ms. In previously developed systems time required for the whole process is compared in Figure 5.8.

D. **Hash bits required:** For key generation, SHA256 and EdDSA is used for this 512bit required. For registration, 544 bits are required for per message. Total four messages are required at maximum for per run of the device, so the minimum messages hash bit costs is 2176 bits.

5.6 CONCLUSION AND FUTURE WORK

In the proposed scheme, a secure messaging system for IoE devices with the blockchain is implemented. These devices have features for inter-device communication and recording of the data in the blockchain public ledger. The key generation process, device registration process, authentication of messages and transactions, and verification of signatures are done using the Ed25519 algorithm. Since our proposed system is novel and based on IoE and blockchain, we have only a few papers for the comparison section. But according to the comparisons, our proposed system is better in terms of device privacy, security, and security of the messages. Our proposed scheme also provides security analysis and verification of the security protocol proposed in this chapter. In future goals, we can use these schemes in the e-governance schemes to control the different services and manage data in the blockchain-based e-governance system.

REFERENCES

1. Shari, Nur Fadhilah Mohd, and Amizah Malip. "State-of-the-art solutions of blockchain technology for data dissemination in smart cities: A comprehensive review." *Computer Communications* (2022).
2. Dewangan, Narendra Kumar, and Preeti Chandrakar. "Patient feedback based physician selection in blockchain healthcare using deep learning." In *Advanced Network Technologies and Intelligent Computing: First International Conference, ANTIC 2021, Varanasi, India, December 17–18, 2021, Proceedings*, pp. 215–228. Cham: Springer International Publishing, 2022.
3. Vivekanandan, Manojkumar, and Srinivasulu Reddy U. "BIDAPSCA5G: Blockchain based Internet of Things (IoT) device to device authentication protocol for smart city applications using 5G technology." *Peer-to-Peer Networking and Applications* 14 (2021): 403–419.

4. Unal, Devrim, Mohammad Hammoudeh, Muhammad Asif Khan, Abdelrahman Abuarqoub, Gregory Epiphaniou, and Ridha Hamila. "Integration of federated machine learning and blockchain for the provision of secure big data analytics for Internet of Things." *Computers & Security* 109 (2021): 102393.
5. Yang, Qing, and Hao Wang. "Privacy-preserving transactive energy management for IoT-aided smart homes via blockchain." *IEEE Internet of Things Journal* 8, no. 14 (2021): 11463–11475.
6. Prakasam, P., Md Shohel Sayeed, and J. Ajayan. "Guest editorials: P2P computing for 5G, beyond 5G (B5G) networks and internet-of-everything (IoE)." *Peer-to-Peer Networking and Applications* 14 (2021): 240–242.
7. Siriweera, Akila, and Keitaro Naruse. "Internet of cross-chains: Model-driven cross-chain as a service platform for the internet of everything in smart city." *IEEE Consumer Electronics Magazine* (2021).
8. Dhiman, Gaurav, Atulya Nagar, S. Vimal, and Seungmin Rho. "Guest Editorial: Cybertwin-Driven 6G for Internet of Everything: Architectures, Challenges, and Industrial Applications." *IEEE Transactions on Industrial Informatics* 18, no. 7 (2022): 4846–4849.
9. Li, Daming, Lianbing Deng, Zhiming Cai, and Alireza Souri. "Blockchain as a service models in the Internet of Things management: Systematic review." *Transactions on Emerging Telecommunications Technologies* 33, no. 4 (2022): e4139.
10. Yang, Chen, Shulin Lan, Zhiheng Zhao, Mengdi Zhang, Wei Wu, and George Q. Huang. "edge-cloud blockchain and IoE enabled quality management platform for perishable supply chain logistics." *IEEE Internet of Things Journal* (2022).
11. Chai, Haoye, Supeng Leng, Jianhua He, Ke Zhang, and Baoyi Cheng. "CyberChain: Cybertwin empowered blockchain for lightweight and privacy-preserving authentication in Internet of Vehicles." *IEEE Transactions on Vehicular Technology* 71, no. 5 (2021): 4620–4631.
12. Jabbar, Rateb, Noora Fetais, Mohamed Kharbeche, Moez Krichen, Kamel Barkaoui, and Mohammed Shinoy. "Blockchain for the Internet of vehicles: how to use blockchain to secure vehicle-to-everything (V2X) communication and payment?." *IEEE Sensors Journal* 21, no. 14 (2021): 15807–15823.
13. Velliangiri, S., Rajesh Manoharan, Sitharthan Ramachandran, and Vani Rajasekar. "Blockchain based privacy preserving framework for emerging 6G wireless communications." *IEEE Transactions on Industrial Informatics* 18, no. 7 (2021): 4868–4874.
14. Hossein, Koosha Mohammad, Mohammad Esmaeil Esmaeili, Tooska Dargahi, Ahmad Khonsari, and Mauro Conti. "BCHealth: A novel blockchain-based privacy-preserving architecture for IoT healthcare applications." *Computer Communications* 180 (2021): 31–47.
15. Da Silva Rodrigues, Carlo Kleber. "Analyzing Blockchain integrated architectures for effective handling of IoT-ecosystem transactions." *Computer Networks* 201 (2021): 108610.
16. Wang, Wenming, Haiping Huang, Lingyan Xue, Qi Li, Reza Malekian, and Youzhi Zhang. "Blockchain-assisted handover authentication for intelligent telehealth in multi-server edge computing environment." *Journal of Systems Architecture* 115 (2021): 102024.
17. Mehdinejad, Mehdi, Heidarali Shayanfar, and Behnam Mohammadi-Ivatloo. "Decentralized blockchain-based peer-to-peer energy-backed token trading for active prosumers." *Energy* 244 (2022): 122713.

18. Lee, Yen-Liang, Hsiu-An Lee, Chien-Yeh Hsu, Hsin-Hua Kung, and Hung-Wen Chiu. "SEMRES-A triple security protected blockchain based medical record exchange structure." *Computer Methods and Programs in Biomedicine* 215 (2022): 106595.
19. Auer, Sophia, Sophia Nagler, Somnath Mazumdar, and Raghava Rao Mukkamala. "Towards blockchain-IoT based shared mobility: Car-sharing and leasing as a case study." *Journal of Network and Computer Applications* 200 (2022): 103316.
20. Ye, Jingwei, Xin Kang, Ying-Chang Liang, and Sumei Sun. "A trust-centric privacy-preserving blockchain for dynamic spectrum management in IoT networks." *IEEE Internet of Things Journal* 9, no. 15 (2022): 13263–13278.
21. Yang, Xiaodong, Tian Tian, Jiaqi Wang, and Caifen Wang. "Blockchain-based multi-user certificateless encryption with keyword search for electronic health record sharing." *Peer-to-Peer Networking and Applications* 15, no. 5 (2022): 2270–2288.

6 Machine Learning (ML) and Blockchain for IoE

A Deep Insight into Framework, Security, and its Applications with Industrial IoE

Avishake Adhikary, Soma Debnath, and Dhrubasish Sarkar

CONTENTS

6.1 INTRODUCTION

The exponential hype in the available information has created an opportunity for industries to capture big data over the past few years. This creates the opportunities to gather high quality data in large volumes, which are raw and sensor specific, and specific to the applications, for example, applications like speech recognition, dialect recognition, sign language detection, and so on.

We know that computers only know and understand numbers and logic, only zeros and ones to be more specific. But in today's world, technology has risen so much it has kept us on the verge of spontaneous development and has led us to the question: "can machines think?" And has also led us to convert straight and simple instructions given to a computer to complex ones in order to make things simple in reality by making machines think like humans by feeding them huge amounts of data and helping them extract useful information from that data and putting it to use to solve a specific type of problem. This process is also known as the technology, machine learning.

Similarly, the rising data has led to a huge hype in transactions of data, which has also led to concerns of data misuse and security. This security concern has created

DOI: 10.1201/9781003366010-8

the possibility to find loopholes and close them through modern technology itself, which is by creating a chain of all the transactions and using good levels of encryption on them to keep them protected. This process is also known as the technology, blockchain.

The rising technologies have opened up the possibilities to convert traditional technology limited to distinct locations into mobile technology, enabling devices to have higher computation power that is not hostile to a single location: these devices were later categorized to be called "things." These things were later connected to the internet to control them. This technology was later termed as Internet of Things, or IoT. These IoT devices were later used to capture huge amounts of data they generated in order to contribute to machine learning or even big data. This IoT technology was later known to be Internet of Everything or IoE, which not only connects the things to internet but also generates huge amounts of data and ensured new ways of communication, which was between machines and humans, humans and humans through assisted technologies, and machines to machines, declared to be the superset of IoT.

These three technologies have been dominating the market ever since and GPUs have been a huge help in training these machine learning models with huge amounts of data within reasonable amounts of time which is way faster than the traditional CPU-assisted training through the CUDA cores, and even Tensor Cores in the modern GPUs. The interest in these technology domains has helped people create and reuse libraries like TensorFlow, Keras, and more, which supports these GPU-assisted training and also provides power to quickly calculate hashes for these blockchain.

But these technologies are completely different in their areas and have completely different features and have never been combined yet, and so the combination of these technologies to build a more secure IoE industrial framework model through the help of ML and blockchain are discussed later in this chapter.

6.2 FUNDAMENTAL CONCEPTS OF MACHINE LEARNING AND BLOCKCHAIN

Blockchain is a chain of fundamental blocks used for decentralizing a system to increase security. Blockchain contains three properties, namely:

- Blocks
- Miners
- Nodes

Blocks in the blockchain are fundamental pieces that contain three basic elements as mentioned in Figure 6.1.

- Data: The transaction data that the programmer wants to store for further processing.
- NONCE: A number used only once that is randomly generated and is used for header hash generation.
- Hash Value: The hash value calculated on the data and connected to the NONCE.

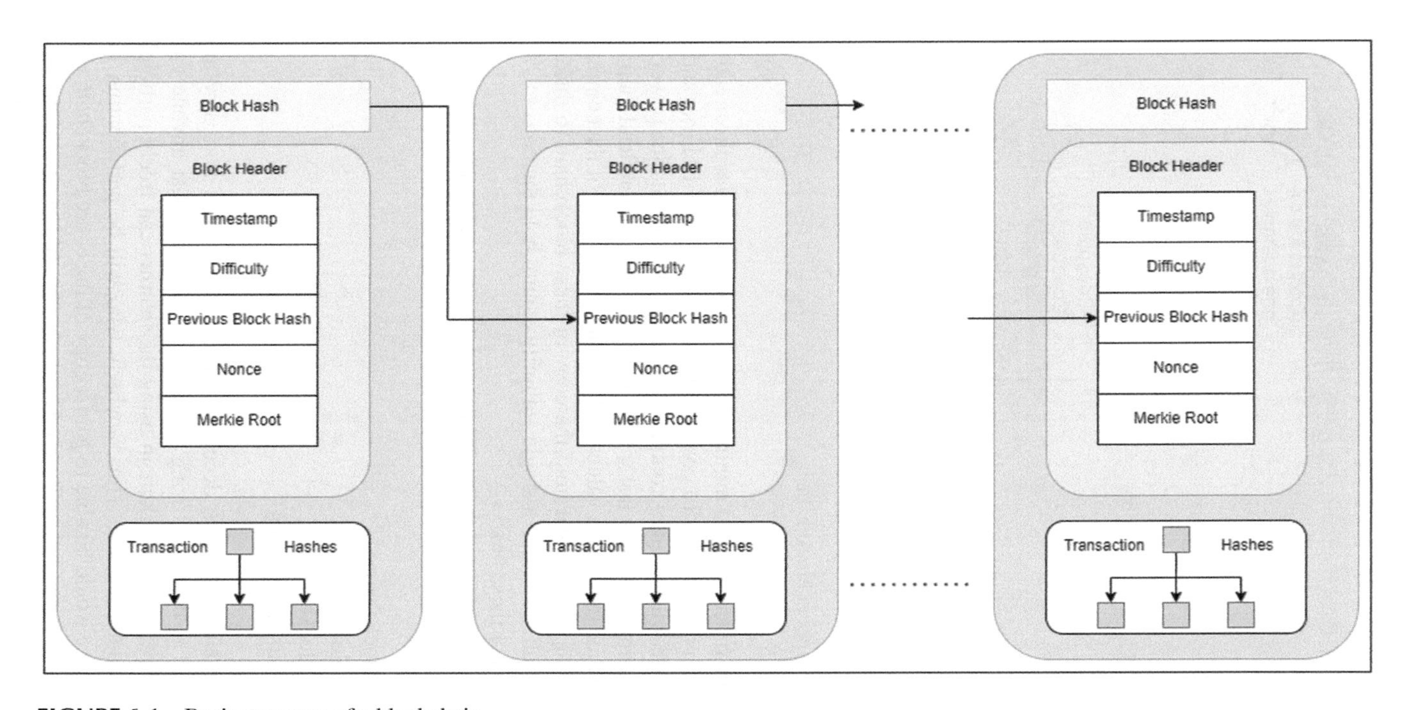

FIGURE 6.1 Basic structure of a blockchain.

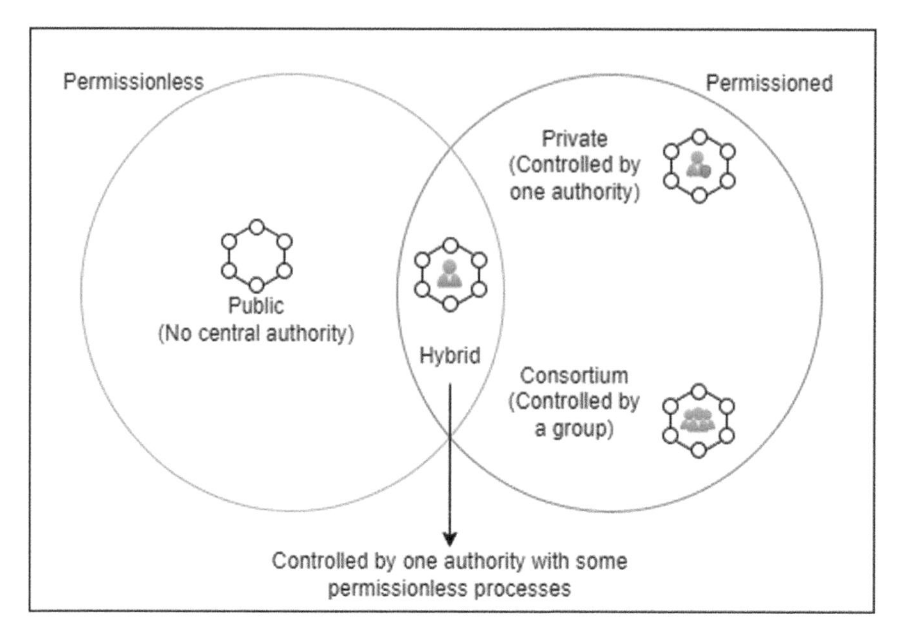

FIGURE 6.2 Types of blockchain.

Miners use the process called mining, which is responsible for the creation of these new blocks of a blockchain. To compute these long chains of blocks and their hash values, it requires quite a bit of complex mathematical computation power, which can be solved by using GPUs for faster computation. These miners lend their computing power to calculate these hash values and income from a part of the computation.

Nodes are the devices that hold the copy of these blockchain to make the entire system decentralized so that a single authority cannot hold the power to control and change or hamper these blocks.

Blockchain can be subdivided into four major categories as mentioned in Figure 6.2, namely:

- Public Blockchain
- Private Blockchain
- Hybrid Blockchain
- Consortium Blockchain

Public blockchain follow the decentralization rule. It does not carry any type of restrictions and is deployed using the internet, meaning that anyone carrying the ability to connect to the blockchain using the internet can freely participate in this type of blockchain network. Thus, anybody who will be participating in this network via a device would be able to gain a copy of the entire blockchain network and verification of the transactional records are also possible to perform. This type of blockchain carries the ability to be trustworthy as there is a large amount of distribution throughout the network and there are no chances of frauds happening in this network. Blockchains are primarily used for domains of cryptographic currencies like mining cryptocurrencies or even exchanging cryptocurrencies. The only problem with this type of network is that it takes a huge amount of computation power to participate in this type of network and a central authority is not present to govern each transaction to facilitate the transactions faster, which results in slower computation and ultimately results in slower transactions.

Private blockchain on the other hand follows the centralized approach. This is open to only selected members to gain authority, and, thus, proves the acronym "private." The speed of the blockchain is greatly increased due to the smaller size of the chain and verification of the transactional records needs only a small amount of time. This type of blockchain is completely restricted and therefore provides a huge amount of security over the previously discussed public blockchain. It also provides higher levels of privacy for blockchain confidentiality as only a few authorized members are given the authority to participate in this network. This is manually scalable to their required size and is customizable at any point of time, meaning that if the organization decides to increase the blockchain network, they surely can do it at any point of time. This comes at a number of risks, like the nodes that are handled by the authority can get manipulated by the authority handling it and makes the blockchain vulnerable by carrying a bias. Thus, building trust in these authorities becomes a huge problem due to their centralized structure. And the biggest disadvantage of this type of blockchain is that because it is authorized for only a few members to participate, if say, these few nodes go offline, the entire network goes down with it.

Coming to consortium blockchain, it is managed by a group of different organizations, unlike one private organization in a private blockchain. This is the extension of the private blockchain and is also known as the Federated blockchain, because multiple organizations tend to manage these blockchains and also tend to have the property of decentralization as a result of larger levels of security. But it requires the cooperation of multiple different organizations and setting them up can be a really fraught process. This blockchain also carries the advantage of making decisions faster than the public blockchain and having the flexibility to scale. It is unknown to the public but anyone having the authority to access may access the blockchain and as multiple organizations are authoring the blockchain, the verification of the blockchain becomes faster than the previously used private blockchain. But taking the approval of the blockchain transaction becomes a hassle as multiple organizations govern the decision and one or more organizations may have differences in vision of interest that might affect the decision.

Hybrid blockchain (Ayush, 2020) on the other hand is the type of blockchain that is the combination of private and public blockchain as mentioned in Figure 6.3. This is managed and controlled by only a single organization but is made open to the public with a number of restrictions and other permissions. Thus, this system of blockchain holds the property of having both permissionless and permission-based systems. Here, even though a single entity holds the blockchain, because they make them open to the public, the number of chances of alteration is zero and makes these systems unhackable. And because only a single entity manages the blockchain, the cost required is much less as they don't need the computational power like the public blockchain. This blockchain is highly customizable in nature and maintains a good amount of

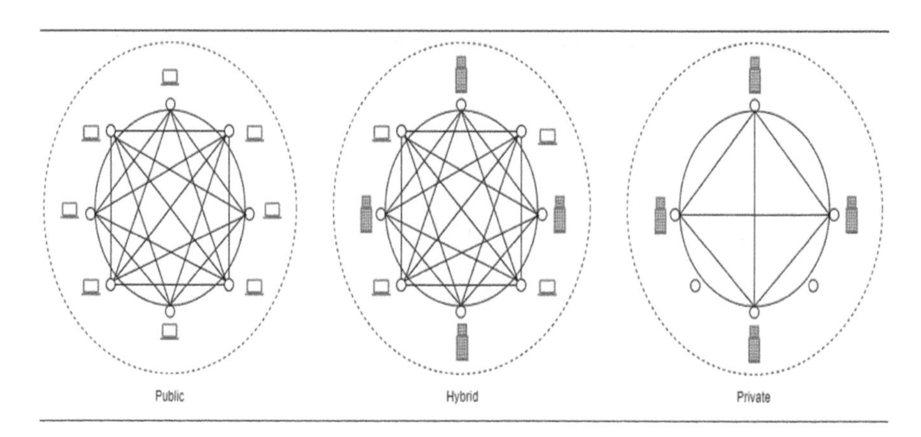

FIGURE 6.3 Hybrid blockchains.

security and transparency to the users along with integrity. But, because of the complex nature of this blockchain, it is very difficult to implement and even difficult to maintain efficiency, and also because a single entity manages this blockchain these organizations can have a bias on whom to give permission to access the blockchain and whom not to give the permission to access the blockchain. Because the organization restricts the access to the blockchain to only specific public users, the ecosystem of the blockchain lacks the incentives for participation in the network. But these blockchains also require advanced hashing power, even if the computational architecture is limited to a single entity or organization. These types of blockchains (Wegrzyn and Wang, 2021) can be very useful for the health care industries and financial and real estate agencies, for example, when we want to share our data with the doctor and life insurance companies, but we also want to rely on methods using better privacy and security, or when we want to use some private payment architecture systems on public networks. These blockchains can also be used in Hybrid IoTs when devices are kept in a private network but gives access to the selective users in public networks.

Machine Learning (ML) on the other hand is the sub-category of Artificial Intelligence (AI) which makes the computerized machines learn distinct human behavior and perform some tasks. Machine Learning uses a mathematical model that defines an algorithm or a set of multiple algorithms that are performed on huge amounts of data collected to improve its accuracy and predict some results.
Machine Learning is based on three general types of learning techniques as mentioned in Figure 6.4, namely:

- Supervised Learning
- Unsupervised Learning
- Reinforcement Learning

Supervised Learning refers to the learning technique of the machines where the machines are fed with labeled data and the outputs are known, which are then mapped together.

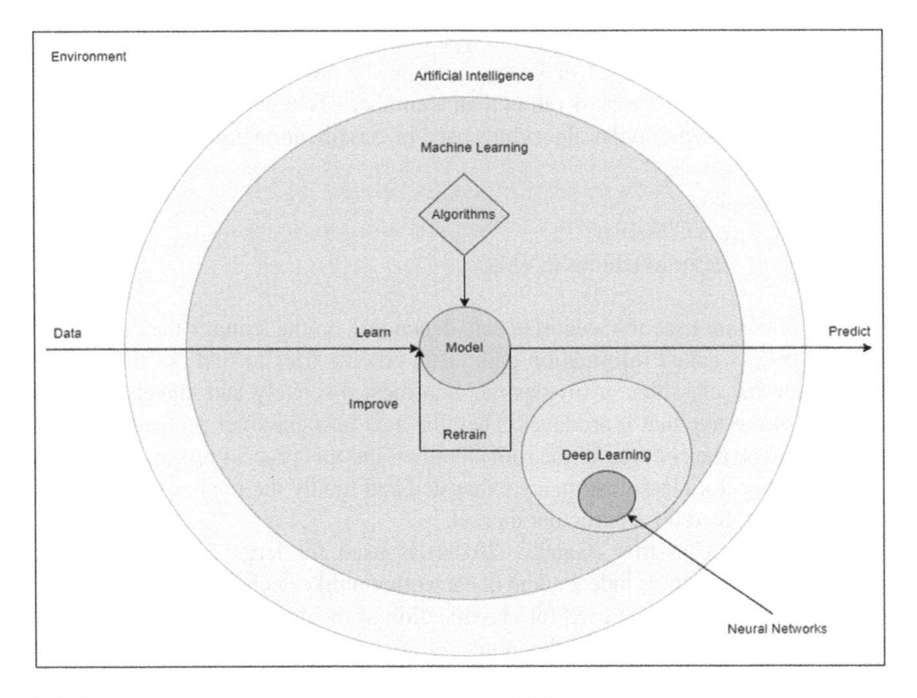

FIGURE 6.4 Basic structure of machine learning (ML).

Supervised Learning can be further subdivided into two major subcategories, namely:

- Regression
- Classification

Regression is the type of Supervised Learning that is trained using labeled datasets and a continuous-value is produced as an output for prediction.

Some of the most popular algorithms used in regression are:

- Linear Regression
- Logistic Regression

Linear Regression maps the continuous-variable from the input variables X and Y which are assumed to be having a linear relationship, and the input variables are known to be the independent variables and the output continuous-variable is known to be dependent on the inputs that are processed (Haridas et al., 2020).

Logistic Regression on the other hand maps discrete-variable values for the name set of X and Y independent variables as linear regression through a logit function that is coded inside the model; the model then produces a prediction in the range between 0 to 1 which tends to be discrete in nature and is the probability of the newly produced data (Webster, 2022).

Classification (Akash, 2020a) is the type of Supervised Learning where the algorithm maps different classes or groups into newly processed data in binary form which is used to choose a class rather than a number like regression.

Some of the most popular algorithms used in classification are:

- Decision Tree
- Naive Bayes Classifier
- Support Vector Machines (SVMs)

Decision trees are used to categorize data depending on the features the data include using a process called information gain. This process tries to find out the property of the data that classifies or divides the data most discretely and makes it the root classifier of the tree that it produces. Then the next best classifier property is chosen and made to be the children of the root classifier and this process repeats until all the properties used for classification are exhausted and finally the final model for classification is produced for a particular dataset.

Naive Bayes Classifier (Gandhi, 2018a) is used for large datasets which are assumed to have features independent of each other unlike decision trees and Directed Acylic Graphs (DAGs) are used for classification of the data.

Naive Bayes Classifier uses the theorem:

$$P(A|B) = P(B|A)P(A)/P(B) \tag{1}$$

According to the theorem mentioned above, the model tends to find the probability of A happening in the future assuming that B has happened before. Here A is known to be the hypothesis and B is known to be the evidence and properties or features do not affect others.

There are further three types of mostly used Naive Bayes Classifiers namely:

- Multinomial Naive Bayes
- Bernoulli Naive Bayes
- Gaussian Naive Bayes

Support Vector Machines (SVMs) (Gandhi, 2018b) use Vap Nik's learning theory of the statistical learning. SVMs Kernel Functions create a hyperplane in an N-dimensional space (or with N number of properties or features) that distinctly classifies two classes from each other. Here the number of features determines the hyperplane dimensions. Say if the number of features are two then the hyperplane is a simple line, and if the number of features are three then the hyperplane is a two-dimensional space or graph; similarly when the number of features are four then the hyperplane is a three-dimensional space, and so on. Here the support vectors are the data points plotted on the hyperplane that are close to the hyperplane and support the location and the orientation of the hyperplane which are then used to enlarge the hyperplane margins to their maximum extent.

Unsupervised learning is a machine learning technique in which the machines are fed with unlabeled data and attempt to extract useful information and features, recognize various patterns, and predict certain outcomes.

Unsupervised Learning (Akash, 2020b) can be further subdivided into two major sub-categories, namely:

- Clustering
- Association

Clustering is the type of unsupervised learning technique where the algorithm finds different patterns on the dataset to classify the data into small groups or clusters based on their properties.

Some of the most popular algorithms used in clustering are:

- Hierarchical Clustering Algorithm
- K-Means Clustering Algorithm
- K-NN Clustering Algorithm

Hierarchical Clustering Algorithm uses the similarities between dataset's plotted data points and builds small groups or clusters using them. This process works until the entire dataset has been classified as a cluster to form a hierarchy to classify the data.

K-Means Clustering Algorithm uses the homogeneous data by fixing a centroid around a cluster and makes sure that the features of the centroid and data points of the dataset are as close as possible to the centroid itself, making it group together and classify itself from other centroids and produces labeled data clusters. Here, the nearest data point from cluster centroids determines which dataset cluster it belongs to.

K-NN (K-Nearest Neighbors) Clustering Algorithm or the lazy learning algorithm learns only when a new data point is assigned to the model and takes a huge amount of time to train and works well with small amounts of datasets. This algorithm classifies data points based on their similarities, meaning that whenever a new data point arrives it is classified into well-suited category easily using this algorithm.

Association is the type of unsupervised learning technique where the algorithm tries to find how a data can be dependent on one another in a way to maximize profits.

Some of the most popular algorithms used in association are:

- Apriori Association Algorithm
- FP-Growth Association Algorithm

Apriori Association Algorithm calculates the support between data items using the most popular breadth first search algorithm (BFS). Here the dependency of one data point is mapped to understand the probability of it affecting the other data points on change. For example, if a person tries to buy a computer monitor as a peripheral device for their computer, there is a great chance that the person would like to buy other peripheral devices such as keyboards and mice too. This algorithm is also therefore used for maximizing the profits of a particular store or organization.

FP-Growth (Frequency Pattern Growth) Association Algorithm is used to identify repeating patterns and creates a table out of the items, then for most reasonable items that are found are added to a tree and the support of these items are calculated and the root tree item creates the rules for association upon completion of all the iterations. All the branches that do not meet the threshold requirements for the support created are pruned on checking. Because the increasing iterations are calculated and others are pruned and checked for support, this algorithm tends to be faster than the Apriori algorithm.

Reinforcement Learning (Keshari, 2019) on the other hand refers to the learning technique of the machines that involves trial and error technique and is based on rewards and punishments. Here, the model incorporates an agent which has information about its environment and learns through feedback which is delayed by interacting with the environment available. The agent is given rewards if the prediction remains accurate and there are punishments if the prediction is not.

Some of the most popular algorithms used in reinforcement are:

- State-Action-Reward-State-Action (SARSA) Reinforcement Learning Algorithm
- Q-Learning Reinforcement Learning Algorithm
- Deep Q-Networks Reinforcement Learning

State-Action-Reward-State-Action (SARSA) Reinforcement Learning Algorithm (Karuppasamy, 2018) is an on-policy reinforcement learning algorithm because it tends to update the policies based on the actions taken. Here, S is a state, A is action taken and the agent handling these actions gets a reward R based on the actions, which updates the next state to be S' and A'. This process repeats itself and the tuple (S, A, R, S', A') stands to be SARSA acronym for the algorithm.

Q-Learning (Quality Learning) (Choudhary, 2019) Reinforcement Learning Algorithm is an off-policy reinforcement learning algorithm that aims to find the best possible action that can be taken from the current state the model stands in. This learning algorithm is considered to be an off-policy because it learns from actions revolving outside of the current policy state, which might take any random action, which results in no policy requirements. This learning algorithm tends to maximize the total reward of the entire model. In this approach, the model constructs a q-table using the state parameter and its values, where the values are initialized with zeros and the q-values are updated with each iteration. Based on the values, the optimum course of action is then chosen.

The q-table updates using the following references:

- Learning Rate – The alpha value (α) which represents the difference between the old values and the new values to identify how much the model learns after each iteration
- Gamma – The discount factor (γ) which is used to balance the current and future receiving rewards.
- Reward – The value that is acquired after each iteration of the model
- Maximum Reward – The maximum reward value that is determined by applying the current reward state to already acquired rewards to see the impact on future reward state.

Deep Q-Networks Reinforcement Learning (Choudhary, 2019) is an extension of the Q-Learning Reinforcement Learning that uses neural networks to map the input states to their actions or q-values.

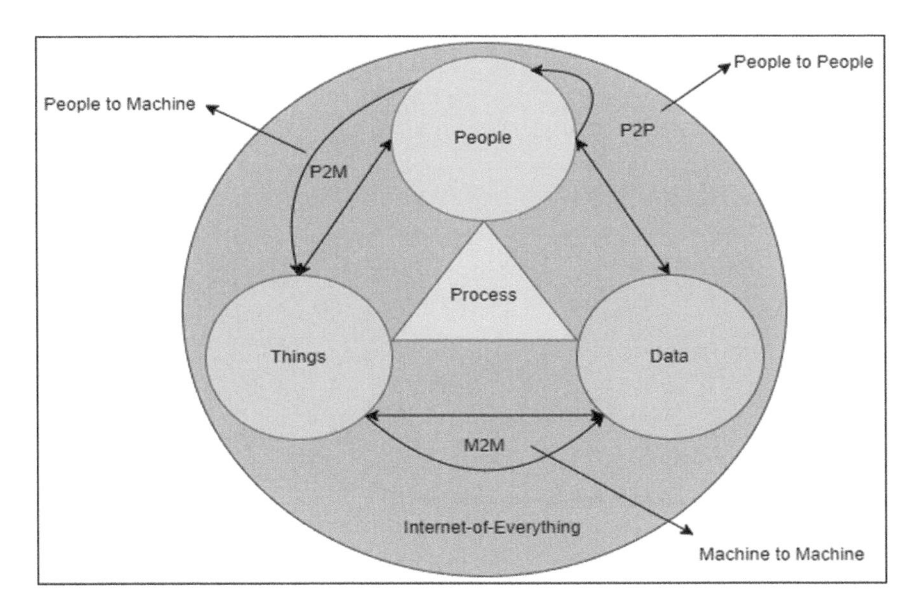

FIGURE 6.5 Basic structure of Internet of Everything (IoE).

The neural network is initialized with weights and an action is chosen using the Epsilon-Greedy Exploration Strategy (Nuer, 2020) and the network is updated using the Bellman Equation.

Internet of Everything (IoE) is the super-category of Internet of Things (IoT) which is based on four main pillars as mentioned in Figure 6.5, namely:

- People
- Process
- Data
- Things

People refer to users that get connected to the internet through devices in order to generate data. People act as nodes that are connected to the internet. The process refers to the IoE process where the correct information gets delivered to the correct people or machines in the correct way at the correct time. Data refers to the raw data which is sent and processed by these connected devices on a higher level to convert the data into intelligence in order to make intelligent decisions.

Things refer to the physical elements on these devices such as sensors, industrial devices, and assets that get connected to the internet which also gather information from their environment. These things are also known as the Internet of Things or IoT.

6.3 RECENT DEVELOPMENTS (AND RELATED WORKS)

With the rising technology, there have been quite a few new recent developments based on the world of Machine Learning and Blockchain. The three most impacted domains by both ML and Blockchain have been:

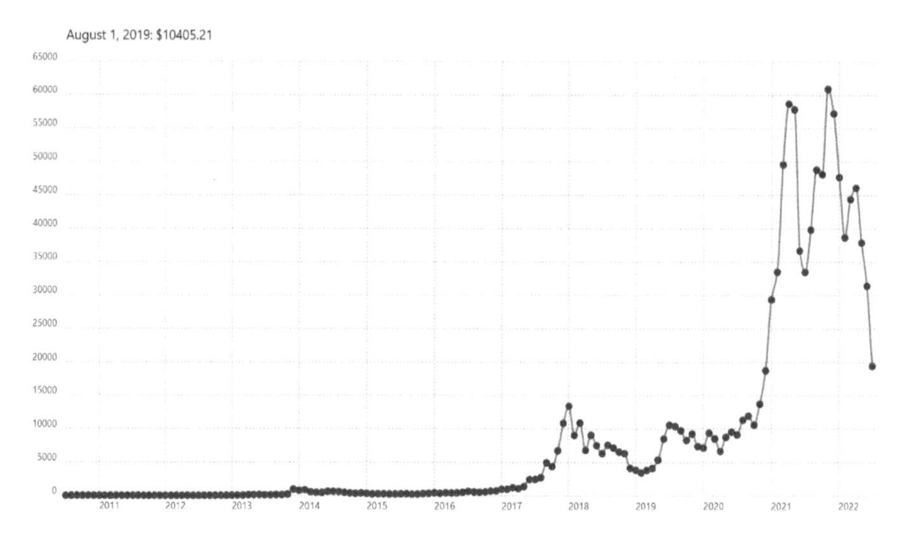

FIGURE 6.6 Trading crypto trends over the past 10 years.

a. Trading Cryptocurrencies (using Reinforcement Learning)
b. Optimizing Crypto-Mining Strategies (using Reinforcement Learning)
c. Tackling Crypto Hijacking (or Crypto-jacking with the help of Deep Learning)

Trading (Koker and Koutmos, 2020) has been one of the most profitable sources of income and has been one of the largest markets of investment all over the world as mentioned in Figure 6.6. Trading cryptocurrencies, in particular, has grown exponentially over the past 5 to 10 years. Currencies like Bitcoin and Ethereum have been one of the most popular cryptocurrencies along with a few others and have now become an activity for daily exchanges, which has attracted thousands of developing markets to develop applications and bots for trading. Traditionally bots were used in stock markets which now come with embedded machine learning powered algorithms as mentioned in Figure 6.7. Reinforcement Learning (RL) is a technology developed in these few years that does not collect feedback from the previous windows of the agent, creating a new sub-model of RL known as Direct Reinforcement Learning or DRL. This new model has created new possibilities through which the traditional RL models can be optimized to achieve better performance and speed up the machines which helped the modern-day trading applications get faster price forecasting models and have created systems that can adapt to specific time intervals on a regular basis and have made the current cryptocurrency trading strategies way more adaptable and profitable.

Crypto-mining (Wang et al., 2021) has been one of the hottest topics for the past five years and has also been one of the most active stages for passive income. In the recent research of 2021 the reinforcement learning techniques have been dynamically extrapolated for different mining strategies which resulted in performing way better than the traditional reinforcement learning methods known as the selfish mining

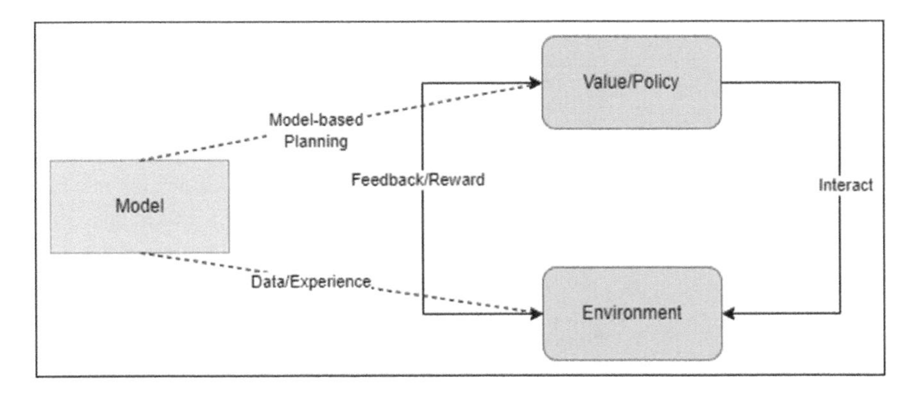

FIGURE 6.7 Reinforcement learning by direct RL, without model-based planning (shown in dash-lines).

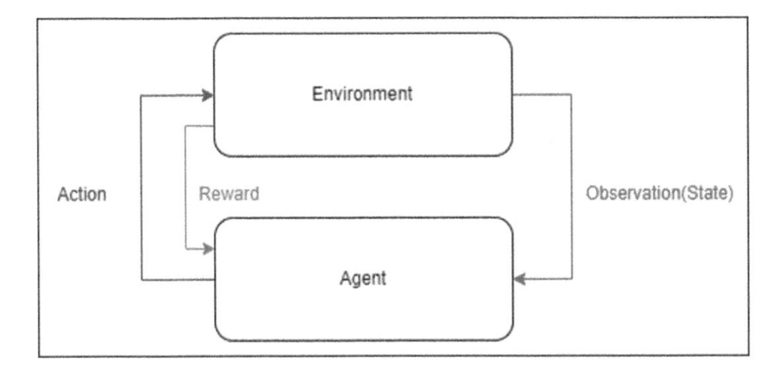

FIGURE 6.8 Q-Learning architecture.

which used agents to maximize rewards in an environment. So researchers have used multidimensional reinforcement learning algorithms that use Q-Learning or model free algorithms to mine crypto that was much more dynamic in nature, as mentioned in Figure 6.8 and Figure 6.9.

Crypto-jacking (Bhatt, 2021) has been the target for most of the hackers in the modern world to acclaim money from the transactions happening spontaneously throughout the world and has been one of the biggest concerns for the national government which provided people with sufficient infrastructure to protect them. Researchers have created a system called SiCaGCN which checks for the similarities between different codes and is based on distance metrics on a control flow graphical representation of the codes and defines a convolutional neural network architecture for Deep Learning (DL) (Amber, 2009).

SiCaGCN system detects bitcoin mining code and graphically analyzes it so that the chain of blocks mined in the blockchain remains the same and performs significantly better than other systems detecting crypto-jacking, as mentioned in Figure 6.10.

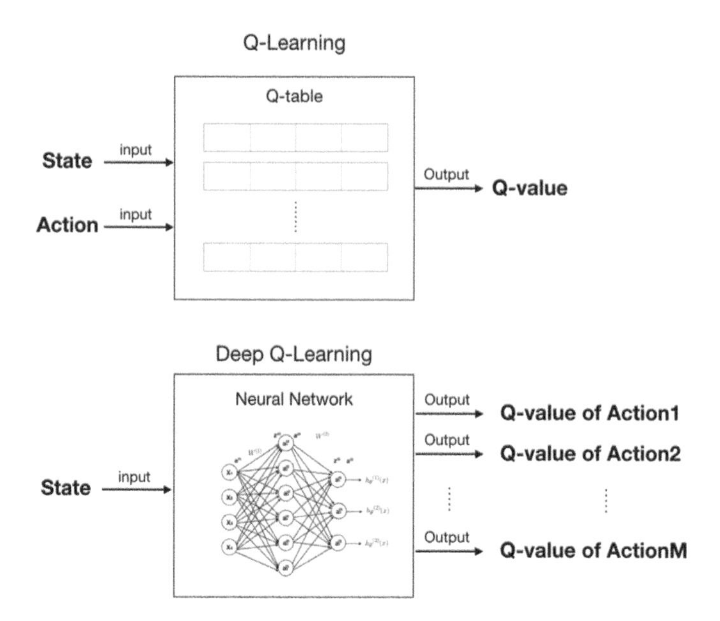

FIGURE 6.9 Q-Learning neural network.

6.4 OVERVIEW OF TENTATIVE INDUSTRIAL IOE FRAMEWORK

In this chapter we discuss how machine learning and blockchains can be added to IoE, to achieve better security and effectiveness. As discussed earlier, IoE is the product of people, process, data, and things. These things that we refer to are none other than the sensors of IoT that take millions of data as input from the environment and generate millions of processed data throughout the world, which are then taken for further computation or outputs which are based on the data. We can take the example of automated cars which are getting more popular every day and are getting recognition for their implementation from the largest MNCs like Google and Tesla (Pisarov and Mester, 2020). These cars in general have many sensors like Lidar, Proximity Sensors, Gyroscopic Sensors which take their respectable inputs from the environment and are stored in a data store which are later used to make other computations, which results in the cars to get knowledge on what to do by making predictions. For example, if somebody wants to steal an automated car, the car sensors keep on checking for unusual activities to prevent casualties that might happen in the future by making predictions and making the car lock itself or even take feasible actions offered by the car technology. Or the car can also detect how it should automate itself by learning from its environment. To do this the car can simply take inputs from its speedometer sensor to check the current speed of the car at every interval of a threshold time and can keep itself automated based on certain parameters. Say the car wants to maintain a speed of 60 kms/hour, the module that processes the data simply uses a model that keeps the car within 0–60kms/hour and keeps on checking so that the car slows down if it tends to exceed 60 kms/hour by making predictions beforehand. The car can also

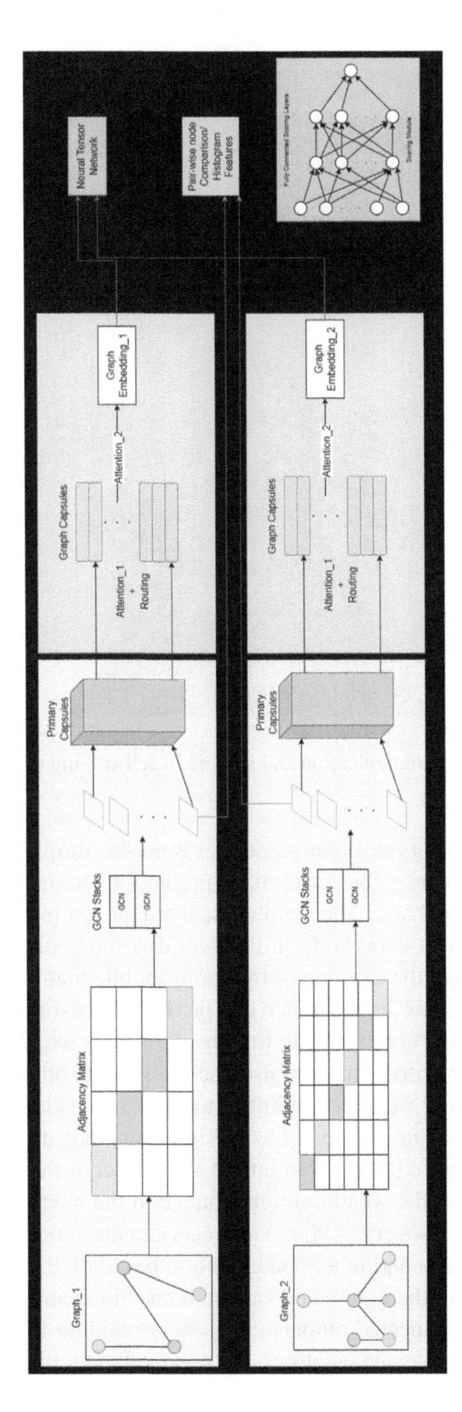

FIGURE 6.10 SiCaGCN neural network.

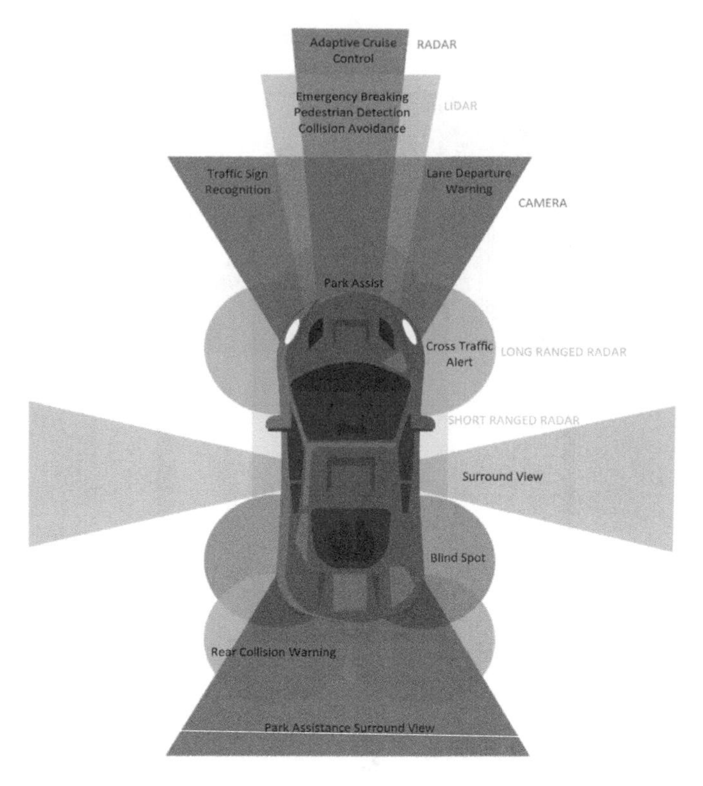

FIGURE 6.11 Wireless communication and sensors in self-driving cars.

prevent accidents by using the Lidar sensors or even the simplest proximity sensors to detect obstructions nearby the car to make predictions based on the sensor inputs and take measures beforehand. The simplest sensor that can make predictions and is used every day by almost everybody in this world is the geolocation sensor which helps the sensor to detect the location of the car or mobile that is later used to predict the movement of the car or device which can predict the best route for the device to react its location by using complex path finding algorithms which tend to be efficient and user-friendly. Automated cars can also use the help of other sensors which can help them detect the outer weather conditions to keep the insides of the car in normal conditions as mentioned in Figure 6.11: say in a very hot day the car can detect the weather which may be hot or even humid and can cool the car down and make predictions of what how the weather might change in the next interval of threshold, or if the outside weather is very cold, modern cars can also cool down the car internally to make the car warm and in a good condition based on the prediction it makes. All of these functions in the automated car use a machine learning model that helps the sensor achieve the respected outputs the users are hoping for or the outputs they require. They make complex calculations and make efficient predictions, which tend to be very accurate. These computations can be very device heavy and need a good amount of modern computing power which grants them a better architecture to make

the predictions faster and gain better accuracy. Therefore, businesses like Google and Tesla perform these calculations and optimize their models to make the most of decent hardware to utilize the processing modules and make better predictions. This helps to automate the car in real-time, as well as provide hospitable testing environments to help test these models and even simulate the automations in a virtual environment or in a game. These tests and simulations defines a scene and different scenarios in which a car can operate and also provides computations and model optimizations to make the best out of decent hardware to utilize the processing modules and make.
We can also take the example of the most common voice assistants like Google Assistant from Google, Siri from Apple, Alexa from Amazon and Bixby from Samsung and so on. These voice assistants are intelligent enough to perform speech recognition of a particular person and take decisions based on that. Most of these models follow a structure that includes the acquisition of raw sounds (or analog sounds) from the environment to the model, which is known as the analog to digital conversion of the sound data. Then the model tries to recognize the voice of the individual from the processed sound data by canceling redundant noise and using various sound frequency detection techniques. Then the prediction is ready for further processing, which may include different types like sentiment analysis, or even taking decisions like voice commands for feasible actions that can be taken by the system. This sentiment analysis can help the system understand what the user wants to say or even by taking decisions like voice commands which can perform several tasks like smart home automation based on the voice which can be done after feature extraction from the voice data. This voice data can also be used to control the security systems of the smart home and other systems like smart cars or even any supporting smart automation devices. These systems go through a translation technique known as Neural Machine Translation (or NMTs) (Vashisht et al., 2021) which predicts the sequence of keywords in an integrated model which might end up making whole sentences as mentioned in Figure 6.12.

All the discussed problems earlier use a machine learning model to facilitate their tasks to either perform them or increase their security using powerful models used

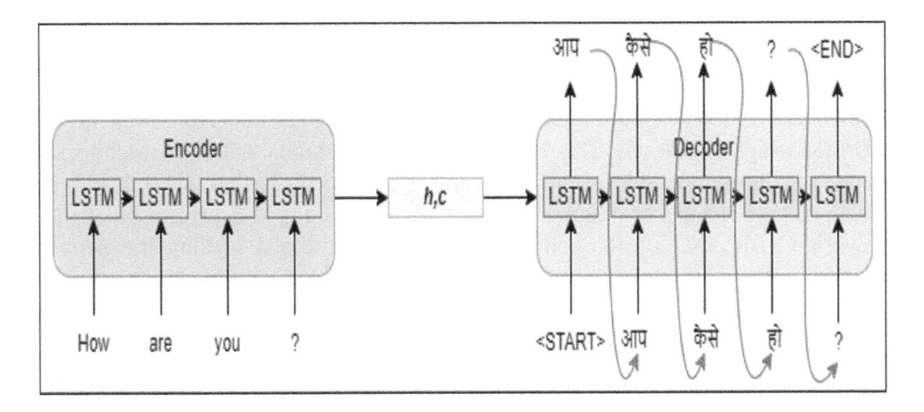

FIGURE 6.12 Encoder-decoder inference phase.

for specific tasks. These devices that collect the raw data from the environment generate a huge amount of useful data that might be used for millions of different tasks that helps human behavior and is known as big data. This huge amount of data can be really useful to these machine learning models. These machine learning models are first trained with a huge amount of data and later tested using a huge amount of data as well, after dividing the data into two segments, namely, training data and testing data. These data are also used by these models as a set that large companies use to sell user specific products. Large companies train a model which checks for user activities, like how a person interacts with other things in daily life, what kind of products the user might be interested in, and so on.

So, in our tentative IoE framework, we try to develop an IoE model where all the people, data, process and things are connected in a network and these things (or device modules) gather data together which can be processed through their own individual processing modules and can be termed as their respective processed data. This data can be trained using the reinforcement learning techniques which can help perform different tasks like smart home automation or smart car automation, by using prediction roles. These predictions are also a form of data that can be later used in back propagation of the neural network of the reinforcement learning models to further train the model in a healthy way to make the model learn about specific users to make the model fit for specific users (Khandelwal, 2020). This tentative model uses a central data store for accessing and storing data and is updated frequently. This model can be monitored initially using a set of stream analytics tools and later deployed in industrial architecture, for companies to use this architecture.

6.5 SECURITY AND PRIVACY ISSUES IN INDUSTRIAL IOE

Even with the most powerful machine learning models used for specific tasks discussed in the earlier section of the chapter we still see that security remains one of the most critical sections of the IoE model and remains to be tweaked. With the knowledge of the blockchains, we can now combine our tentative industrial IoE model to work with blockchains. Let's say we term each event in our architecture as transactions that happen throughout the network. These transactions can now access the frequent file store system that comes with our tentative model to store consecutive blocks inside the database. And the model's data and events both can now be secured using good hashing functions like SHA1(Secure Hashing Algorithm v1) and MD5 (Message Digest v5). This hashing of events and data makes the architecture secure and "un-tamperable" as all the events are chained together and a third-party device cannot change or manage the blockchain. Making it a private blockchain network for the user customizing the smart home for and making the network scalable so that whenever the user wants to add a new device module to the network can access and given the authority to change the blockchain but nobody else. Additionally, the design will become less expensive as a result of the user's ability to lower the computational resources required to centralize and protect the system, creating a Chain-of-Things (CoT) and Chain-of-Security (CoS). This system can, thus, result in a better IoE framework ready for deployment in industrial architecture and is highly customizable to fit almost any industry type already using the IoE framework for daily tasks.

6.6 FUTURE SCOPE

The tentative industrial IoE architecture discussed earlier for making the system work with both machine learning and blockchains can be a hassle because the devices need to coordinate with each other and use a single file store. Since the entire network will be connected using a single file store and each of the things or devices connected to the network will be generating huge amounts of processed data and will be updated more frequently. One of the crucial tasks that must be implemented in the future is to implement a file store that can be accessed more frequently. Then connecting the architecture to a faster file service system, and also a file system that can store enormous amounts of data. This architecture also craves the need to have proper computation power for each thing or device to preprocess the data into processed data, so, modern hardware is a must implementation for the future for each device module. As this architecture uses high amounts of data transfer, the network it is connected to tends to slow down on older hardware bandwidth network, so modern network bandwidth support is also a must on future implementation of the architecture which might result in smoother data transfer and better bandwidth that reduces the network load from each thing or device.

6.7 CONCLUSION AND FUTURE SCOPE

Combining technologies like Internet of Everything, Machine Learning and Blockchain in industry application can be a really complex task because each of these domains carry their own properties and features. With the increase of modern technologies, modern architecture implementation becomes necessary for performing complex tasks. In this chapter we discussed what each of these domains include and gave an overview insight over where each of these domains work. We saw how machine learning algorithms perform and their subcategories, we saw how a blockchain performs and also discussed its categories. Then we dove into the domain of Internet of Things (IoTs) and how they are extended to become Internet of Everything (IoE). We discussed how each of these devices generate huge amounts of data to contribute into something called a bigdata. Then we discussed various scenarios of how these machine learning models and IoE framework can be used for user specific applications and industry applications and came to a conclusion that reinforcement learning technique and IoE result in a better fitting of these applications and produce the most accurate predictions while being time efficient. We then saw how we can combine machine learning technologies and IoE framework together to contribute to big data and further make the machine learning model better using back propagation to use its own output to further improve the model. Then we analyzed how we can add the blockchain technology to further improve the security in the same model to make the data and the transactions in the model "untamperable." This resulted in an architecture that combined these very different domains to make the IoE framework industry-ready. More importantly, it resulted in an architecture capable of dealing with complex data and different scenarios so that this same model can be implemented anywhere in the industry. Then we discussed what should be our future intake in the model and what are the things that need to be improved and further implemented to make it totally industry-ready.

REFERENCES

Akash, (2020a). "What is Supervised Learning and Its Different Types?", Edureka, November 25, 2020, www.edureka.co/blog/supervised-learning/

Akash, (2020b). "What is Unsupervised Learning and Its Different Types?", Edureka, July 21, 2020, www.edureka.co/blog/unsupervised-learning/

Amber, "(Deep) Q-learning, Part1: Basic Introduction and Implementation", Medium, April 9, 2019, https://medium.com/@qempsil0914/zero-to-one-deep-q-learning-part1-basic-introduction-and-implementation-bb7602b55a2c

Ayush, (2021). "Deciphering Hybrid Blockchain Technology and Its Use Cases", Oodles Blockchain, April 28, 2021, https://blockchain.oodles.io/dev-blog/hybrid-blockchain/

Bhatt, Shweta (2021). "Introduction to Direct Reinforcement Learning by Example", Towards Data Science, August 28, 2021, https://towardsdatascience.com/introduction-to-direct-reinforcement-learning-by-example-3f69af9353b2

Choudhary, Ankit (2019). "A Hands-On Introduction to Deep Q-Learning Using OpenAI Gym in Python", Analytics Vidhya, April 18, 2019, www.analyticsvidhya.com/blog/2019/04/introduction-deep-q-learning-python/

Gandhi, Rohith (2018a). "Naive Bayes Classifier", Towards Data Science, May 5, 2018, https://towardsdatascience.com/naive-bayes-classifier-81d512f50a7c

Gandhi, Rohith (2018). "Support Vector Machine—Introduction to Machine Learning Algorithms", Towards Data Science, June 7, 2018, https://towardsdatascience.com/support-vector-machine-introduction-to-machine-learning-algorithms-934a444fca47

Haridas, Poornima, Chennupati, Gopinath, Santhi, Nandakishore, Romero, Phillip, and Eidenbenz, Stephan (2020). "Code Characterization with Graph Convolutions and Capsule Networks." IEEE Access: 1–1.

Karuppasamy, Anuradha (2018). "Introduction to Reinforcement Learning (Coding SARSA)", Medium, July 23, 2018, https://medium.com/swlh/introduction-to-reinforcement-learning-coding-sarsa-part-4-2d64d6e37617

Keshari, Kislay (2019). "Q Learning: All You Need to Know about Reinforcement Learning", Edureka, June 12, 2019, www.edureka.co/blog/q-learning/

Khandelwal, Renu (2020). "Intuitive Explanation of Neural Machine Translation", Towards Data Science, January 13, 2020, https://towardsdatascience.com/intuitive-explanation-of-neural-machine-translation-129789e3c59f

Koker, Thomas E. and Koutmos, Dimitrios (2020). "Cryptocurrency Trading Using Machine Learning," in Machine Learning Applications in Finance, August 10, 2020.

Nuer, Jeremi (2020). "A Practical Guide to Deep Q-Networks", Medium, November 18, 2020, https://towardsdatascience.com/deep-q-learning-tutorial-mindqn-2a4c855abffc

Pisarov, Jelena and Mester, Gyula (2020). "The Future of Autonomous Vehicles." *FME Transactions*. 49: 29–35.

Vashisht, Vineet, Pandey, Aditya, and Yadav, Satya (2021). "Speech Recognition using Machine Learning." *IEIE Transactions on Smart Processing & Computing*. 10. 233–239.

Wang, Taotao, Liew, Soung Chang, and Zhang, Shengli (2021). "When Blockchain Meets AI: Optimal Mining Strategy Achieved by Machine Learning," *International Journal of Intelligent Systems*, January 6, 2021.

Webb, G.I. (2011). Naïve Bayes. In: Sammut, C. and Webb, G.I. (eds) *Encyclopedia of Machine Learning*. Springer, Boston, MA. https://doi.org/10.1007/978-0-387-30164-8_576

Webster, Ian (2022)."Bitcoin Historical Prices," U.S. Finance Reference, July 1, 2022, www.in2013dollars.com/bitcoin-price#:~:text=Growth%20%C2%B7%202016%20and%202017%20saw,2020%20and%202021%20increased%20dramatically

Wegrzyn, Kathleen E. and Wang, Eugenia (2021). "Types of Blockchain: Public, Private, or Something in Between", Foley, August 19, 2021, www.foley.com/en/insights/publications/2021/08/types-of-blockchain-public-private-between

7 Biometric Authentication in Internet of Everything

P. Gayathiri

CONTENTS

7.1 INTRODUCTION

IoE can be integrated with people, processes, and data in several applications [1-3]. Despite this, a huge number of linked systems and the high volume of data flow make it difficult to provide the necessary Quality-of-Services, because IoE devices have limited computing, storage, and bandwidth. Blockchain encryption keys are used in IoE applications to make changes [4-7]. A blockchain block is composed of data that can be uniquely identified by encryption keys. In blockchain technology, each block contains a hash value in the previous block. A decentralized system with anonymous and trustworthy transactions exists within the block. The previous

DOI: 10.1201/9781003366010-9

block's identification and the proof of work are all entered in the header. IoE systems integrated with blockchain technology, including lower operational costs and protection from threats.

The combination of IoE with blockchain technology intends to address the major difficulties in the implementation of the IoE platform. Initially designed for cryptocurrency, blockchain is a distributed ledger technique. Blockchain technology was first established in the year of 2008 by Satoshi Nakamoto [8], and it has a peer-to-peer network that has incorporated cryptographic technology and lacks centralized data storage, making it resistant to attacks that seek to take over the system. To create blockchain technology, a transaction-management platform called Ethereum was later released to the general public in 2013. The user signs in to the transaction via a private key. A block is created by a miner from all legitimate transactions and transmitted back to the network. The end user's private key is used to verify a transaction before it is broadcast to the network. After its validity is confirmed, the transaction is completed over the network.

Consensus agreements are considered complete when they have been agreed upon by all parties. Its distinctive and appealing characteristics are privacy, security, immutability, integrity, authorization, and transparency. Blockchain is used in several cryptocurrencies, including Ripple, Litecoin, Swift coin, Peercoin, and Bitcoin, due to its immutability, privacy, security, and transparency. The block is then added to the blockchain after it has been hashed against the previous block and validated with the transaction. Identity management [9], intelligent transportation [10], mobile crowd sensing [11-16], supply-chain management [17], agriculture [18], Industry [19], Internet of energy [20-22], and security [23]. The hash values link the blocks together to form the blockchain's structure. The users' digitally verified transactions are stored in the distributed ledger of the blockchain network. A user normally possesses two keys: a private key for decrypting communications sent to them and a public key for encrypting messages sent to them. According to blocks, the private key is used for blockchain transactions, while the public key encryptions the private address. Once a public key has been used to encrypt communication, asymmetric cryptography is used to decrypt it. In this process, the user validates a transaction with their private key and transmits it to their network. The network verifies the transaction once it has been received and then sends it over the network. During the transaction, both parties validate one another's transactions in order to agree. Once a distributed network has been reached, the node, as the miner, verifies the transaction. Before being affixed to the chain, each block, which contains the transaction, is compared with the previous block in the blockchain. The transmission block, which contains the transaction, is validated with hash-matching technology whether the transaction is valid or invalid. Data can be encrypted or decrypted using a private or public key, depending on how the data is managed and the types of apps used. Blockchain technology is shielded against inaccurate ledger users.

Using the private and public keys, the ledger is maintained and the network is approved by encrypting and decrypting the data. In [24], these classes are demonstrated through concrete examples. In the IoE context, blockchains can be grouped according to authorization and authentication. In a private blockchain, the miners are chosen by the central trusted authority that controls the authentication and authorization mechanism, as shown in Figure 7.3. In blockchain technology the public key does not allow third parties, to make the decision. Blockchain technologies have been addressing major challenges in research for the last few years [25, 26]. Consensus protocols, for example, are fundamental to blockchain technology, hence attacks against them have gained prominence as a subject of study for blockchain academics. Blockchain splits the consensus protocols that control the blockchain. The vulnerability of a recently developed blockchain to cyberattacks has also been discovered [27]. A lot of power must be consumed to maintain multiple blockchains at once [28].

7.2 EVOLUTION OF IOE

Today, the Internet of Everything is an advanced technology that can be connected all over the world. By bringing people and things together in novel ways, building new enterprises and industries, tackling pressing environmental problems, stabilizing unstable countries' economies, reshaping how people are educated, and supporting a wide range of other public and private sector initiatives. To combine these technologies in novel ways to address these kinds of regional, global, and national concerns while promoting corporate innovation and growth. The development of technology, business plans, and people interaction in ways that redefine how businesses produce value is what Mach Nation refers to as the Internet of Everything. A major component of the Internet of Everything is the Internet of Things, as shown in Figure 7.1. Eventually, public and private organizations will eventually implement various business and technology-related measures for economic and societal benefits. Organizations should be aware of the IoE's growing future potential and design their strategies to maximize their success. Using the networks/connectivity, hardware/devices, applications, and platforms depicted in Figure 7.1, this whitepaper

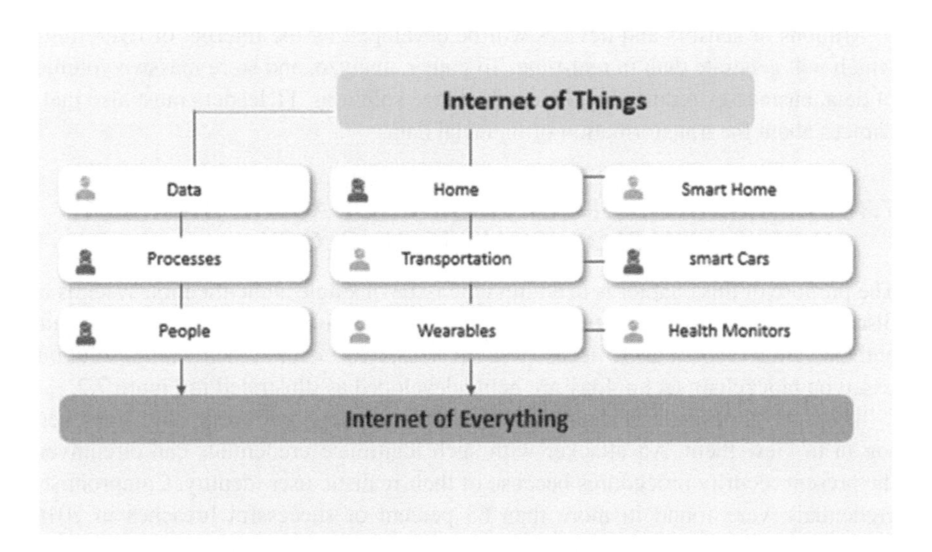

FIGURE 7.1 Concepts of IoT and IoE.

investigates the effects of the IoE's development on business strategy, human interaction with technology, and technology adoption.

The Internet of Everything is a collection of interconnected objects or gadgets. Electronics, sensors, software, and network connectivity are all included in Internet of Things (IoT) devices. These components allow IoT devices and other networked devices to share and exchange data. Current wired and wireless technology is utilized by the Internet of Everything, which also provides a basis for network creation. As a result, for the Internet of Things to work, other systems are required. Before the Internet was created, M2M was the industry standard measurement system since there was human involvement; it was just an interaction between two or more machines. In the Russian army in 1845, the first M2M systems with data transmission devices were developed. It served as the wired data network's brain. When GSM data connections started to be a feature of cell phones in 1995, Siemens offered these systems of M2M technology. By the early 1990s, vending machines were utilizing wireless connections, which had evolved. After developing analog-based M2M technology, the first sensor-based Internet of Things goods, including thermostats and house lighting, was introduced in 2013 and 2014. Significant technological companies like Apple (which launched HomeKit in 2014), Google, and Microsoft are all contributing to the Internet of Things, which is encouraging the future of M2M systems.

A brand-new field called the Internet of Everything has recently emerged, building on the Internet of Things, where objects will gain context awareness, more potent computation, and more potent sensing capabilities. The inclusion of people and information creates previously unthinkable opportunities in a network with billions or trillions of connections. As the number of people, data, and communications has grown, the Internet, which is simply a network of networks have considerably expanded their possibilities. The extent of digital investment is driven by unparalleled speed and flexibility.

Millions of sensors and devices will be developed by the Internet of Everything, which will generate data in real-time. To gather, analyze, and store massive volumes of data, businesses require big data and storage solutions. IT leaders must also make choices about the transformation of financial data.

7.3 THE ARCHITECTURE OF TWO-FACTOR AUTHENTICATION WORKS WITH BLOCKCHAIN TECHNOLOGY

The purpose of this chapter is to address the issue of single authentication systems by discussing two-factor authentication. As more Internet of Everything devices require authentication credentials to access web services, two-factor authentication solutions based on blockchain technology are being developed as illustrated in Figure 7.2.

There are thousands of databases in every company, and every day, employees log in to view them. An attacker with such legitimate credentials can circumvent the present security procedures because of their realistic user identity. Compromised credentials were found in more than 63 percent of successful breaches in 2016, according to the Verizon Data Breach report. Many data breaches involve social media and professional websites being hacked. Two-factor authentication adds an

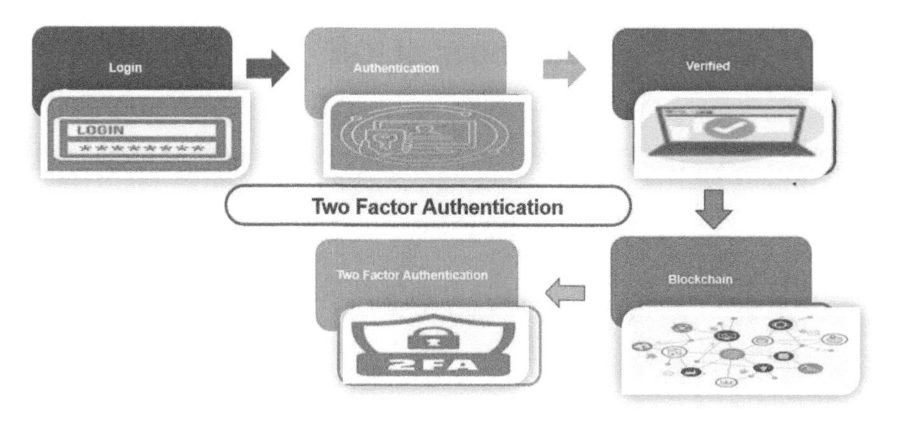

FIGURE 7.2 Two-factor Authentication.

additional layer of security to the existing credential-based system. Something the consumer is aware of is information like a password, the response to a secret question, or perhaps a personal identification number.

- This technique uses a software token, smartphones, other devices, or the second level of authentication based on card information.
- Second part of the process uses biometric information, such as keystroke dynamics and mouse behavior, to validate the user in one of the most efficient methods possible.

As a result of calculating hashes, the blocks that make up the blockchain are connected. Users' digitally verified transactions from a peer-to-peer network are kept in the blockchain network's public ledger. A student used a private key and a public key to decrypt messages for themselves and other users. The key was also provided as a public key to users. Similarly, blockchain theory specifies that a private key is required to verify whether the transaction is valid or not. When a message is encrypted using a public key, cryptography is used to decrypt it. Using their private key, a user verifies and transmits their first transaction to their peers. Once the signed transaction has been received by the peers, they confirm it before sending it over the network. To reach a mutually beneficial agreement, the parties to the transaction mutually approve one another's position in the transaction. Once a distributed ledger has been reached, the miner checks the valid transaction using the hash-matching method to verify the transactions of each block in the blockchain. It depends on how data is handled, what applications are supported, and whether a blockchain can be considered private. The decentralized nature of the blocks provides a higher level of security from malware attacks. The methods are illustrated with specific examples in [24]. According to Figure 7.3, private blockchains are selected by a central trusted authority that controls the authentication and authorization mechanisms in the IoE.

The public blockchain doesn't allow third parties to access the chain network. Recent years have witnessed significant commercial investment [25, 26]

FIGURE 7.3 Challenges of IoE.

as well as considerable academic interest in tackling the main research issues in blockchain technologies. The key elements of blockchain technology include the risks associated with consensus protocols, for instance, which have become a complex area of research in the field. Blockchain splits also pose a challenge to protocols. [27]. To sustain several blockchains, a lot of electricity must be used concurrently [28].

7.4 SECURITY AND PRIVACY CHALLENGES OF IOE

The Internet of Everything is described as a multi-domain environment. IoE is made up of a huge number of services and devices that communicate with one another. Each domain has its own set of security, privacy, and trust criteria. Figure 7.3 depicts some of the challenges associated with IoE securities.

7.4.1 User Privacy and Data Protection in IoE

Privacy is a major issue with the Internet of Everything. Data is also shared and moved online, allowing the Internet of Everything to connect people, processes, things, and data. Within the IoE network, data gathering, sharing, management, and security must all be done confidentially.

7.4.2 Authentication and Identity Management

In the Internet of Everything, identity and authentication are a combination of technologies and processes. The purpose is to secure and control information. Objects can be identified by two communicating parties using identification and authentication.

7.4.3 Trust Management and Policy Integration

In an IoE environment where numerous things are communicating with one another, trust is essential for secure communication. A reliable process is necessary for gaining the trust of users in the IoE context.

7.4.4 Authorization and Access Control

Authorization determines if someone or something is allowed to access the resource once it has been identified. Access to resources is provided or refused based on an extensive list of criteria. Through access controls, authorization is implemented.

7.4.5 End-to-End Security

Security at the endpoints where IoE devices connect to Internet hosts is highly significant.

Session keys and algorithms must be implemented securely for comprehensive end-to-end security.

7.4.6 Attack Resistant Security Solution

There are many types of devices connected to the Internet of Everything. The devices may be attacked in various ways, including denial-of-service attacks, flood attacks, etc.

7.5 APPLICATIONS OF IOE

A growing number of connections will be enabled by the "Internet of Everything," which will link peer-to-peer networks as shown in Figure 7.4. Currently, 1% of the physical world is connected, but that number will grow dramatically in the future. As connections become smarter, faster, and more perceptive, the Internet of Everything will revolutionize the world.

Healthcare: Cisco says secure and reliable communications will allow people to manage their health conditions, and medical practitioners to monitor recuperation remotely.

Retail: In spite of Cisco's claim that stores are too busy to make the high street the center of their operations, the Internet of Everything is expected to bring people back to the high street. Furthermore, retailers can advertise and sell their goods to local customers based on customer behavior, at precise times.

Transport: The smart city will be revolutionized by connectivity and intelligent engineering on highways, railroads, and stations. With the help of digitally interconnected rails and highways, smart cities and Internet of Everything technologies will be able to collect real-time data for travel-planning applications and monitor infrastructure problems remotely.

Energy: Cisco says our thermostats will be connected to sensors that monitor our homes in the future, allowing the UK to conserve more energy.

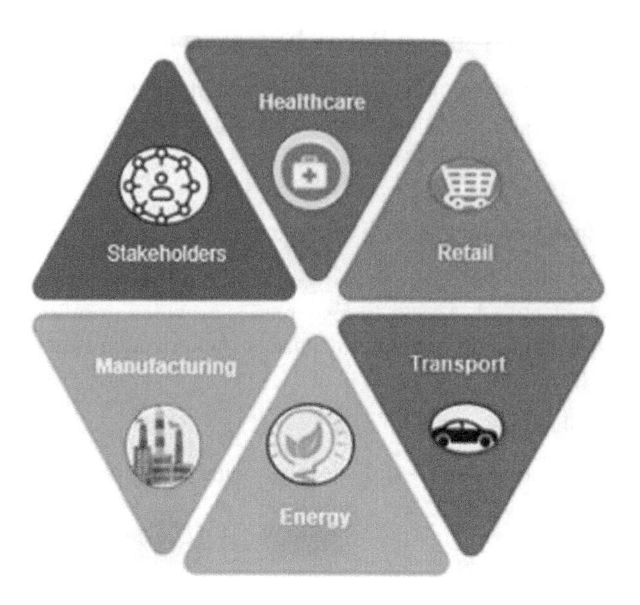

FIGURE 7.4 Applications of IoE.

Manufacturing: To turn disposable products into renewable energy sources of the future, Cisco recommends IoE technology. In order to achieve this, they apply the same efficiency to production as they do to sales. UK organizations in both the public and private sectors must examine their business models closely to take advantage of the IoE's potential. Three factors must be present to effectively utilize these opportunities. Understanding how IoE relates to the corporate environment. In the second step, effective management skills must be invested in and implemented. The third step is to determine which of the groups will enable ground-breaking initiatives that will benefit all parties.

7.6 USE CASES OF IOE

IoE devices now have the ability to increase security and provide transparency to IoT networks thanks to blockchain. For IoT platforms, gadgets, and applications, blockchain provides a decentralized environment. Banks of Deutsche Bank, and HSBC are conducting Proof of Concept to certify the blockchain technology illustrated in Figure 7.5. Many other businesses plan to research the potential of the blockchain. In addition to financial institutions, the Internet of Things offers organizations a vast array of options for carrying out intelligent operations. In the modern world, everything we come into contact with has sensors that transmit data. By combining these two technologies, systems can become more effective. There is a significant chance that the convergence of IoE and blockchain will have an impact on numerous industries. These industries include the pharmaceutical sector, agriculture, water management, supply chain and logistics, and the automotive sectors.

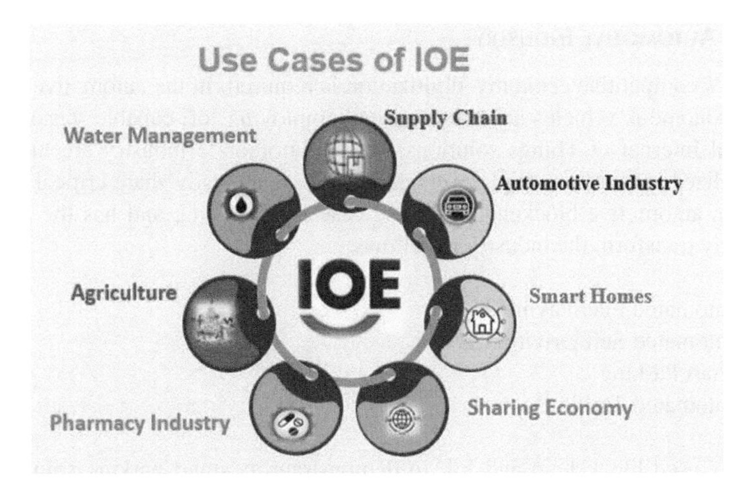

FIGURE 7.5 Use cases of IoE.

7.6.1 Supply Chain

Blockchain includes a wide network of stakeholders, including brokers and raw material suppliers. The process of seeing everything from beginning to end can be challenging. Payments and invoices for the supply chain may take several months to process. A number of stakeholders are involved, which has resulted in delivery delays being the main issue. Therefore, businesses are seeking IoT-enabled trucks so they can track movements throughout the shipment process. Combining IoT and blockchain can improve supply chain and logistics traceability and reliability. A wide range of businesses can benefit from IoE and block chain integration, including:

- Sensors
- Motion Sensors
- GPS
- Temperature Sensors
- Vehicle Information
- Connected Devices

The data is accessible in real-time to the parties listed in the smart contracts after it is saved on the blockchain. Following that, parties in the supply chain can arrange transshipments and conduct worldwide business. Golden State Foods, a supplier known for developing and selling food goods, offers a wide range of items. GSF's goal is to develop and distribute premium things to its over 125,000 restaurant clients. IBM and GSF are collaborating to improve business processes by utilizing blockchain and IoE. To ensure that no severe issues arise, faults are automatically detected and reported using sensor data collected on the blockchain. To improve responsibility and transparency, GSF can use blockchain to build a secure, public, and accessible ledger that is available to all participants.

7.6.2 Automotive Industry

In today's competitive economy, digitization is required. In the automotive industry, fully autonomous vehicles are being created employing IoE-capable sensors. When industrial Internet of Things solutions for the automobile industry are linked to a decentralized network, multiple users can quickly and easily share critical information. The automotive blockchain IoT use case is interesting and has the ability to drastically transform the industry as follows:

- Automated Fuel Payment
- Automated Self Driving Cars
- Smart Parking
- Automated Traffic Control

NetObjex used blockchain and IoE to demonstrate its smart parking solution. The link automates payments made with bitcoin wallets and speeds up the process of obtaining a free parking spot. The company has collaborated with parking sensor startup PNI to identify cars in real-time and assess parking space availability. The cost of parking is computed by IoT sensors, and payment is processed directly with a cryptocurrency wallet.

7.6.3 Smart Homes

IoT-enabled smart gadgets are essential in our daily lives. The home security system may now be controlled remotely from a smartphone thanks to IoT blockchain technology. The typical centralized way of sending IoT device-generated data, on the other hand, lacks data ownership and security standards. Blockchain has the potential to improve the concept of the smart home by overcoming security concerns and decentralizing infrastructure. Telstra, an Australian communications and media business, provides smart home services. To prevent fraud with data collected by smart devices, the company deployed biometric security and blockchain technology. Sensitive user data, including biometric data, is kept on the blockchain to boost security. Voice and facial recognition data saved to the blockchain may only be accessed by the appropriate user and cannot be modified.

7.6.4 Sharing Economy

The sharing economy is now widely accepted throughout the world. Blockchain could aid in the development of decentralized, shared economy systems capable of efficiently distributing things and producing enormous sums of money. Stock uses blockchain technology to distribute IoE-capable items or equipment. The stock market sought to build a Universal Sharing Network in order to develop a secure online marketplace for connected items. Without the requirement of a middleman, any object can be safely rented, acquired, or shared with USN. Any object could be boarded into the USN without permission from the owner or any other third party. Smart contracts provide data privacy and transparency by limiting data access.

7.6.5 Pharmacy Industry

The problem of counterfeit drugs is growing in the pharmaceutical business. The pharmacy industry is in charge of drug development, production, and distribution. As a result, it is difficult to follow the medical course from beginning to end. Using the openness and traceability provided by blockchain technology, medication distribution may be monitored from its origin to its final destination in the supply chain. The ownership of prescription medications is tracked using Medi ledger, a blockchain IoE use case. Transparency and traceability are important when it comes to sensitive medical products. The distributed ledger's unchangeable, time-stamped data is available to:

- Manufacturers
- Wholesalers
- Dispensers
- End-customers

7.6.6 Agriculture

Transparency throughout the supply chain is essential for achieving the highest levels of customer satisfaction, as is growing more food to feed the rising population while minimizing environmental impacts. As a result of blockchain and IoT, the food production sector includes farms, supermarkets, and residences. Installing IoT sensors in farms and instantly uploading data to the blockchain can dramatically improve the agricultural supply chain. Let us look at how IoT blockchain can be used to improve the agricultural supply chain. Farmers may approach agriculture with a new level of knowledge and transparency due to blockchain IoE. The blockchain stores the data collected by IoT equipments installed in fields. Farmers can enhance their farming operations by analyzing the data collected. Distributors, merchants, and customers may all make informed decisions when purchasing a certain crop or food product. Farmers may now use smart contracts on the Pavo marketplace to presell crops instead of waiting for payment after harvest.

7.6.7 Water Management

One trillion gallons of water might be lost annually in the USA due to leaking plumbing fixtures. The Aqua Smart Water Sensor Puck can do the following things:

- keeps tabs on your water consumption;
- automates water shutoff in the event of a leak.

It can monitor and save data using NetObjex's IoT and blockchain technology. Another planned initiative is the use of blockchain and IoT to measure river pollutants. Drone on the Volga is the result of a collaboration between Libelium and Airalab. A drone outfitted with IoT sensors and blockchain technologies is used to collect data on water pollution levels autonomously. The drone captures data from the Volga River's Kuybyshev Reservoir and broadcasts it in real-time on the Ethereum Blockchain.

Scientists can pinpoint the source of pollution thanks to the drone's ability to understand where and when IoT readings were obtained.

7.7 CHALLENGES OF IOE

This research focused on two-factor authentication, which is based on the use of OTP SMS. The proposed architecture overcomes the two biggest drawbacks in the OTP-SMS two-factor authentication technique: the Man in the Middle Attack and third-party assaults. An attack of this type is rendered impossible, and the data is kept secure by encapsulating the OTP message and encrypting it using the user's public key. Furthermore, using a hash value to authenticate users instead of the original OTP is an innovative method that, to our knowledge, has not been researched. In this section, Homelin et al. [29] and Park, Hwang, and Kim [30] will be compared to our proposed framework. To generate the OTP during authentication, send it in plain text to the smart contract for usage with blockchain technology's pseudo-random approach. At this point, an attacker can carry out his MITM attack using the OTP non-encrypted network transfer. The root value computation, which occurs later in the authentication process, also takes a significant amount of time and effort. As a result, many OTPs must be generated from the first one, and their hash values must be calculated. Although the OTP is sent in plain text, this architecture is vulnerable to attacks due to the use of hash values, which makes changing the OTP difficult. This methodology addressed the issue of third-authority attacks by generating the OTP using the blockchain's membership function. However, because the OTP is sent as a text message, it is susceptible to MITM attacks. There is no way to ensure that the OTP is the same as it was when it was formed and has not been altered, which is another attack vector against the malware. In this chapter, the OTP-SMS framework was introduced as a two-factor authentication mechanism utilizing blockchain technology to improve security and overcome login problems. Instead of depending on a third party, we generated an OTP using an Ethereum smart contract, encrypted it with the user's public key, and delivered the encrypted OTP along with the hash value to the requested URL. The user will compute the hash value and send it to the application after decrypting the received OTP with his private key. The website will authenticate the user once it has been received from both entities. An MITM attack cannot access or change the OTP's value in these operations. Furthermore, the blockchain addresses the issue of third-party attacks. Our framework improved security in less time and with less computational power.

7.8 CONCLUSION

The Internet of Everything (IoE) has replaced the IoT. Nowadays, IoE network security is an issue. The key challenges in IoE are security and privacy concerns. Home automation, smart cities, smart agriculture, and smart finance all benefit from IoE security. IoE technology has the potential to revolutionize the planet. Because of you, we are buying and using more sophisticated things. These are stimulated by technological advances in big data, artificial intelligence, semantic operability, and networking. Biometrics is a cutting-edge technology for system authentication. With the advancement of technology, the results have shown a better security standard implemented by combining the Internet of Things with biometrics. It enables better encryption standards to raise the level of security for these fingerprint-scanning systems.

Traditional password authentication systems will have security issues. The IoE integration is intended to simplify authentication by incorporating two-step authentication to access, eliminating the traditional method of using login/passwords, which allows for sharing, forgetting, or guessing. Public and private organizations alike require two-factor authentication based on OTP-SMS technology in the event that the passwords are same. This is due to how the Internet of Everything will impact our societies.

The OTP cannot be accessed or changed by an MITM attack. The issue of a third-party attack is also addressed by implementing blockchain technology to add a layer of protection and handle issues with the login process. Biometrics is a cutting-edge technology for system authentication. With the advancement of technology, the results have proven a better security standard created by merging the Internet of Everything with Biometrics. It enables stronger encryption standards to raise the level of security for these fingerprint-scanning technologies. Traditional methods of authentication on a smart node, such as login/password, do not complement IoT technologies, much less provide adequate security. When it comes to IoT efficiency, traditional authentication methods become a bottleneck. Passwords can be misplaced, guessed, or leaked, jeopardizing security and exposing the endpoint. If the same password is used, the security of additional IoT devices may be jeopardized. Combining passwords with a second factor of authentication to enable two-factor authentication improves security while reducing convenience and user experience. The focus of future work would be on banking and e-payment.

It helps with online payment solutions, blockchain systems, E-tracking systems, and so forth. Businesses and customers will have a more secure purchasing experience as a result of recent advancements in token technology. They also create enormous prospects in eCommerce by connecting channels and speeding up the payment process. It integrates payment data, customer data, and other network tokens into a unified super token. This provides businesses with a comprehensive view of their clients' purchase behavior across channels and card types, while also offering customers a more seamless and secure shopping experience. Visa maintains the highest level of security to protect customer payment information and boost consumer loyalty by making the payment process better.

REFERENCES

[1] IDC. (2020) Worldwide Internet of Things Forecast, 2015–2020, IDC #256397.

[2] IDC. (2019). Worldwide Internet of Things Forecast Update 2015–2019, Feb. 2016, Doc #US40983216.

[3] Miorandi, S. Sicari, F. De Pellegrini, and I. Chlamtac (2012). Internet of Things: Vision, applications and research challenges, Ad Hoc Networks, vol. 10, no. 7, pp. 1497–1516, Sept. 2012.

[4] Puthal, N. Malik, and S. P. Mohanty (2018). Everything you wanted to know about the blockchain: Its promise, components and problems, IEEE Consumer Electronics Mag. vol. 7, no. 4, pp. 6–14, July 2018.
[5] Mukherjee, R. Matam, L. Shu, L. Maglaras, M. A. Ferran, N. Choudhury, and V. Kumar (2017). Security and privacy in fog computing: Challenges, IEEE Access, vol. 5, pp. 19293–19304, 2017.
[6] Swan. (2015). *Blockchain: Blueprint for a new economy*. O'Reilly Media, Inc.
[7] Tschorsch, F. and Scheuermann, B. (2016). Bitcoin and beyond: A technical survey on decentralized digital currencies, IEEE Commun. Surveys and Tutorials, vol. 18, no. 3, pp. 2084–2123.
[8] Nakamoto, S. (2008). Bitcoin: A peer-to-peer electronic cash system. *Decentralized business review*, 21260.
[9] Wilson, D. and Ateniese, G. (2015). From pretty good to great: Enhancing PGP using bitcoin and the blockchain, in Network and System Security. Springer International Publishing, pp. 368–375.
[10] Huang, C., Xu, P., Wang, and Liu, H. (2018). LNSC: A security model for electric vehicle and charging pile management based on blockchain ecosystem, IEEE Access, vol. 6, pp. 13565–13574.
[11] Dorri, A., Steger, M., Kanhere, S. S., and Jurdak, R. (2017). Blockchain: A distributed solution to automotive security and privacy. *IEEE Communications Magazine*, vol. 55, no. 12, pp. 119–125.
[12] Lei, H., Cruickshank, Y., Cao, P., Asuquo, C. P., A. Ogah, A., and Sun, Z. (2017). Blockchain-based dynamic key management for heterogeneous intelligent transportation systems, IEEE Internet Things J., vol. 4, no. 6, pp. 1832–1843.
[13] Kang, J., Yu, R., Huang, X., Maharjan, S., Zhang, Y., and Hossain, E. (2017). Enabling localized peer-to-peer electricity trading among plug-in hybrid electric vehicles using consortium blockchains. *IEEE Transactions on Industrial Informatics*, vol. 13, no. 6, pp. 3154–3164.
[14] Li, L., Liu, J., Cheng, L., Qiu, S., Wang, W., Zhang, X., and Zhang, Z. (2018). Creditcoin: A privacy-preserving blockchain-based incentive announcement network for communications of smart vehicles. *IEEE Transactions on Intelligent Transportation Systems*, vol. 19, no. 7, pp. 2204–2220.
[15] Yang, Z., Zheng, K., Yang, K., and Leung, V. C. (2017, October). A blockchain-based reputation system for data credibility assessment in vehicular networks. In *2017 IEEE 28th annual international symposium on personal, indoor, and mobile radio communications (PIMRC)* (pp. 1–5). IEEE.
[16] Wang, C. (2018). A blockchain-based privacy preserving incentive mechanism in crowdsensing applications. IEEE Access, vol. 6, pp. 17545–17556.
[17] Tian (2016). An agri-food supply chain traceability system for China based on RFID and blockchain technology. IEEE 13th Int. Conf. on Service Systems and Service Management (ICSSSM).
[18] Kang, Z. (2017). Blockchain for secure energy trading in the Internet of Things. IEEE Informatics, pp. 1–1.
[19] Ahram, T., Sargolzaei, A., Sargolzaei, S., Daniels, J., and Amaba, B. (2017, June). Blockchain technology innovations. In *2017 IEEE technology & engineering management conference (TEMSCON)* (pp. 137–141). IEEE.
[20] Gao, J., Asamoah, K. O., Sifah, E. B., Smahi, A., Xia, Q., Xia, H., Zhang, X., and Dong, G. (2018). Grid monitoring-secured sovereign blockchain-based monitoring on smart grid. IEEE Access, vol. 6, pp. 9917–9925.

[21] Liang, G., Weller, S. R., Luo, F., Zhao, J., and Dong, Z. Y. (2018). Distributed blockchain-based data protection framework for modern power systems against cyber attacks. *IEEE Transactions on Smart Grid*, vol. 10, no. 3, pp. 3162–3173.

[22] Aitzhan, N. Z. and Svetinovic, D. (2016). Security and privacy in decentralized energy trading through multi-signatures, blockchain and anonymous messaging streams. *IEEE Transactions on Dependable and Secure Computing*, vol. 15, no. 5, pp. 840–852.

[23] Kshetri (2017). Blockchain's roles in strengthening cybersecurity and protecting privacy, Telecommunications Policy, vol. 41, no. 10, pp. 1027–1038.

[24] Fernández, FragaLamas (2018). A review on the use of blockchain for the internet of things. IEEE Access, pp. 1–23.

[25] Crypto-currency market capitalizations were accessed on 15 June (2018). [Online]. Available: https://coinmarketcap.comhttps://coinmarketcap. com

[26] Blockchain technology report to the US federal committee on insurance (2018). accessed on 15 June 2018. [Online]. Available: www.treasury.gov/initiatives/fio/ Documents/ McKinsey_FACI_Blockchain_in_Insurance.pdf

[27] Bahack (2013). Theoretical Bitcoin attacks with less than half of the computational power. [Online]. Available: https://arxiv.org/pdf/1312.7013.pdf

[28] Ali, J. Nelson, R. Shea, and M. J. Freedman. (2016). A global naming and storage system secured by blockchains. Annual Technical Conference (USENIX ATC), June 2016, pp. 181–194.

[29] Homelink, D. Breitenbacher, A. Binder, and P. Szalachowski (2018). An Air-Gapped 2-Factor Authentication for Smart-Contract Wallets. Dec. 2018.

[30] Park, D.-Y. Hwang, and K.-H. Kim. (2018). A TOTP-based two factor authentication for hyperledger fabric blockchain. Tenth International Conference on Ubiquitous and Future Networks (ICUFN), Prague, pp. 817–81.

8 High Data Priority Endorsement and Profile Overhaul Using Blockchain against Remapping Attack in MANET-IoT

Gaurav Soni, Kamlesh Chandrawanshi, Ravi Verma, and Devdas Saraswat

CONTENTS

DOI: 10.1201/9781003366010-10

8.1 INTRODUCTION

The nodes in the MANET structure provide dynamic connectivity between the source and receiver. Wireless mobile ad hoc networks (MANETs) have emerged as the most prominent area of research in recent years [1] [2]. Mobile ad hoc network is an emerging technology that allows users to communicate without any physical infrastructure, irrespective of their topographical location. As a result, it is sometimes stated as an "infrastructure-less network" but supports a hybrid solution to the problem [3]. MANET is the fastest-growing network because more affordable, smaller, and more powerful devices are becoming available. Self-organizing and adaptable are two characteristics of an ad hoc network, but in terms of security, it will be more vulnerable to attacks. Blockchain is a powerful concept, and by that, we can enhance the security of the route during data forwarding between the sender and receiver. A device that is part of a mobile ad hoc network should have the ability to detect the presence of other devices and carry out the necessary setup in order to facilitate the communication process as well as the sharing of data and services. This will make the mobile ad hoc network more user-friendly [2] [3]. In a decentralized environment with a constantly shifting topology, the problem of message routing becomes more difficult. In a static network, the best path from a source to a destination based on a given cost function is usually the shortest path between two nodes in the network. However, it is hard to apply this idea to a MANET for strong route establishment. A sample diagram for a mobile ad hoc network is shown in Figure 8.1 and with IoT nodes mentioned in Figure 8.2.

Over the course of the last few decades, wireless networks have gained an ever-increasing amount of popularity, particularly in the 1990s, when they were first changed to make them portable and when wireless devices started to become more popular. As a result of the significant rise in reputation of mobile devices (MDs)

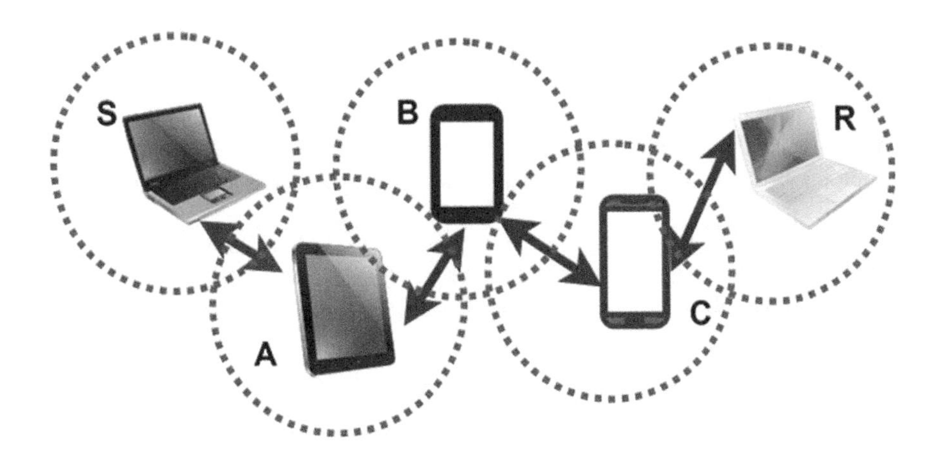

FIGURE 8.1 Example of mobile ad hoc network.

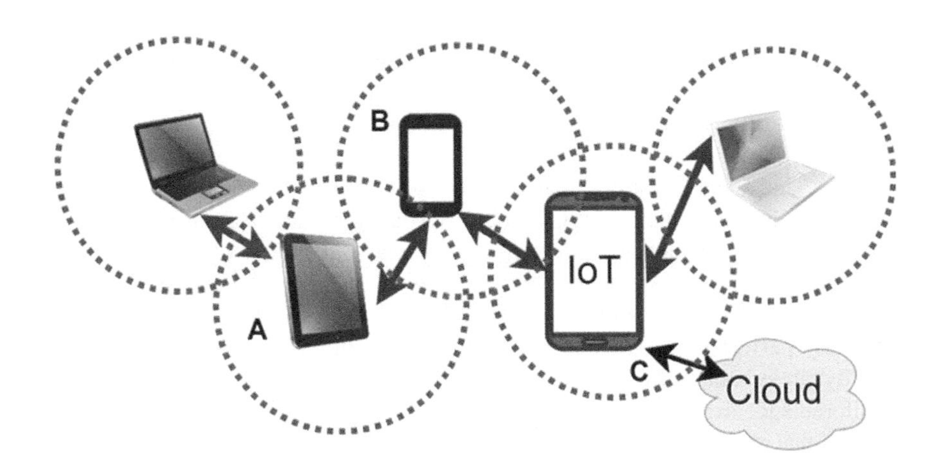

FIGURE 8.2 MANET with IoT.

and wireless networks over the past few years, research into wireless mobile ad hoc networks has evolved into one of the liveliest and research activities in the most active domains of communication and networking. Through the use of ad hoc networking, devices are able to keep their connections to the network intact while also having the ability to easily add and remove devices from the network. Because of the mobility of the nodes, the topology of the network may undergo sudden and unpredictable shifts over the course of its lifetime. The Internet of Things (IoT) plays an important role to control the functions of other IoT devices or other connected devices [4] [5]. The function of the IoT sensor is only to monitor suspicious predictions. The C node is the normal node, and when data is forwarded by the sender or any other node, the IoT node's role is to check the normal and abnormal functioning, and after that, the proposed scheme will activate to capture the attacker presence. The scenario of MANET with IoT is mentioned in Figure 8.2.

In IoT, connecting the devices or things by internet is required for transferring the information in limited areas or transferring the data anywhere in the world. It's possible to connect the mobile node to control the communication or detect the attacker symptoms by applying a reliable security scheme against the attacker in the network [6].

The network is decentralized, which means that the individual nodes are responsible for both the organization of the network and the delivery of messages. Attackers can simply affect the routing performance by injecting unwanted packets or dropping the data packets after connection establishment. There are many types of attackers in MANET and IoT and only they differ in malicious functionality. The aim of some attackers is to consume resources such as bandwidth and energy, and some attacker's aim is to damage data packets [6]. Blockchain technology provides a permanent solution to attackers by using a strong hash code [7].

In a blockchain, the size of each block is the same and all the blocks are interconnected with each other. If the data packets are continuously received by their destination, it means a continuous change in the hash code [7] [8]. Once the hash code is generated, then decryption of the hash code is not possible. When it comes to securing routing in regular and high-speed networks, MANETs are a particular sort of sensor network that has the potential to be employed in a wide range of novel strategies [6] [9]. However, there are still a number of significant obstacles and open questions that need to be answered [4]. The mobile ad hoc network strategy, on the other hand, does not incorporate any fixed infrastructure into any of its components in any way. Every node in a mobile ad hoc network has complete freedom of movement and the ability to make a dynamic connection to any other node in the network at any point in time. Each individual node in the network also performs the duties of a host computer and a router [1]. Because of the very dynamic nature of ad hoc networks, when developing a routing protocol that is tailored specifically for these networks, there are a great deal of unique considerations that need to be taken into consideration. The topology of the network, the routing path, the routing overhead, and the requirement to locate a path quickly and effectively are some of these factors. Ad hoc networks often have a smaller pool of available resources, whereas infrastructure networks typically have a larger pool. Because the transmission range of each node is limited in wireless ad hoc networks, it is not possible for all of the nodes to directly communicate with one another. However, certain nodes may be able to do this. It is frequently necessary for one node to transfer data packets to another node in order to carry out communication across the network. This can be accomplished in a number of ways. An ad hoc routing system must dynamically construct and maintain routes between source and destination nodes in order to function correctly. This is because there is no static network topology and no fixed routes. This research in the field of MANET-IoT is primarily focused on ensuring the safety of the system against remapping attacks by utilizing the blockchain approach.

The remapping attack changed the priority of sending data across the network. This attacker itself gives first priority, and the rest of the senders wait for their chance.

The proposed blockchain method only does the change in the hash code if there is any change observed in the data packets. If only one sender identity is observed during a flood of unwanted data, then the hash code will be the same. The proposed approach only changes the hash code when that contains actual data packets and has different and same priority with limited flooding of routing packets. The rest of the sections of chapters describe the flow of work, such as Section 2 describes local repair concept and section 3 describes important MANET challenges and Section 4 describes routing protocols in MANET-IoT. Section 5 describes block chain security considerations in MANET-IoT and Section 6 describes types of attacks in MANET-IoT. Section 7 describes work done by researchers in the field of security with its Gap and Section 8 describes Problem statement. Section 9 describes Proposed Block Chain security scheme for Remapping attacks, and the simulation tool, parameters, and performance metrics are mentioned in Section 10. Section 11 describes simulation results comparison and finally, section 12 concludes chapter with future work.

8.2 LOCAL REPAIR

In the event that a route is changed, a signaling protocol ought to make use of a mechanism known as "Local Repair" to accomplish the goal of achieving speedy and effective flow restoration [10]. When we talk about something being "local," we mean that the signaling is confined to a small area, and end-devices must not be involved [10] [11].

Consider the following scenario in a network with four intermediary nodes, A, B, C, and D, situated between the source (s) and destination (R). Node S sends data to node R via nodes A and C, the intermediary nodes shown in Figure 8.3.

Along this route, it is presumed that a service of an accurate and high quality is already being provided. Now, at a specific point in time, the node C starts to move, and it finally moves out of the transmission range of the node S. The routing process

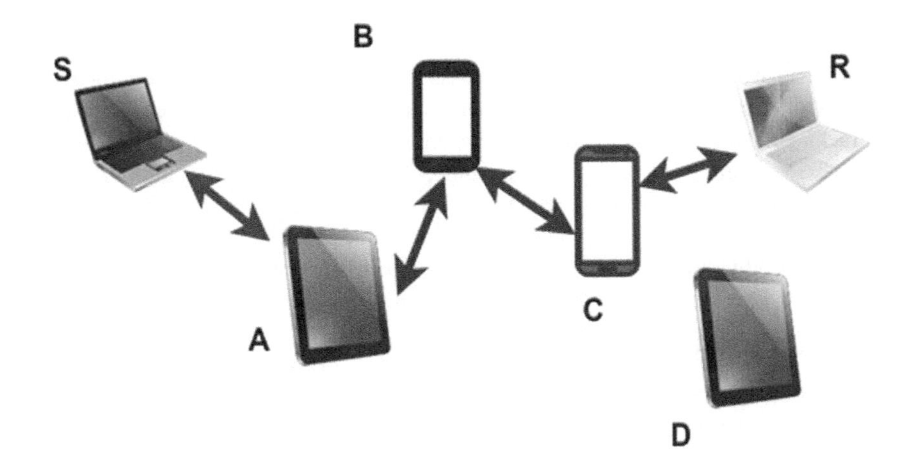

FIGURE 8.3 Route selected S-A-B-C-R.

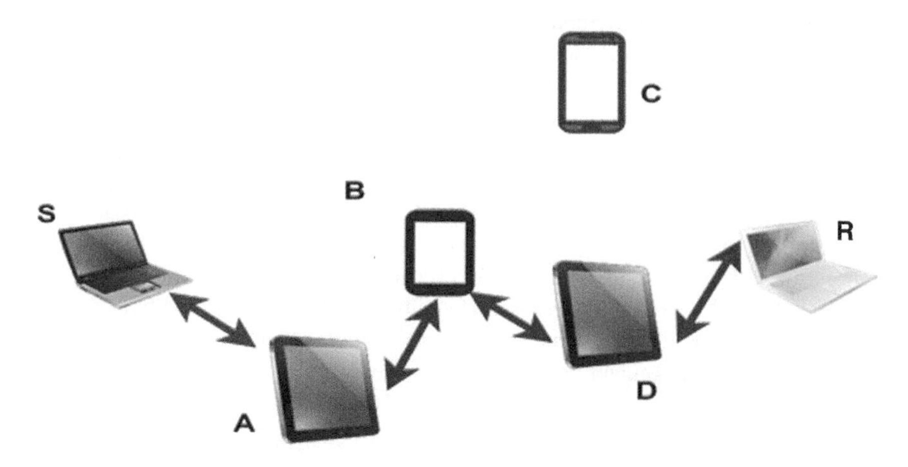

FIGURE 8.4 New Route from S to R.

then identifies a new route from S to R, and it uses node D as the connection point between A and B, as shown in Figure 8.4. Local repair is now in charge of figuring out if the route has changed, making sure that the highest level of service is kept along the new route, and canceling any reservations that have already been made. The objective is to reestablish the connection as rapidly as humanly possible while at the same time sustaining a low level of signaling overhead. There are a few different ways to accomplish this, the most general of which are known as "Proactive" and "Reactive" methods.

8.2.1 Proactive Approach

Local repair mechanisms that take a proactive approach make an effort to maintain the required quality of service for a particular flow in advance, which means before the old path is severed and a new one is constructed. This is done in order to prevent disruptions to the flow and in an effort to prevent disruptions to the flow of data and can be accomplished in a few different ways like by trying to anticipate potential route changes in advance, by tracking the movement of nodes, by measuring transmission quality, or by allocating an excessive number of resources. Signaling, in the latter scenario, simply reserves resources on every feasible path, which can ultimately be determined by inspecting the routing forwarding table. The problem with both of these methods, though, is that they can waste bandwidth by allocating and reserving too many resources.

8.2.2 Re-Activate Approach

No resources are reserved in advance for the reactive signaling protocols to use; rather, they make it a priority to respond as quickly as possible to any changes in the route. Being triggered by the routing layer whenever a route change is made is the most straightforward method for accomplishing this objective. Sending monitoring messages at regular intervals is an additional approach that could be taken to solve the problem. This would allow for the detection of any changes in the flow's route. Probably, the frequency with which monitoring signals are issued has little to do with the efficacy of this method.

8.3 IMPORTANT CHALLENGES IN MANET

Despite the fact that MANET has a number of challenges that are appealing to the researchers to do continuous work on it, the challenges are affected by the performance of the network [12] [13]. The some of the challenges are as follows.

8.3.1 Dynamic Topologies

Since nodes are free to move wherever they want, the network topology, which is usually multi-hop, can change quickly and in unpredictable ways. Depending on the situation, the network may have both one-way and two-way links.

8.3.2 Routing

The topology of the network is in a state of constant change and, because of that, the task of routing packets between any pair of nodes is one that presents a great deal of difficulty. The majority of protocols ought to be constructed on top of reactive routing rather than proactive routing. An additional obstacle is presented by multicast routing, which arises from the fact that the multicast tree is no longer static as a result of the random movement of nodes within the network. Communication between nodes that takes place over just one hop is simpler than communication that takes place over multiple hops between the same nodes. It is necessary to identify relevant newly moved nodes and inform them about their existence in order to facilitate automatic route selection that is optimal for the device. Linking with fluctuating capacities and limited bandwidth, the capacity of wireless links will always be far lower than that of their cable counterparts.

8.3.3 Power-constrained and Operational

It's possible that some or all of the nodes in a MANET get their energy from batteries or some other type of resource that depletes over time. When it comes to these nodes, energy conservation may be the single most important design criterion for the system that needs to be optimized. Communication-related functions should ideally be optimized for low power consumption in the majority of lightweight mobile terminals. As well as power-aware routing, it is important to think about ways to save power.

8.3.4 Security and Reliability

In addition to the vulnerabilities that are common to wireless connections, an ad hoc network has its own unique security issues. For example, a malicious neighbor could relay packets, which could compromise the network's safety. The capability of distributed operations necessitates the utilization of several distinct authentication and key management systems. The limited wireless transmission range, the broadcast nature of the wireless medium (such as the hidden terminal problem), mobility-induced packet losses, and data transmission errors are all problems that are introduced by the characteristics of wireless links. In addition, wireless link characteristics also introduce reliability issues. In general, mobile wireless networks are more susceptible to threats to their physical security than fixed-cable internet systems are. The increased likelihood of attacks like eavesdropping, spoofing, and denial of service is something that must be carefully thought through.

8.3.5 Quality of Service (QoS)

It will be difficult to provide different quality of service levels in an environment that is constantly changing. Because the quality of communications in a MANET is inherently stochastic, it is challenging to provide a device with fixed guarantees regarding the services that are being provided to it. If multimedia service support is going to be added to traditional resource reservation, there needs to be an adaptive quality of service (QoS) implementation.

8.3.6 Inter-networking

In addition to communication taking place within an ad hoc network, inter-networking between MANET and fixed networks, the majority of which are based on IP, is typically something that is anticipated in many situations. The problem with mobility management is that a mobile device like this must support multiple routing protocols at the same time. In order to support wireless communications between multiple parties, multicast is a desirable protocol. Now that the multicast tree is no longer static (leave and join), the multicast routing protocol must be able to handle mobility, including changes in which is a part of a multicast group.

8.3.7 Mobile Routing at the IP or Routing Layer

An improved routing capability with mobile nodes at the IP layer can give a benefit that is comparable to the intention behind the creation of the original internet, which was to provide "a felicitous internetworking ability over a heterogeneous and homogeneous infrastructure for communication in a network." The problem with the diffusion hole is that nodes that are located on the boundaries of holes may suffer from excessive energy consumption. This is due to the fact that geographic routing has a tendency to deliver data packets along the boundaries of holes using perimeter routing if it needs to circumvent the hole. Because the nodes at the edges of the node require a lot of energy, this could make the hole a lot bigger.

8.4 ROUTING PROTOCOLS OVERVIEW

There are three main categories of routing protocols in ad hoc networks, each of which is determined by the other two [14] [15]. The routing protocols of wireless

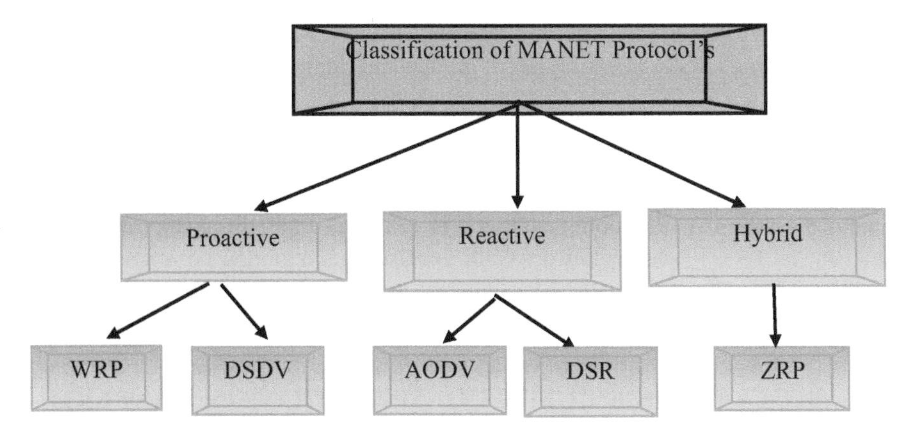

FIGURE 8.5 Classification of routing protocols.

network are different and the routing protocols for dynamic network are different. Each node functions as a sender, router, and receiver, making rapid routing only possible through protocol. The classification of routing protocol mentioned in Figure 8.5.

8.4.1 Proactive Routing Protocol (Table-driven)

In this active routing environment, wireless mobile nodes at various distances from one another will exchange data via a multi-hop process. Route tables are dynamic and updated based on the actions of each mobile node. All nodes will be informed of the path's current status when a change in the network's topology renders the previously established path invalid or when a new path is established. The route is constantly maintained so that the node may always be reached using its own routing tables, even when it is not actively used. There is a significant waste of wireless bandwidth and wireless node power because of these agreements, but we can mitigate this by increasing the interval between broadcasts. This, however, will cause the path table to no longer accurately reflect changes in the network's structure. Preventive procedures include Destination-Sequenced Distance Vector Routing (DSDV) and Wireless Routing Protocol (WRP).

8.4.2 Reactive Routing (On-demand)

When it became necessary to send packets exclusively, preparations were made to transmit the routing table. In order to communicate data to another wireless node, the sending node will initiate a route discovery procedure and save knowledge about the paths between the nodes. The first phase of a ratio of such agreements in each node must have passed before the path may be considered genuine. Less information is required because you won't be storing the full topology of your network. This agreement's primary benefit is its reduced bandwidth consumption; nevertheless, not all wireless nodes that transfer packets can reliably do so in a short amount of time. It takes longer on average if symmetric links are unavailable, as the path discovery process can create delays. Ad hoc On-Demand Distance Vector (AODV) and Dynamic Source Routing (DSR) are two types of on-demand routing protocols.

8.4.3 Hybrid Routing Protocol

This is a hybridization of the best aspects of the two protocols discussed above. It is said that a node is in the routing zone if it is within a certain distance of the node in question or if it is located within a specific geographical region. A proactive approach is utilized for routing within the zone, whereas a proactive routing protocol is utilized for routing outside of the zone. ZRP is the example of hybrid routing protocol.

8.5 BLOCK CHAIN SECURITY CONSIDERATIONS IN MANET-IOT

Generally, wireless mobile ad hoc networks are more vulnerable to physical security risks than static networks that rely on hardwired connections. Within wireless network, the existing link layer security mechanisms, such as encryption, are frequently utilized in order to mitigate the effects of these threats [16] [17]. The most important problem at the network layer is authentication between routers before they may communicate network configuration data. This is because link-level encryption is not present at this level. The group will investigate various different authentication levels, from completely unsecure through at all (which is always an option) and straightforward shared-key methods, all the way up to comprehensive authentication mechanisms on the bases of public key mechanism. The efforts of the working groups may be supplemented by the standardization of several different optional authentication modes that can be used in MANETs-IoT. Specifications demanded for an ad hoc network's safety. The subsequent requirements for the security of an ad hoc network are:

- Unforgeable route signaling is required.
- No fictitious routing messages may be inserted into the network.
- Changing the routing of messages during transmission is not possible.
- It is impossible for malicious users to create routing loops through their actions.
- Malicious users are unable to deviate in any way from the shortest possible path.
- Unauthorized nodes should be kept out of route computation and new route creation.

Every routing protocol needs to incorporate a fundamental collection of security mechanisms into its design. The aforementioned are examples of mechanisms that aid in the prevention, detection, and response to security. Five key security goals that are needed to be addressed in order to secure an ad hoc network environment are:

1) Confidentiality
   Preventing sensitive information from falling into the hands of unauthorized parties. This is much harder to do with ad hoc networks because intermediate nodes (nodes that act as routers) get packets that are meant for other people, which means they can listen in on the information being sent.

2) Availability
   Whenever there is a need for them, services should be accessible. Even in the face of a denial of service (DoS) attack, there ought to be some guarantees of continued operation. A high priority capturing technique can be utilized by an attacker on both the physical and the media access control layers in order to disrupt communication on the physical channel. An attacker can cause disruptions in the routing protocol at the network layer. An attacker could take down higher level services, such as the key management service, if they gained access to higher layers of the network.
3) Authentication
   It is a verification that an organization of concern or the purported source of a communication is in fact who or where they claim to be. Without it, an adversary would be able to impersonate a node, gaining unauthorized access to resources and sensitive information as well as interfering with the operation of other nodes.
4) Integrity
   The message being transmitted is never altered. During the course of the communication, the data packets are protected from any malicious users. The malicious user did not change the content of the actual information that was sent by the sender in network, nor did they delete it.
5) Non-repudiation
   This eliminates the possibility of either the sender or the recipient denying that they ever sent or received the message. In this case, the data was only received by the real sender. If any other bad nodes had tried to get the data, the security system would have been able to find them quickly.

All of the aforementioned security mechanisms have to be incorporated into any and all ad hoc networks in order to guarantee the safety of any transmissions that take place over those networks. So, whenever we think about the security of a network, we should always make sure that the five security goals we talked about earlier have been met and that none of them (or at least the vast majority of them) are wrong.

8.6 TYPES OF ATTACK IN MANET-IOT

The communication protocol that allows for the delivery of data between wireless devices is provided by routing, which is an important operation. It is possible to provide a secure system by either preventing attacks or detecting them and then providing a mechanism to recover from those attacks. Both of these approaches are viable options. There are different types of attacks in MANET-IoT [18] [19]. Attacks in MANET-IoT are:

8.6.1 The Active Attack

An intruder engages in a passive attack when they observe the data being exchanged without attempting to change it. The attacker does not knowingly engage in intentionally malicious behavior in order to deceive other hosts. The objective of the attacker is to gain access to information that is being transmitted, which would constitute a breach of the confidentiality of the message. Because these attackers do not prevent the network from functioning normally, it is difficult to find them.

8.6.2 Passive Attack

Active attacks are those in which the person conducting the attack takes an active role in disrupting the normal operation of the network services. It messes up the routing procedures and brings down the performance of the network. Internal and external attacks are both types of active attacks that can be carried out.

8.6.2.1 External Attack

Nodes that are not authorized to be a part of the network are the ones responsible for carrying out attacks from the outside. In the event that an organization is attacked from the outside, it is possible for the parking lot in front of the office to become a source of communication disruption.

8.6.2.2 Internal Attack

Attacks from within the network come from compromised nodes that were once fully functional components of the network. When compared to attacks from the outside, those that originate from authorized nodes within ad hoc wireless networks are significantly more damaging and difficult to identify.

8.6.3 Security Threats in MANET

The performance of MANET has been impacted by some flaws and vulnerabilities that have allowed malevolent and undesired nodes to attack and take advantage of it. The following is a list of some of the most frequent attacks in MANETs are mentioned in Figure 8.6.

1) **Remapping Attack**
   The goal of this attack is to ensure that the malicious node is included in the path taken by the route. During the phase of route discovery, malicious nodes will send RREQs to the neighbors of the target node. These RREQs will be sent by the malicious nodes. The behavior of the attacker is consistent with what is expected in the network; however, it is attempting to gain the opportunity to route on first priority. It is also known as a "high priority capture attacker," and its purpose is to devour all of the valuable resources in the target. Because the activities of a remapping attacker appear to be normal at first, it can be challenging to identify one. However, after a period of time, unwanted packets can distract the entire routing procedure.
2) **Location Disclosing Attack**
   The privacy requirements of the ad hoc network are the focus of the attack that discloses location information. The aim of the attacker is to find the location of the destination node in the network by performing traffic analysis or using

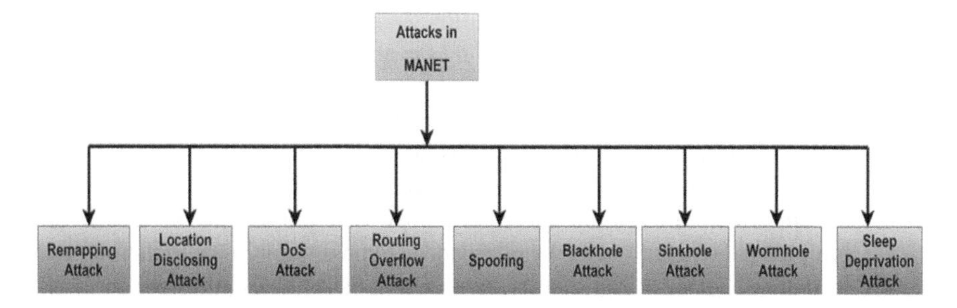

FIGURE 8.6 Attacks in MANET.

simple monitoring. Because the attacker is familiar with the intermediary nodes, they are able to locate the node of concern and acquire information regarding the structure and topology of the network.

3) **Denial of Service (DoS) Attacker**
The uninterrupted operation of the network is severely hampered as a result of this attack. This attack is carried out by continuously sending packets into the network, which forces the targeted node in the network to process those packets and keep its resources occupied, which ultimately results in the targeted node crashing. The attacker prevents the legitimate RREQs from being dropped by the targeted node by using this attack, which keeps the targeted node busy processing the attacker's fabricated packets. This attack has the potential to bring down the infrastructure of the network.

4) **Routing Overflow Attacker**
The goal of this attack is to use up all of the available space in the routing table buffer. The traffic that is introduced into the network by the attacker is fabricated and false. It makes modifications to the packets and adds in the routes to nodes that do not exist. This attack floods the routing table buffer with false routing information, which causes it to become overwhelmed. Because of this, processing and storing valid routes becomes impossible because the routing table is already at capacity.

5) **Spoofing or Impersonation**
The true identity of the one carrying out the attack is meant to remain a secret throughout the course of this assault. During this type of attack, the adversary takes on the persona of a node in the network that has a higher level of trust. If this is done, the other nodes will include the malicious node in their routing path. Once this occurs, the malicious node will be able to interfere with the normal operation of the network without drawing attention to itself.

6) **Black Hole Attack**
This attack is carried out with the intention of making the network more congested. During this type of attack, the malicious node does not forward any of the packets that have been forwarded to it; rather, it discards all of them. Because of this attack, the packets that are being forwarded by the nodes do

not make it to their intended destination, and as a result, the congestion that is already present in the network becomes even worse.

7) **Sinkhole Attack**
The goal of the attacker in this particular attack is to direct all of the traffic on the network towards themselves. This attack is carried out by the adversary by deceiving the neighboring nodes into thinking that the quickest route to the target lies through the adversary's node. The purpose of this attack is to force all of the traffic from the other nodes to pass through the malicious node so that the attacker can modify, fabricate, or simply listen to the incoming packets.

8) **Wormhole Attack**
The replaying of the packet on the opposite side of the network is the primary objective of the wormhole attack. This attack is carried out by two nodes working together to create a wormhole in the network. The adversary on one side tricks the nodes into thinking that the distance to the destination is only one hop when, in reality, the distance is greater than one hop. The result of this is that the attacker pulls in all of the traffic from one side of the network and relays it through the wormhole. Meanwhile, the attacker on the other side of the network replays the same packet. When the attacker does this, they are able to steal packets or gain unauthorized access to any service.

9) **Sleep Deprivation Attack**
The attacker's objective in this type of attack is to maintain a high level of activity at the node they are trying to compromise. The first step in this attack is to flood the network with routing traffic, which causes the node to use up all of its available computing power as well as its battery power. This attack causes the targeted node to waste its battery, network bandwidth, and computing power by sending spurious requests to destination nodes that either exist or do not exist. As a result, the targeted node is unable to process the legitimate requests that are being sent to it.

8.7 LITERATURE REVIEW

In this section we discuss the previous efforts of the authors to secure the network from the remapping attack. The remapping attacker alters the priority, but the malicious behavior of the node remains unchanged. Some of the author's contributions are mentioned here.

In Saxena and Khule [20], an algorithm for detecting and protecting against traffic remapping attacks on mobile ad hoc network communication was designed. First, they set all of the variables to their default values and observed how the remapping attacker behaves. If any node in the network is observed to be sending an undefined type of packet at an extremely rapid rate, this shows that the remapping attack shifted the priority of network traffic and harmed the network's protocols. Analyzing the behavior of the traffic and the traffic's data sending rate-based methodology was what led to the discovery of the traffic remapping attack. They are not working to identify the data lost owing to congestion or attackers, which is a research gap. Various attackers and various congestion symptoms. So, how can you differentiate it?

Konorski and Szott [21] offered a game-theoretic approach, which means that the selfish stations are viewed as (bounded) rational players striving to maximize some payoff in a game that does not entail cooperation. They inquire as to whether or not the possibility of detection can act as an incentive for a selfish station to continue to be truthful when the TRA has the potential to harm the service levels obtained by other honest stations. While this is going on, traffic remapping needs to be allowed in situations where it won't negatively impact other stations because it has the potential to boost the service level that the attacker provides. Although they focus research on EDCA because it is a practical means of carrying out a TRA, the conclusion they reached from this research is that they can apply security scheme to other distributed networks in which similar attacks are a risk. The lack of research on the likelihood that an attacker may affect routing performance is a research gap. Although the existence of an attacker affects network performance, the amount of bogus or pointless packets that attackers overwhelm the system with is not known.

Konorski and Szott [22] proposed a MAWiN-oriented defense mechanism that could be justified by noncooperative game-theoretic considerations. The goal of this mechanism was to make TRAs unbeneficial to attackers in terms of perceived quality of service. In order to generate closed-form solutions and, by extension, helpful payoff functions for emerging games, it is necessary to use straightforward MAWiN performance models subject to TRAs. There are a few models of this kind, but none of them can capture on a macroscopic level the complex interplay of channel access, node mobility, and intraflow competition caused by multi-hop forwarding when hidden nodes are present. The fact that they designed a wireless network scheme and made the assumption that performance would improve with IoT represents the primary limitation of their research. The research gap measures the loss of data percentage in the attacker presence on edge, interior, and corner. There is no need to identify the attacker in all locations. The only focus is on the symptoms and infection by attacker.

Abdel-Sattar and Azer's [23] work integrates blockchain technology in ad hoc networks was proposed, along with the drawbacks that come with it. In addition to this, we discuss the most effective methods, from our point of view, for overcoming these limitations and achieving the greatest possible success with blockchain technology in MANETs. The assumption that a node or user has a sufficient stake in the system underpins this approach, which can be thought of as a technique. The primary limitation of the research is that the results are not evaluated to demonstrate that the proposed scheme is able to protect users from malicious nodes. This is the most significant gap in research.

Thura Lwin, Yim, and Ko [24] proposed a block chain-based trust system that gives nodes in MANETs trust that is distributed, consistent, and can't be changed. It has been studied in a lot of different areas, such as wireless networks and the Internet of Things (IoT), [25] and it has the potential to be a solution for trust management. A trust system needs to take into account the limited resources of MANETs in order to make the most of the fact that blockchain technology is decentralized. The study gap relates to network performance based on trust value, yet trust value alone does not measure the existence of attackers. Measurement of trust value with consideration for other nodes.

Goka and Shigeno [26] proposed a system for managing trust and rewards that is spread out (DMTR). The DMTR uses a block chain to improve on the traditional reputation system, which gives each node a score based on how it acts, and the price-based system, in which a sender node gives points to relay nodes. The DMTR does this by using a price-based system. Specific mining nodes are responsible for managing the blockchain, and general nodes are responsible for reporting information regarding neighboring nodes in the network to specific mining nodes. The application of PoW has been removed from MANETs, which has allowed the mining nodes to work together to produce a single target block. Despite the fact that the authors of this work have proposed a decentralized management system, the management of a blockchain relies on a centralized authority in conjunction with some trusted nodes that have already been established. This is due to the fact that the processing power required for blockchain operations is beyond the capabilities of MANET routing nodes.

Soni and Sudhakar [27] proposed a L-IDS scheme security mechanism against black hole attack in WSN-aided IoT. Through the use of the wireless link, the sensor nodes are able to establish connections with other nodes and engage in the exchange of data routing as well as data packets. IPv6 addresses the issue of routing by utilizing the RPL routing protocol as its primary means of communication. The proposed intrusion detection system was able to verify the presence of the blackhole attacker and put a stop to the latter's malicious activity within the blackhole network. The fact that they are not working on TCP and UDP end-communication is the most significant limitation of the research.

Szott and Konorski [28] have demonstrated the potentially harmful effects of TRAs in single-hop wireless LANs and provided some illumination on those effects as well as potential preventative measures. To begin, it is not entirely clear how far a TRA's effects will reach because of the intricate relationship between MAC contention, interference from hidden stations, and transport-layer flow control. This interaction also causes a distortion in quality of service perception and makes it difficult for honest stations (stations that are not the attackers) to determine with absolute certainty that a TRA is occurring. A TRA operation carried out on a victim flow with a long path may not be significant given that the throughput of the victim flow is already quite low. The biggest problem with research is that it can only explain both the bad and good effects of flooding in terms of congestion and the person who caused it.

The research gap in recent research [20] [21] [22] [23] [24] [25] is mentioned at the end of work done by the respective authors. The main gap is only to not measure the data dropping or flooding and malicious behavior of attacker only.

8.8 PROBLEM STATEMENT

Ad hoc wireless network with IoT is stimulating research in order to allow portability and availability everywhere, but number of challenges, like security, power consumption, resource access, and so on. In order to detect and protect MANET-IoT against Remapping attack, this work will create a system for safe and efficient power utilization. Remapping attack is a resource-consumption attack where the attacker node changes the communication priority and uses the power of a mobile node that is within range of the attacker's network. Attacker nodes overwhelm the network with unnecessary packets, shortening the network's lifespan.

The remapping attack can be performed locally or remotely, and it is one of the most popular types of security assaults because it takes only ordinary, affordable resources and does not require a high level of technical expertise. Multiple tactics, including direct active attacks, contribute to the rapid growth of packet flooding and packet complexity. A security issue like this prevents authorized users from accessing the wireless channel by interfering with network operations, affecting network connectivity and availability. In mobile ad hoc networks, misbehavior in routing is the primary issue. Here routing misbehavior occurs due to the presence of an attacker. The behavior of the jamming attacker is to inject a huge number of packets that are received or forwarded and routed through that node and this degrades the network performance. Due to heavy packets jamming, not only is the routing affected, but also other protocols (like other layer protocols) that function in network, for example:

- Packet dropping increases
- Packet Delivery Ratio decreases
- Packet receiving decreases
- Overhead increases
- End-to-End delay increases.

8.9 PROPOSED BLOCK CHAIN TECHNIQUE FOR REMAPPING ATTACK

Security is an important parameter for data communication. A freestanding system, which is more secure than nodes connected to the network, is the target of an attack when a node communicates with other nodes in a network. In this proposal, we take the network architecture as mobile ad hoc IoT devices, where each node is capable of performing routing without the need for any other networking devices. But the nature of mobile ad hoc devices is self-configured, dynamic, and movable, which creates more vulnerability against secure communication. The network is less reliable as compared to static networks or wired-based networks due to their nature.

In a MANET-IoT scenario, when any device needs to send data to a receiver node, other mobile nodes provide intercommunication in a multi-hop manner, increasing the possibility of vulnerability because all intermediate nodes are free for centralized control, requiring self-decision to provide communication support. The abovementioned nature of MANET-IoT devices makes the network more vulnerable during communication, resulting in more unsecured communication. The chances of attack in a network are data link layer attack, wormhole, blackhole, grey hole, denial of service, or remapping attack during communication. In this study, we take vulnerability type as the remapping attack because very few researchers have done work in the field of remapping attack security. A remapping attack is a type of priority inversion attack where an attacker node increases the priority of taking network resources, that is, channel, intermediate devices for unwanted message flooding, and so on. Detection and prevention of remapping attacks is a challenging task because no centralized controller exists in the network to monitor the activity of participating nodes in the network and all devices are unreliable when moving from one location to another.

This article proposes a block chain-based security technique that provides high security as compared to other techniques because it uses the hashing technique for communication security. The term "block chain" refers to a specific type of DLT in which a growing list of records, termed "blocks," are dispersed across a network and cryptographically linked together. The proposed security technique uses the hashing function of SHA1 and creates block chain security. SHA1 is applied to every mobile node that participates in communication. Initially, the source node starts the execution because it is a message generator. While input text comes into the network layering architecture, it converts into a 448-bit message. After that, we include the 64-bit message length field and create the 512-bit long packet. This is called the "final message" (FM). The next step is to divide a 512-bit message into 16 words, each with 32 bits, and perform 80 iterations for each word using the SHA1 algorithm, which uses the XOR and circular shift operations defined in the algorithm section of this article, to generate the five hash code values, which are then sent to the next connected node that participated in the data communication or within the source-to-destination route. The message digest is received by the next connected mobile device, and in the same way applies SHA1-based message digest generation and forwards the message to the end of the receiver where the message is received by the receiver node which decrypts it and also gets the hash value as 0. This means the message is authorized and concludes that no one can modify the message or any mis-activity can be performed from the data.

The data is forwarded to the intended receiver using the same SHA1-based algorithm, but when the message reaches the receiver node and is decrypted, the hash value is not obtained as 0 (since any node is treated as a remapping attacker), wants to change the priority of data transmission, wants to change the route, and so on. This means someone can modify the message, which is identified by the source id contained in the packet. The decrypted message is further analyzed by the receiver node and if the message does not match the criteria of the network, it is treated as an unwanted message by the receiver and immediately sent to the attacker information into the network to block the node from further communication with the attacker node.

8.9.1 Proposed Algorithm to Hash Generation for Block Chain-based Security

Input:

M = Input text

Initialize:

Random string of hex characters

$H_{[0-4]}$ = 32bits * 5 (hash characters)

Step1:

Message padding converts into 448 bits

Message Length is represented by 64 bits which is added to the end

Final Message (FM) = 512 bits long

FM is divided into 32 bit each chunks

Step2:

$Wt_{[0-15]}$ = sixteen word each 32 bits

Step4:

Perform 80 iteration for each $Wt_{(i)}$

for i 16 to ≤ 79

$Wt_{(i)} = CS^1 (Wt(i-3) \oplus Wt(i-8) \oplus Wt(i-14) \oplus Wt(i-16))$

End for

$CS^n(x)$ = The operation of circular shift on word x by n bits (in between 0 to 32

$CS^n(x) = (x << n) \| (x >> 32 - n)$

Step5:

Store (A, B, C, D) = $H_{[0-4]}$ respective in increasing order

Step6:

for i 0 to ≤ 79

temp = CS^5 * (A) + f(i, B, C, D) + E + Wt(i) + K(i)

End for

Where

for i 0 ≥ to ≥ 19

$f(i, B, C, D) = (B \wedge C) \vee ((\neg B) \wedge D)$

K(i) = hex (32bits)

End for

for (i 20≥ to ≥ 39)

$f(i, B, C, D) = B \oplus C \oplus D$

K(i) = hex (32bits)

End for

for i 40≥ to ≥ 59

$f(i, B, C, D) = (B \wedge C) \vee (B \wedge D) \vee (C \wedge D)$

K(i) = hex (32bits)

End for

for (i 60≥ to ≥ 79)

$f(i, B, C, D) = B \oplus C \oplus D$

K(i) = hex (32bits)

End for

Step7:

Reassign Variables:

E = D

D = C

C = CS^{30} (B)

B = A

A = TEMP

Step8:

Store result has value of all chunk

$$H_0 = H_0 + A$$
$$H_1 = H_1 + B$$
$$H_2 = H_2 + C$$
$$H_3 = H_3 + D$$
$$H_4 = H_4 + E$$

Step9:

Final Digest message as 160 bit string

$$HH = CS^{128}(H_0) \vee CS^{96}(H_1) \vee CS^{64}(H_2) \vee CS^{32}(H_3) \vee H_4$$

Hash digest message send to receiver node

Step10:

Receiver receives digest message and decrypt by inverse process
If hash value ==0 **then**
    Authorize Message
Else
    Not authorize
End if

In this technique, every node contains their previous hash code and develops a chain link while any one makes an update to generate a hash or insert unwanted data is detected through its hash value, which is not found as zero at the receiver side.

8.9.2 Proposed Working Architecture

In this section we describe the working architecture of block chain security against a remapping attack as mentioned in Figure 8.7. The architecture shows the communication from source to receiver which includes the hash code generation and remapping attacker node detection as well as prevention. The key plan is to visualize the attack effect on a network and apply the novel hash function to recognize the device or nodes that are affecting the performance. The role of IoT nodes is to control the routing performance after getting malicious information. The IoT nodes only measure the normal routing information by making a profile of successful data reception and flooding of data by high priority. Remapping attack is given a high priority for data transmission before launching a flooding attack to impede network performance. High priority values are unfairly received by senders, and the primary goal of this kind of sender is to negatively impact normal routing performance. Blockchain determines the high data profile recommendation by examining each node's profile and generating a hash code for each legitimate and fraudulent user.

The SHA1 algorithm is used for the development of the next hash code block as the source node starts to generate the input text that will be passed to the SHA1 module and generate the hash code in accordance. When the packet reaches the receiver's end, it is checked to see if the message is genuine or not. If it is not, the packet is discarded, and a thorough internal packet analysis is used to determine the attacker node's identity. When a node receives information from a packet that is undesirable and changes its priority, the hash value is not equal to zero, thus it is discarded. Instead, the node broadcasts the attacker's information into the network so that it can no longer

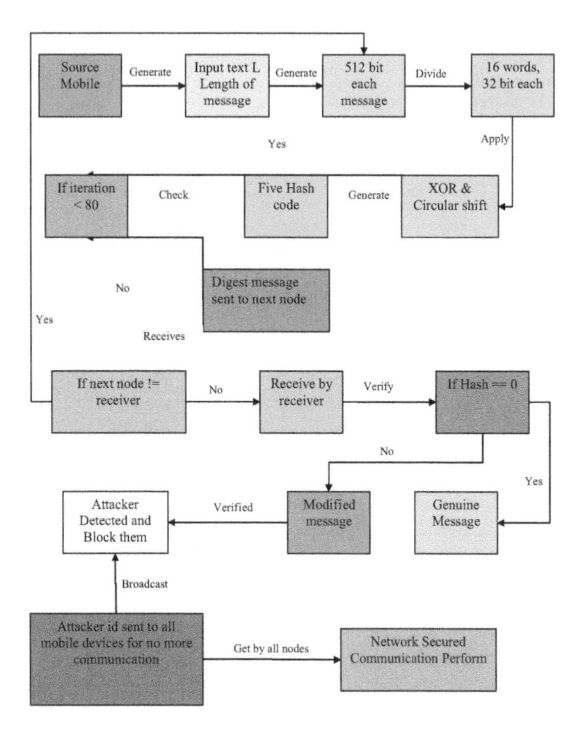

FIGURE 8.7 Architecture of proposed security scheme.

communicate with other nodes. This approach employs the idea of a blockchain to construct a hash code in each node and conduct distributed security to mobile ad hoc networks, making it more secure than any other security system now in use.

8.10 SIMULATION TOOL, PARAMETERS AND PERFORMANCE METRICS

The C++ object library has been heavily integrated into Network Simulator-2 (NS-2) [29]. It is recommended that a TCL (Tool Command Language) simulation script be used in conjunction with these C++ objects in order to set up a simulation. However, advanced users might consider these objects to be lacking in some way. They are responsible for developing their own C++ objects and utilizing an OTcl configuration interface in order to put these objects together. After the simulation is complete, NS-2 will output the results of the simulation using either text or animation. Tools such as NAM (Network AniMator) and XGraph are used in order to interpret these results graphically and in an interactive manner, respectively. Users are able to extract a pertinent subset of text-based data and transform it to a more comprehensible presentation so that they can conduct an analysis of a specific behavior exhibited by the network. This simulation is carried out using network simulator 2.31 [29], which is a simulator for wireless mobile ad hoc networks. Table 8.1 contains information regarding the simulation's parameters. The random waypoint movement model is one

TABLE 8.1
Considered simulation parameters

Network Protocol	AODV
Number of Nodes	Fifty Nodes
Terrain Size	800×600
Simulation Completion Time (seconds)	Fifty Seconds
Range of Radio Signal	550m
Application Layer Data	CBR, 3pkts/s
Size of packet (bytes)	512Byte
Transport Layer Protocol	UDP, TCP
Node Speed Maximum (m/s)	30
Movement of Nodes	random
Attack Types	Remapping Attack
Number of Attacker Nodes	5
Number of Preventer Node	1

that we implement for the simulation. In this model, a node begins the simulation at a random position, the time for the simulation is set to 100, and then the nodes move to another random position at a speed that is assigned to them. Up to 30 meters per second is the maximum speed that can be assigned to nodes. Packets of 512 bytes in size, with a rate of transmission of four packets per second have been considered.

The primary objective of this project is to protect the MANET-IoT network from priority-based flooding of attackers and to develop a profile-based technique for network security against attack. Both of these objectives will be accomplished through the completion of this title. This scheme is able to identify the perpetrator of the attack based on their profile, which has been analyzed and compared to the routing and delivery of packets.

8.10.1 Simulation Parameters to be Used in the Case Study

The simulation parameters considered for simulation are mentioned in Table 8.1. When simulating ramping attacks and the suggested security measure against traffic remapping attacks in MANET-IoT, the same parameters are taken into account.

8.10.2 Performance Metrics

In our simulations, we use a variety of performance metrics to make a comparison between the previously mentioned priority and attack, and the proposed prevention protocol. During the comparison, the following metrics were taken into consideration:

- **Packet delivery Fraction (PDF)**
  As a percentage, the number of packets accurately received by the nodes that correspond to the packets' PDR and the number of packets accurately transmitted by source nodes are compared.

$$PDR = (\Sigma Pr / \Sigma Ps)*100$$

Where Pr is total packet received & Ps is the total packet sent.

- **Average End-to-end delay**
  End-to-end delay is the average time it takes for data packets to travel from source to destination. It is measured from source to destination. The number of senders and receivers in the network is taken into account when calculating the average delay, as well as the total delay. However, the end-to-end delay will be calculated by the given formula.

$$\text{E-to-E Delay} = (Tr+Tp)-Tt)/\ Cn$$

Where Tr is receiving time, Tp is processing time and Tt is transmission time, Cn is number of connections

When there is a high volume of traffic, an insufficient number of two-way links or an attacker present in the network, the end-to-end delay will be increased.

- **Normalized Routing Load**
  The routing load is determined by tallying the number of route packets transmitted for each successfully delivered data packet. The performance of the network is negatively impacted when there is a greater quantity of route packets. Senders flood the network with routing packets, which are separate from the ones that attackers who are already present in the network send out to perform traffic remapping.

$$NRL = \Sigma Rp\ /\ \Sigma Pr$$

Where Pd is total packets drop & Pr is the total data packets received.

- **UDP Packets Analysis**
  In order to verify the dependability of the network in the event of a traffic ramping attack and the proposed security scheme in MANET, an evaluation of the UDP packets is performed. After discarding the data confirmation, the UDP protocol does not forward to the sender, which makes it an unreliable transport protocol.

$$UDP(Pt,Pr,Pl) = (\Sigma Pt,\ Pr,Pl)$$

  Pt = Transmit, Pr = receive, Pl = loss

- **TCP Packets Analysis**
  In the event of an attack utilizing the suggested security approach, the number of TCP connection packets received equals the total number of packets received. The Transmission Control Protocol (TCP) is a dependable protocol, and after successfully receiving data, the receiver is required to send a reply to the sender.

$$\text{TCP(Pt,Pr,Pl)} = (\Sigma\text{Pt, Pr,Pl})\text{// Based on Acknowledgment}$$

8.11 SIMULATION OUTCOMES

In this section, the analysis of simulation results is discussed in relation to three different scenarios: the normal routing scenario, the ramping attack scenario, and the protection PPS scheme scenario. Following implementation of the Proposed Prevention Scheme (PPS) scheme, the results unequivocally demonstrate the expected behavior. There is overlap between the graphs representing the normal case and the PPS case, but the performance of the attack case can be seen clearly. The complete findings of the investigation are presented and discussed below.

8.11.1 Analysis of the UDP End Data Stream

The only connectionless protocol in a transport layer is the UDP protocol, which stands for user datagram protocol. The UDP packet analysis in the normal priority-based routing, TPA, and proposed prevention scheme is depicted in Figure 8.8. The connectionless nature of the UDP protocol makes it unreliable for communication; however, if the network conditions are satisfactory, the UDP protocol will also provide high-quality performance as mentioned in Figure 8.8. In this particular scenario, the quantity of UDP packets received in the attack and preventive scenarios, namely, roughly 930 and 900, respectively, is almost precisely the same; however, in the assault case, the destination only receives 600 packets. It suggests that the level of performance is inadequate.

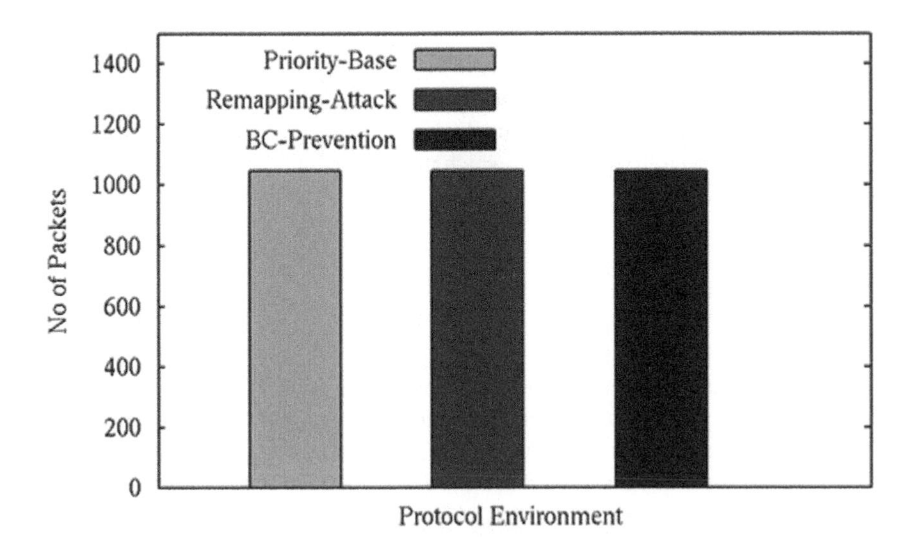

FIGURE 8.8 UDP data receiving analysis.

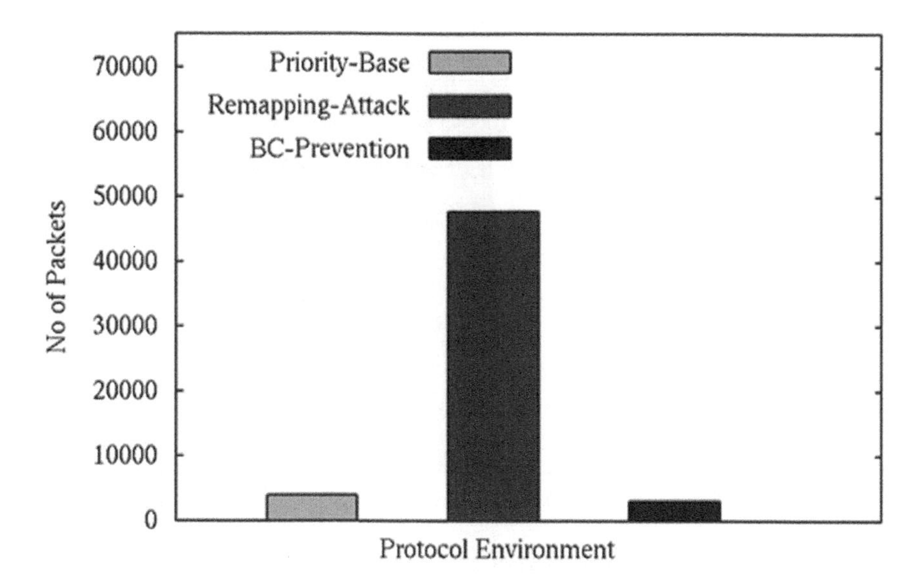

FIGURE 8.9 Overhead analysis.

8.11.2 Analysis of the Routing Packets Flooding

Information pertaining to the receiver is included in the routing packets. It is necessary to conduct a routing load analysis in order to connect the sender and the receiver. When it comes to the amount of routing packets that are sent across the network, or the routing load. In Figure 8.9, the overhead comparison is depicted. Around 48,000 routing packets are sent during a ramping attack, 4,000 are sent during regular routing, and the fewest routing packets are sent during network flood prevention.

8.11.3 TCP PDR Analysis

The performance of the PDF is directly proportional to the ratio of the output value to the input value. The packets that were received are of higher quality compared to those that were sent. PDF performance will suffer if there is a significant disparity between the number of packets received and sent. The newly proposed security scheme not only boosts performance but also delivers the most effective PDR available across the network. Because there are only two graphs that can be visualized equivalently in the network, the PDR value is the same in both the case of normal routing and after the block chain security scheme has been applied mentioned in Figure 8.10. Although the received packets in the attack case are very small, virtually all of them have been obtained, and the PDR is approximately 90%.

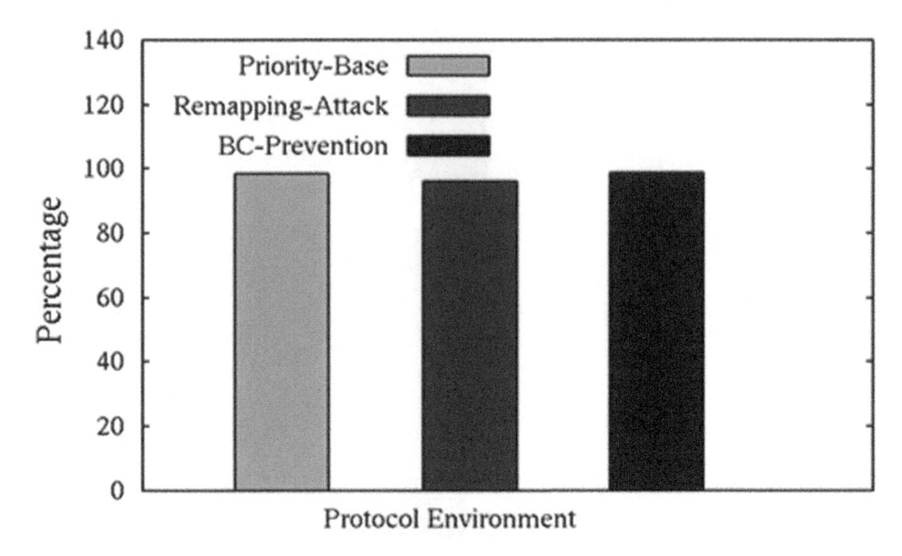

FIGURE 8.10 TCP packets percentage analysis.

8.11.4 End-to-End Delay Analysis

When determining the average delay in all connections within a dynamic network, the average delay in network measures is utilized as the metric of choice. Collisions, congestion, and link time outs are all caused by the frequently shifting topology of dynamic networks, which is also the cause of the delay that occurs in these types of networks. The delay performance of priority-based routing and attack prevention schemes is shown to be nearly as good as, or even better than, that of the attacker mentioned in Figure 8.11. On the other hand, as a result of the significant delay, the packets that are received during an attack are of poor quality. This demonstrates that the performance has declined.

8.11.5 Routing Load Analysis

The TRA attackers are continuously flooding the network with a huge number of packets (about 50,000). Consuming the network's bandwidth in such a way that the nodes are unable to communicate with one another about such inappropriate behavior is what this phrase refers to. There is a heavy flood occurring right now caused by approximately 14% of unwanted messages that are flooding into the network. This graph shows the percentage of the route load that would be affected in the event of an attack, and it shows a very high value. This is the primary cause of the bandwidth bottleneck that currently exists in the network. After applying the block chain security, the network load is now under control and has returned to its normal level of normal routing behavior by about 5% mentioned in Figure 8.12.

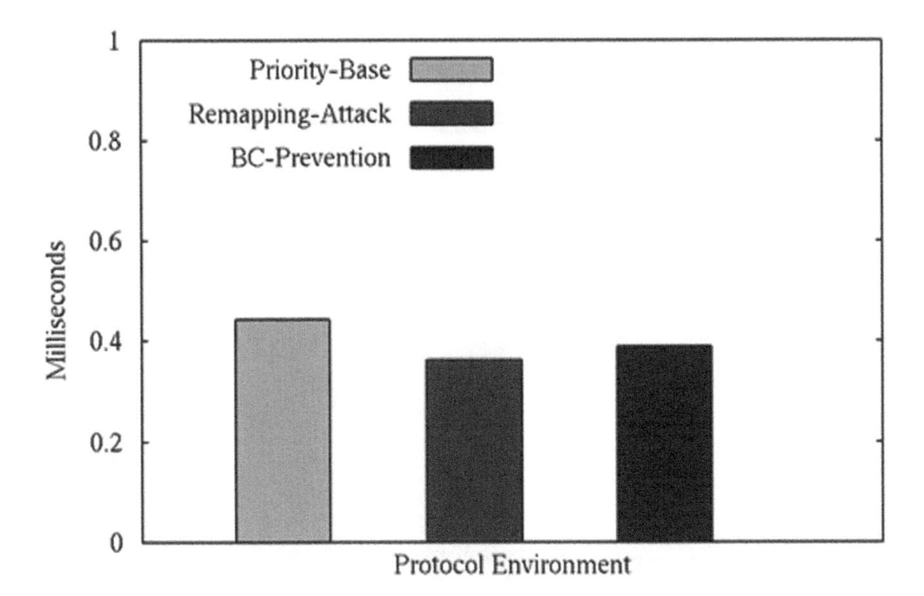

FIGURE 8.11 Delay analyses.

8.11.6 An Examination of the Flooding of Attackers

The traffic ramping attack floods a large number of packets in the network, which has the effect of reducing the network's overall performance. In the event of a ramping attack, this graph depicts the analysis of the attacker flooding the system. Here, we are able to see very clearly that there are approximately 70,000 unwanted packets flooding the network. However, after putting the suggested prevention plan into action, the attacker infection has been reduced to zero despite the ongoing attack mentioned in Figure 8.13. This indicates that the security system is capable of preventing any and all misbehavior on the part of attackers.

8.11.7 Analysis of Drops Caused by Congestion

The graph in Figure 8.13 includes a discussion of the analysis of drop rates as a result of contention, queue, and link timeout. This graph illustrates the performance of a traffic ramping attack, traditional priority-based routing, and the proposed security scheme. These reasons for dropping packets are evaluated here in the normal priority and prevention scheme; however, in the event of a traffic ramping attack, only a small number of packets are dropped for those reasons mentioned in Figure 8.14. This indicates that the primary reason why these factors are not considered in the attacker's scenario is due to the presence of the attacker.

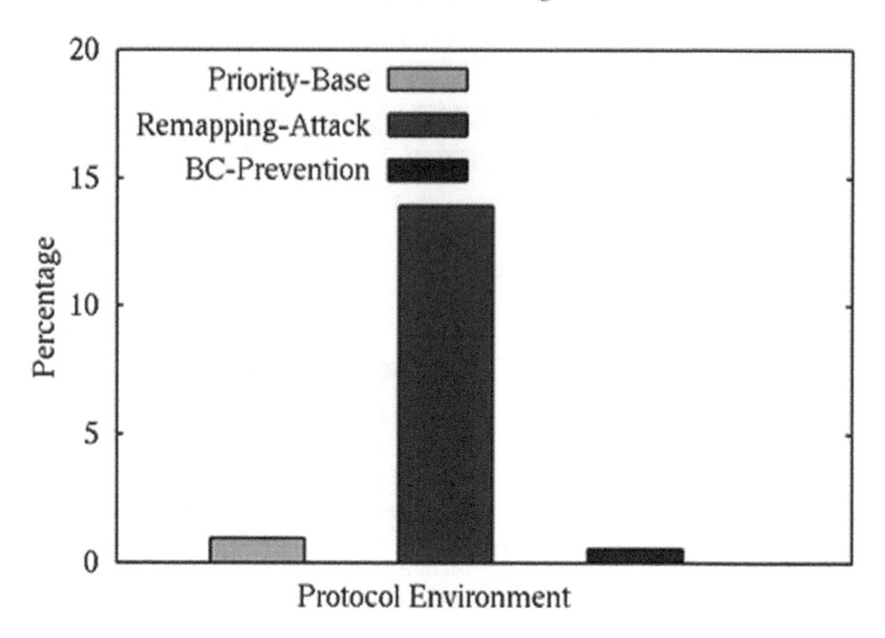

FIGURE 8.12 Routing load percentage analysis.

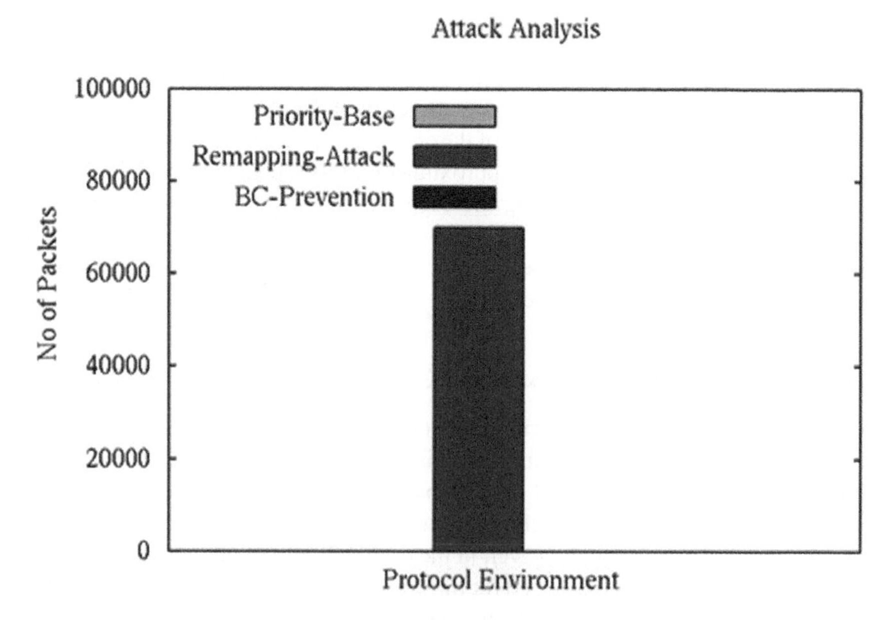

FIGURE 8.13 Attacker analysis.

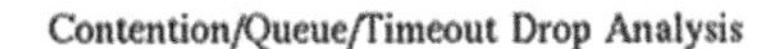

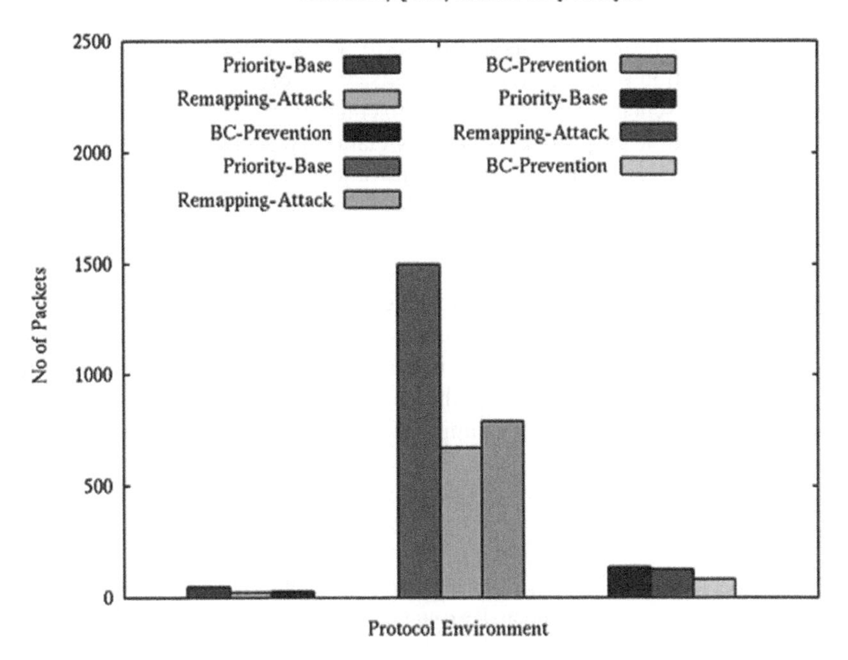

FIGURE 8.14 Drop analysis due to contention, queue, and timeout.

8.12 CONCLUSION AND FUTURE WORK

The nodes transfer information and control the functions of devices based on user requirements in the Internet of Things. Due to the absence of any central authority, the security in mobile ad hoc networks is significantly less robust than that of MANET. The MANET-IoT is the dynamic network and mobile node's actual behavior observed by IoT nodes. Within the network, the mobile nodes are constantly engaged in the exchange of information. On the other hand, if the information is muddled up with many other packets that are flooding the network, then the network is vulnerable to TRA (Traffic Remapping Attack). The block chain method generates the hash code based on the number of packets flooding into the network. The PPS scheme is reliable and efficient to secure MANET-IoT networks from traffic remapping attacks. The findings indicate that the presence of a TRA results in a significant increase in the amount of routing load and packet loss that occurs within the network. The network is prevented by the proposed PPS mechanism, which utilizes a process that is completely decentralized, self-organized, and localized. The performance of the network has been negatively impacted because of the attacker, which in turn has a knock-on effect on the overall performance of the network. The hash code in the presence of an attacker always shows the same value, however, actual data packet forwarding shows the network changes. As per usual, the network overhead is well under control, and the PDF value has improved. The fact that it improves performance by 98% in comparison to the standard behavior of the network's routing protocol is the most significant benefit of this scheme. The overhead is only 0.5 less as compared to the previous security scheme. The proposed prevention scheme shows 0.1 millisecond less delay as compared to the previous priority scheme, and the drop due to contention, queue, and timeout is also less in the case of the proposed scheme. In situations where an adversary is present, the PPS security scheme produces more favorable results. The remainder of the performance of the TCP and UDP analysis is also satisfactory, and after applying the proposed prevention scheme, there is no evidence of an attacker's presence in the performance.

In future, we will apply the PPS scheme on vampire attacks in MANET-IoT. A vampire attack consumes the communication capability of mobile nodes, which means the battery or energy of mobile nodes, which is limited and efficiently used for communication. In other words, this attack is a type of energy consumption attack. After studying how the vampire attack works, the new plan is put into action.

REFERENCES

[1] C. Siva Ram Murthy and B. S. Manoj, "Ad Hoc Wireless Networks, Architectures and Protocols", Low Price Edition, Pearson Education, pp. 521, 2007.

[2] Mohammed Abdulhakim Al-Absi, Ahmed Abdulhakim Al-Absi, Mangal Sain, and Hoonjae Lee, "Moving Ad Hoc Networks—A Comparative Study," Sustainability, pp. 2–31, 2021.

[3] G. Soni, M. K. Jhariya, K. Chandravanshi, and D. Tomar, "A Multipath Location-based Hybrid DMR Protocol in MANET," IEEE 3rd International Conference on Emerging Technologies in Computer Engineering: Machine Learning and Internet of Things (ICETCE), pp. 191–196, 2020.

[4] Hamed HaddadPajouh, Ali Dehghantanha, Reza M. Parizi, Mohammed Aledhari, and Hadis Karimipour, "A Survey on Internet of Things Security: Requirements, Challenges, and Solutions," Internet of Things, Vol. 14, 2021.

[5] Gaurav Soni, R. Sudhakar, and Krishna Pathak, "Cluster based Techniques for Eradicating Congestion in WSN Aided IoT: A Survey," International Journal of Emerging Technology and Advanced Engineering, Vol. 8(12), pp. 166–172, December 2018.

[6] X. Liang and Y. Kim, "A Survey on Security Attacks and Solutions in the IoT Network," 2021 IEEE 11th Annual Computing and Communication Workshop and Conference (CCWC), pp. 0853–0859, 2021.

[7] Yourong Chenad, Hao Chenb, Yang Zhangb, Meng Hancd, Madhuri Siddulae, and Zhipeng Caif, "A Survey on Blockchain Systems: Attacks, Defenses, and Privacy Preservation," High-Confidence Computing Vol. 2(2), June 2022.

[8] Emanuel Ferreira Jesus, Vanessa R. L. Chicarino, Célio V. N. de Albuquerque, and Antônio A. de A. Rocha, "A Survey of How to Use Blockchain to Secure Internet of Things and the Stalker Attack," Hindawi, Security and Communication Networks, pp. 1–27, 2018.

[9] G. Soni and K. Chandravanshi, "Security Scheme to Identify Malicious Maneuver of Flooding Attack for WSN in 6G," 2021 8th International Conference on Signal Processing and Integrated Networks (SPIN), pp. 124–129, 2021.

[10] C. E. Perkins, E. M. Royer, and S. R. Das, "Ad hoc On-Demand Distance Vector Routing," IETF RFC 3561, July 2003.

[11] G. Soni and R. Sudhakar, "AODV Local Route Innovation (AODV_LRI) to Increase the Network Performance by Reducing Congestion Problem on MANET," Journal of Emerging Technologies and Innovative Research, Vol. 5(5), pp 937–942, 2018.

[12] G. Mamatha and D.S. Sharma, "Analyzing the MANET Variations, Challenges, Capacity and Protocol Issues", International Journal of Computer Science & Engineering Survey, pp. 14–21, 2010.

[13] P. Goyal, V. Parmar, R. Rishi, "MANET: Vulnerabilities, Challenges, Attacks, Application," International Journal of Computational Engineering & Management, 11, pp. 32–37, 2011.

[14] Srivarshini S. and Jaya Vignesh Thyagarajan, "Survey of Routing Protocols for Low Power and Lossy Networks (LLN)," International Journal of Pure and Applied Mathematics Vol. 120(7), pp. 375–382, 2018.

[15] Pathan, Al-Sakib Khan, and Choong Seon Hong. "Routing in Mobile ad hoc Networks." *Guide to Wireless Ad Hoc Networks* (2009), pp. 59–96.

[16] Ahmed Said Abdel-Sattar and Marianne A. Azer, "Using Blockchain Technology in MANETs Security," Conference: 2022 2nd International Mobile, Intelligent, and Ubiquitous Computing Conference (MIUCC), May 2022.

[17] Mohammad O. Pervaiz, Mihaela Cardei, and Jie Wu, "Routing Security in Ad Hoc Wireless Networks," Chapter 5, Network Security, Springer, 2005.

[18] Sengupta, Jayasree, Sushmita Ruj, and Sipra Das Bit, "A Comprehensive Survey on Attacks, Security Issues and Blockchain Solutions for IoT and IIoT," *Journal of Network and Computer Applications*, 149 (2020): 102481.

[19] Vikas Hassija, Vinay Chamola, Vikas Saxena, Divyansh Jain, Pranav Goyal, and Biplab Sikdar, "A Survey on IoT Security: Application Areas, Security Threats, and Solution Architectures," IEEE Access 2019.

[20] Aradhana Saxena and Manish Khule, "Remapping Attack Detection and Prevention for reliable data service in MANET." *Proceedings of International Conference on Recent Advancement on Computer and Communication: ICRAC 2017*. Springer Singapore, 2018.

[21] Jerzy Konorski and Szymon Szott, "Discouraging Traffic Remapping Attacks in Local Ad Hoc Networks," IEEE Transactions on Wireless Communications, Vol. 13(7), July 2014.

[22] Jerzy Konorski and Szymon Szott, "Mitigating Traffic Remapping Attacks in Autonomous Multihop Wireless Networks," IEEE Internet Of Things Journal, Vol. 9(15), 1 August 2022.

[23] Ahmed Said Abdel-Sattar and Marianne A. Azer, "Using Blockchain Technology in MANETs Security," IEEE 2nd International Mobile, Intelligent, and Ubiquitous Computing Conference (MIUCC), 2022.

[24] Lwin, May Thura, Jinhyuk Yim, and Young-Bae Ko. "Blockchain-based Lightweight Trust Management in Mobile ad-hoc Networks." *Sensors* Vol. 20(3), 2020, p. 698.

[25] Yang, Jidian et al. "A Trusted Routing Scheme Using Blockchain and Reinforcement Learning for Wireless Sensor Networks," *Sensors* Vol. 19(4), 2019, p. 970.

[26] S. Goka and H. Shigeno, "Distributed Management System for Trust and Reward in Mobile Ad Hoc Networks", in Proceedings of the 2018 15th IEEE Annual Consumer Communications & Networking Conference (CCNC), pp. 1–6, 2018.

[27] G. Soni and R. Sudhakar, "A L-IDS against Dropping Attack to Secure and Improve RPL Performance in WSN Aided IoT," IEEE 7th International Conference on Signal Processing and Integrated Networks (SPIN), pp. 377–383, 2020.
[28] Szymon Szott and Jerzy Konorski, "Traffic Remapping Attacks in Ad Hoc Networks," IEEE Communications Magazine, pp. 1–8, April 2018.
[29] Teerawat Issariyakul and Ekram Hossain, "Introduction to Network Simulator NS-2," Springer Science and Business Media, LLC, 2009.

Part III

Application Areas Integrating Blockchain and IoE

Part III

9 Block Chain Applications and Case Studies for IoE

Roheen Qamar and Baqar Ali Zardari

CONTENTS

DOI: 10.1201/9781003366010-12

9.1 INTRODUCTION

In 1991, Stuart Haber and W. Scott Stornetta released their research on cryptographically secured blockchains, which is when the concept of blockchain technology first emerged. Merkle trees were integrated into the project in 1992, enabling the collection of several documents into a single block (Zheng et al., 2018).

9.1.1 Block

This is a list of fixed-price bills. Every node in the system has access to the information when it is stored in blocks. The size, duration, and triggering circumstances for blocks are based on the type of blockchain.

9.1.2 Chain

This connects a collection of blocks. The concept of connecting every block is the foundation of the blockchain (Wang et al., 2018).

9.1.3 System

A system is a collection of neurons that are linked together. Nodes are viewed as channels in a conventional network, whereas blocks are viewed as nodes in a blockchain network. The blockchain model has been demonstrated here. IoT and computer-based physical systems, crowdsourcing and crowdsensing, trust management, edge and cloud computing, blockchain, and the Internet of Things in 5G, as well as access control and lightweight blockchain-based data structures for IoT data, are some of the topics covered. The blockchain block is the data storage component of the distributed ledger of the blockchain. Each block of the blockchain contains a variety of transactions that contribute to the shared state of the blockchain network. The image above depicts a simplified representation of several different blockchain blocks. The blocks are split into two sections: the header (which runs across the top of the image) and the body.

Block chain connections and distributed storage architecture, by utilizing the "waterfall effect," block chain cryptography, consensus methods, smart contracts, and other technologies, can address the issue of information traceability in the process of information collection, circulation, and sharing. New management options for the Internet of Things platform, administration, and information security have emerged as a result of its emergence. To improve the security of the automotive ecosystem and protect user privacy, a blockchain-based architecture has been proposed by qualitatively proving the architecture's adaptability for typical security assaults (Deepa, 2022).

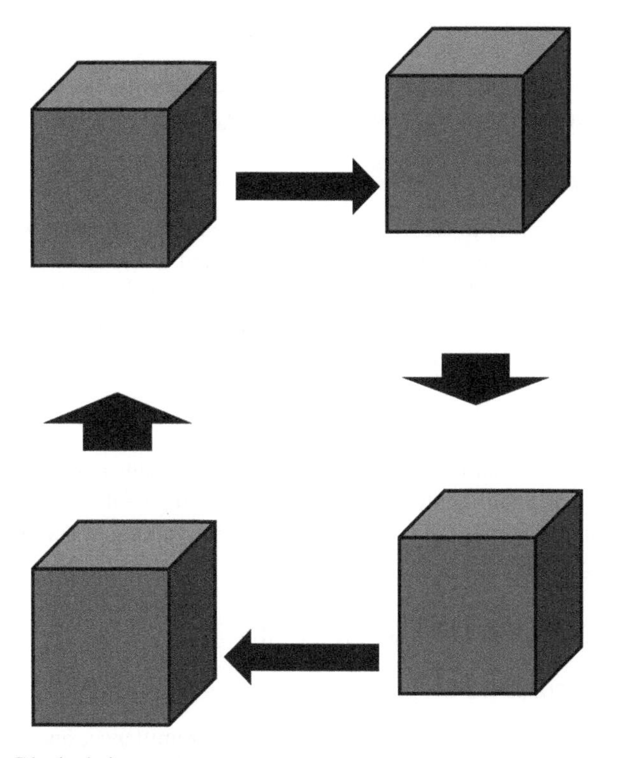

FIGURE 9.1 Block chain structure.

9.2 BACKGROUND

Since the creation of Bit coin in 2008, blockchain has been acknowledged as one of the most promising technologies. Data is contained in blocks on a blockchain, and these blocks link together in the order determined by a distributed consensus method to build a linked list. Blockchain has a lot of advantages due to its chain structure and decentralized nature, including transparency, traceability, and immutability (Nakamoto, 2008). A blockchain is a global ledger that makes it possible for many people to share data. As already mentioned, it is regarded as Bit coin's most significant contribution because it resolved the double-spend problem, a persistent financial problem. By seeking the agreement of the vast majority of mining nodes, Bit coin's approach allowed for the addition of valid transactions to the network (Tariq, 2019). Blockchain technology combines a number of ICTS is given below.

1. **P2P Networks**

The P2P network uses the gossip protocol as the fundamental architecture for information exchange in order to spread messages across the network.

2. **Cryptology**

Numerous cryptographic techniques ensure the security of message delivery. Additionally, peer privacy and anonymity are safeguarded (Yli-Huumo et al., 2016).

3. **General Agreement**

Each peer has the ability to create new blocks. Once it has received widespread consensus approval from peers, new blocks will be added to the chain. Peers attempt to agree on a shared choice during the consensus process, such as whether to approve a freshly generated block. If the majority can come to an understanding, consensus has been attained (Yli-Huumo et al., 2016).

4. **Smart Contract**

When certain trigger conditions are met, smart contracts launch themselves as executable programmers. Smart contracts allow for the editing of its terms by anyone, but any changes must first have consensus approval before taking effect (Mitridati, 2021) as shown in Figure 9.2.

9.3 APPLICATIONS OF BLOCK CHAIN

9.3.1 Cryptocurrency

Shock wave is a decentralized currency exchange, remittance, and real-time gross settlement system (RTGS) that focuses on the banking business. It leverages the ripple protocol through a peer-to-peer network. Coinbase, BitPesa, Billion, Stellar, Kraken, and Crypto Sigma are some more well-known currency exchange and remittance services (Ripple, 2022).

9.3.2 Being Aware

Block chain is a cutting-edge technology used mostly in the financial industry (Bit coin being the most popular application). The adoption of block chain technology is being hampered by a general lack of information about it (Tse et al., 2017).

9.3.3 Strategy

There are still legal barriers to the general implementation of block chain, despite the fact that it does away with the requirement for a centralized authority or a reliable intermediary to confirm transactions. Decentralized systems like blockchain demand new business and governmental standards. Smart contracts must also be legally binding in order to prevent disputes between parties to transactions (Tse et al., 2017).

9.3.4 Food Industry

Product life cycle visibility can be enhanced by BCoT, particularly in the food industry. Food product traceability is especially important for food safety. Nevertheless, food traceability throughout the food supply chain cannot be guaranteed by the current Internet of Things (Wang et al., 2016).

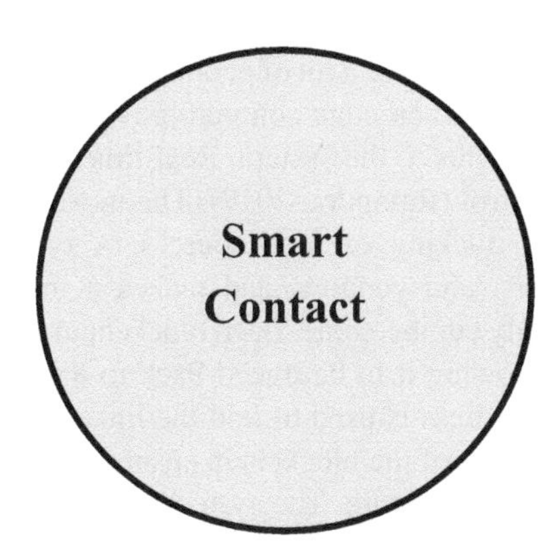

FIGURE 9.2 Smart contracts.

9.3.5 Mobile Commerce

With the growing popularity of mobile commerce (MC), data security issues are becoming increasingly widespread and must be addressed. Block chain was initially offered to safeguard mobile node transactions as a distributed database and to provide m-commerce data transmission and sharing directly from device (Tian, 2017).

9.3.6 Big Data

Because of intelligent production, human genetic data is more important than ever before in the IIoT. Big data and related analytical capabilities are being added to the blockchain ledger to fulfill the business aims of blockchain financial services. Ripple, a blockchain technology startup, has reached a partnership with more than

40 Japanese banks to use the blockchain to make it easier to transfer money between bank accounts and make low-cost real-time payments (Novo, 2018).

9.4 BITCOIN FOR IOT

An innovative idea in IoT key management has emerged due to the introduction of blockchain technology Bot. Second, because there are so many different types of devices in the IoT network, there are a lot of weak nodes. These weak nodes are easily breached by intruders, who then use them for illicit purposes. A blockchain-based intrusion detection system is necessary for IoT security because it can effectively identify intrusions. Third, there are a lot of people using IoT. The network level is also rather complex, especially when edge computing is involved, and IoT makes it difficult to manage access rights to the system. Real-time dynamic problems exist with conventional access control (Ravindra, 2018). The most common use of blockchain to improve IoT is the tracking of food sources. A typical food chain connects manufacturers, suppliers, and vendors, and makes it more difficult to trace food origins and is exceedingly cumbersome. Each blockchain transaction is time-stamped and digitally signed, allowing it to be traced back to a specific time period. On the blockchain, the public address is used to find the linked person. This is because the non-repudiation mechanism of the blockchain ensures that no one can independently verify the validity of their signature. The system gains credibility when the author's identity is linked to the transaction they initiated (Ravindra, 2018).

9.5 INTERNET OF EVERYTHING (IOE)

The idea that the Internet of Things (IoT) can now encompass people, processes, data, and things has recently been coined as the IoE (Vaya and Hadpawat, 2020; Yu et al., 2020). Sensors are typically incorporated with "everything" in the IoE's central processing unit in order to monitor, determine status, and take intelligent action to create new opportunities for society. Numerous sensors, including those for temperature, pressure, biosensors, light, position, velocity, and so on, are available for use (Shilpa et al., 2019) and shown in Figure 9.3.

The IoT, the IoE, and the Internet of Nano-Things (IoNT) are innovative strategies for incorporating the Internet into a broader context of personal, professional, and societal life as well as the impersonal world of quasi-intelligent devices. This chapter looks at the relevant literature to see where these technologies are now and how they can be used in many different ways. In addition, the chapter assesses the various potential future applications of these technologies. In order to support these and other applications in industries, traffic, smart cities, and healthcare systems (Sirma et al., 2019; Alsuwaidan, 2019), a significant number of sensors are anticipated to be placed everywhere. By expanding on the IoT emphasis on machine-to-machine (M2M) communications, the concept of the IoE describes a more complex system that includes people and processes. The Internet of Things is depicted in Figure 9.4.

The problems with sensor connectivity, data collection, and processing have been studied in some IoE initiatives. Utilized Cloud as a Service (CaaS) to address the challenge of integrating, storing, and transporting remote data. By utilizing Software

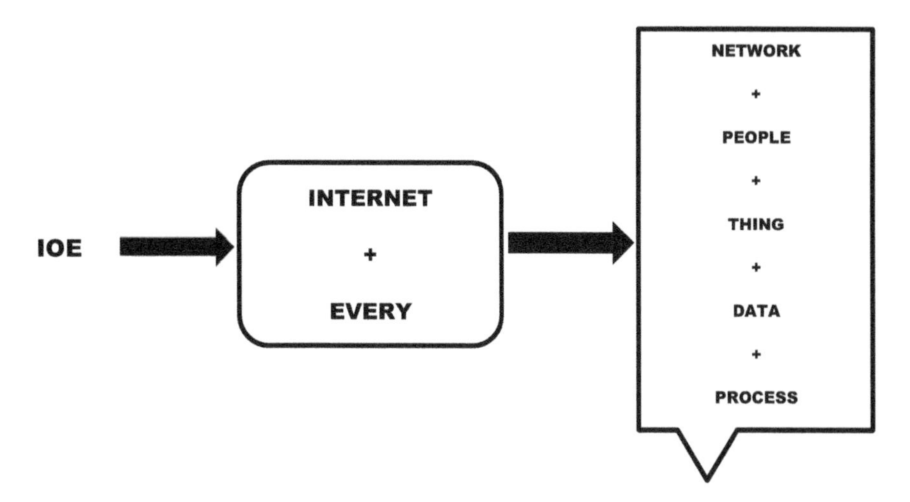

FIGURE 9.3 Internet of Everything structure.

Defined Network (SDN) technology, control over sensors in the 6G/IoE network was improved. The IoE service that makes use of self-context differentiates between the logical and physical contexts. Others, for instance, concentrated on energy harvesting while addressing IoE security and privacy issues (Ryoo et al., 2017).

A paradigm shift that brought data processing closer to the sources of the data was required as a result of the network's ongoing addition of IoE sensors and edge devices (Badr et al., 2019). Two architectures, edge computing and fog computing, strive to bring processing closer to users at the edges of the network. Fog computing, on the other hand, is an intermediate layer that connects users to cloud computing features and extends the cloud layer (Ryoo et al., 2017). Fog nodes are devices that can use resources to provide services. Edge computing localizes processes in edge devices to produce instant results, whereas many authors in the literature do not differentiate between fog and edge (Aiello et al., 2019). They can be devices with limited resources like access points, routers, switches, and base stations or machines with a lot of resources like Cloudlet and IOx (Sunyaev, 2020).

9.6 APPLICATIONS OF THE INTERNET OF EVERYTHING

As a concept, the Internet of Everything has numerous applications and has been implemented in several applications. Let's take a look at the main IoE application fields:

9.6.1 IoE in Smart Airports

The number of passengers will more than double to 8.2 billion by 2037, according to the International Air Transport Association (IATA) (Muhammed et al., 2018). The aviation sector, particularly the existing infrastructure, will experience significant strain as a result of this anticipated increase in passenger numbers. Additionally, IoE

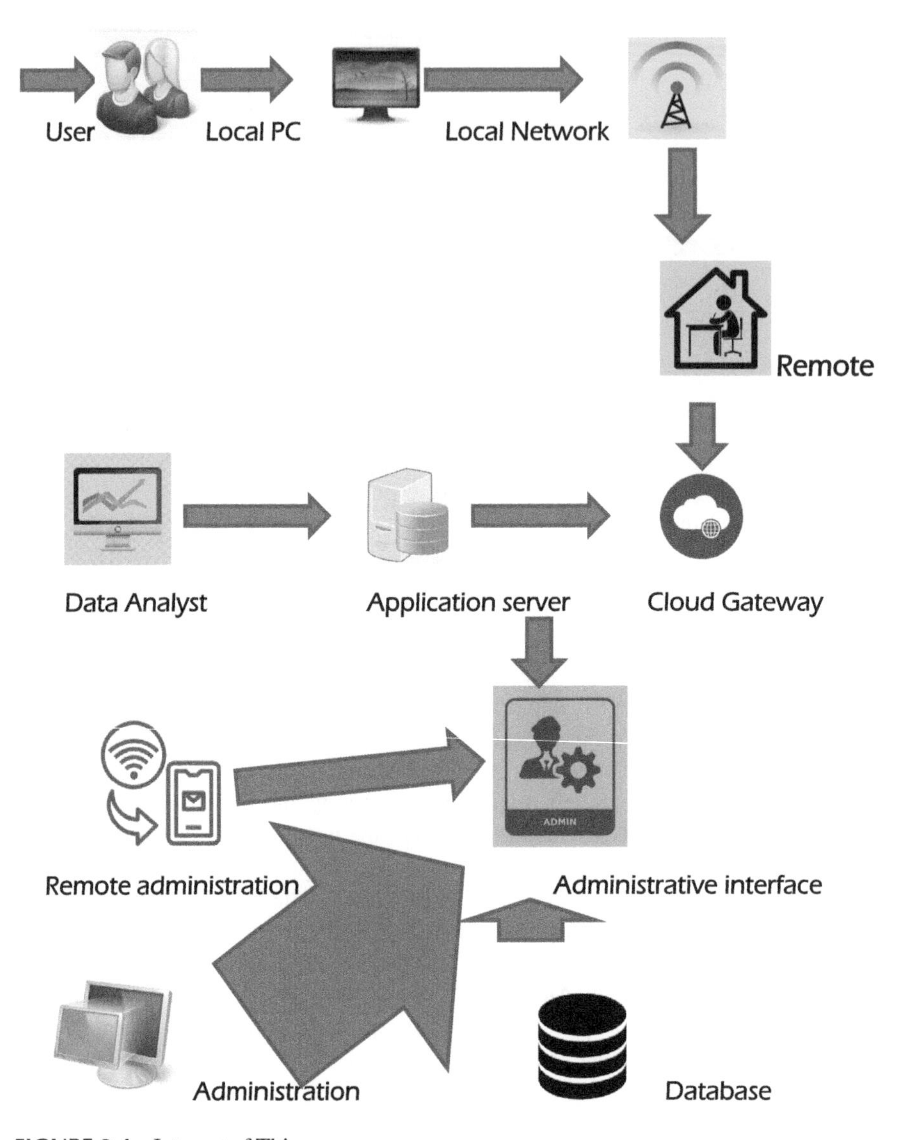

FIGURE 9.4 Internet of Thing.

will provide a fantastic opportunity for both airlines and airports, enhancing the passenger experience (Khan et al., 2020). Surveillance cameras, radio-frequency identification (RFID), various sensors (like an air quality sensor), wearable devices (like watches), avionics devices (like flight recorders), biometric devices, and/or digital regulators can all be deployed to support airport smartness (like electricity) utilizing the information gathered by these devices. The router is connected to all edge devices in the region. There are three types of simulated edge devices: bar code readers, smart cameras, and counter devices, and each have a different sensor or actuator. There are three on display applications: Smart Surveillance, Smart Gate Control, and a Smart

Counter. Despite having the same physical infrastructure, both scenarios have distinct placements for application modules. While Smart Surveillance was covered in the previous section (Lazaroiu and Roscia, 2017), Smart Gate Control and Smart Counter will be discussed in the following sections.

9.6.2 IoE in Smart District

In a smart ecosystem, numerous issues, such as the provision of energy and utilities, healthcare, education, transportation, waste management, and the environment, must be addressed effectively and efficiently. The smart district serves as the foundation for smart cities, and a variety of applications can be incorporated into it to provide the intelligence that will enhance people's quality of life Smart sensors and IoE devices continuously monitor the district while acting absurdly, as shown in the Figure.9.5; parking guidance systems, parking lot monitoring, parking lot reservation, parking entrance, and security management utilize sensors like infrared sensors, ultrasonic sensors, inductive loop detectors, cameras, RFID, magnetometers, and/or microwave radar (Shahinzadeh et al., 2019).

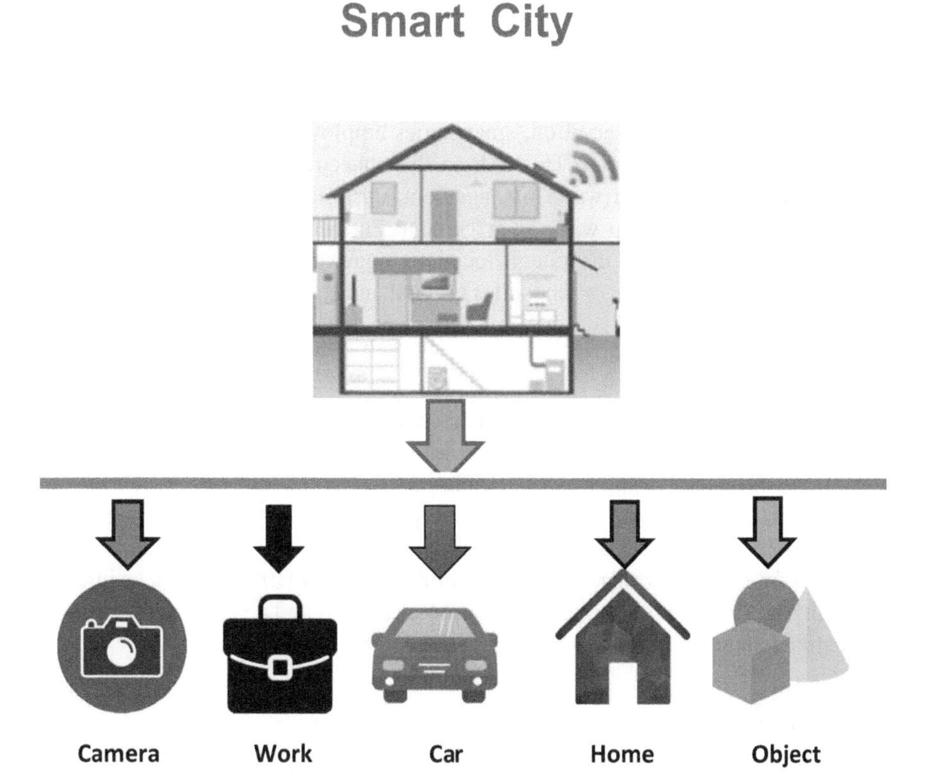

FIGURE 9.5 IoE in smart district [28].

9.6.3 Environmental Surveillance

A. IoE tracks and collects weather data across seasons using a network of sensors. Temperature, humidity, wind speed, rainfall, pressure, air quality, soil conditions, water level, and other variables are included in weather data. When these weather parameters are collected, the data is analyzed and processed to record the events and changes in the surrounding conditions. It aids in the detection of anomalies in real-time, allowing people to take immediate action before the weather disrupts their plans. Smart environmental data is also sent to other applications, such as Air traffic management.

B. Farmers, for agricultural practices industries, because they need to know the environmental impact of their plant while ensuring regulatory compliance and worker safety.

C. An IoE ecosystem is a network of all these applications (Snyder and Byrd, 2017).

9.6.4 Smart Cities

A typical smart city model is powered by IoE solutions. A smart city's goal is to improve its citizens' quality of life, spur economic growth, and organize processes to ensure cities run smoothly.

Automation, AI, machine learning, and IoT technologies are being combined for a variety of applications, such as smart parking systems that assist drivers in managing their parking space and enabling digital payment. Other applications, such as smart traffic management, aid in traffic flow control and congestion reduction.

In terms of energy conservation, smart cities employ streetlights that dim when there is no traffic on the road. This aids in the upkeep and optimization of power supplies. In terms of smart waste management, dustbins and trash collection units are Internet-enabled to help waste management. Furthermore, for the safety of city dwellers, sensors are placed in strategic locations to provide early warning of incidents such as earthquakes, floods, or landslides (Anand and Ramesh, 2021).

9.6.5 Energy Sector

Monitoring energy consumption by industries, communities, and individual households is one application of IoE in the smart energy sector. IoE networks process data collected from energy production sources such as solar, wind, and thermal, both renewable and non-renewable. Smart meters are being used to improve energy management. Users of these smart meters have access to a variety of features. These include instant bill generation for consumed energy units, the ability to display tariff changes, an interface displaying stats related to supplied and consumed energy, and a visual alert to indicate an identified anomaly in the energy system (Di Martino et al., 2018).

9.6.6 Blockchain in the IoE

The current IoE employs a centralized communication model. The central cloud servers validate the IoE devices. As a result, existing IoE solutions for smart cities rely on network and cloud computing resources, raising infrastructure and maintenance costs. The size of devices in smart cities with scalable environments changes frequently. It implies that ad hoc networks of smart sensors are increasingly being introduced into smart city infrastructure. The current system does not support large IoE devices due to scalability issues. As the volume of resources increases, so does the interaction between servers and devices. Another issue that cloud servers face is single points of failure. A centralized system is required for a smart city (Mohanty et al., 2020).

9.6.7 Smart Water Management

Water management encompasses a wide range of issues, including administration, the management of environmental resources in the ecosystem, and the preservation of environmental balance and stability. IoE solutions make it easier to manage real-time processes like monitoring water supply, determining whether the water is fit for consumption, managing water storage systems, tracking water consumption by end users (organizations and individuals), and calculating the cost of water supply to remote-located commercial units (Adenugba et al., 2019).

9.6.8 Smart Apartments

Smart apartments in smart buildings contain a variety of household appliances and devices that are connected to the IoE network. Refrigerators, thermostats, air conditioners, televisions, washing machines, cookers, and other devices that generate raw data fall into this category. Data from each device is combined, analyzed, and processed to allow for informed decisions about how to use them (Zhan et al., 2022).

9.7 COMPONENTS TO AN IOE ENVIRONMENT

An IoE environment is made up of four main parts (see Figure 9.3: (1) Humans; (2) Data; (3) Things and a process).

9.7.1 People

In an IoE environment, individuals serve as nodes in a network. People have traditionally relied on handheld or desktop electronic devices to connect to the Internet and gain access to the rest of the world. The introduction of the Internet of Everything means that people now have access to a wide range of brand-new means of communication. Implantable medical devices (IMDs), such as pacemakers, send data to a server so that doctors can make diagnoses. WMDs are comparable in that they can be worn on the body for a variety of purposes, like monitoring heart rate. These are what are known as implantable and wearable medical devices (IWMDs) (Nozari et al., 2021).

9.7.2 Processes

The processes that make up the Internet of Everything are the "connections" to the network and the real-time data and information that flow between IoE nodes.

Consequently, smartness, intelligence, and real-time insights work together to satisfy demands from management and society for more data and actionable intelligence that go beyond disruptions in IoT context. In a transcendent process in which entities (people, things, and data) acquire new knowledge and new interactions are created in knowledge-creation cycles, actions and interactions generate and expand knowledge of the environment of the IoE. The transformation of data into knowledge provides crucial insights and a wide range of potential IoE applications (Farias et al., 2021).

9.7.3 Data

Data by themselves cannot have any meaning. We evaluate and examine it so that it is useful and meaningful. Then and only then will we have the data at hand. IoE can examine and understand a large amount of customer data. Companies cannot produce what customers want if they do not comprehend user data. As a result, data is a foundational element of IoE. IoE can be used to generate accurate and trustworthy insights. Critical insights and a wide range of potential IoE applications are provided by the transformation of data into knowledge (Farias et al., 2021).

9.7.4 Things

Devices, consumer goods, gadgets, enterprise machines, and assets are all examples of things, which are physical objects with sensors and actuators that can communicate over a network. In addition to acquiring information from their surroundings, these devices generate data. Intelligence, context awareness, and cognitive ability all rise as a result. These actual things are referred to as the "Internet of Things" (Alkhabbas et al., 2019). The May 2022 report from IoT Analytics estimates that there will be approximately 12.2 billion IoT connections worldwide in 2021. By 2022, this number is expected to reach around 14.4 billion connections. It is anticipated that these devices will generate data and transmit it to servers for analysis, enabling better business decisions.

9.8 CHARACTERISTICS OF INTERNET OF EVERYTHING

9.8.1 Connectivity

Connectivity, which links everyday objects, is what makes the Internet of Things possible. The various levels of IoT hardware and devices must be connected, such as sensors and other electronic components and connected hardware and control systems. Smart appliance networking has the potential to open up new market opportunities for the purpose of things and facilitates hardware compatibility and accessibility through this connectivity (Zhang, 2022).

9.8.2 Intelligence

Software and hardware, algorithms and computation, and IoT provide an intelligent combination, intelligence derived from big data analytics and the sensing capabilities of IoT devices. While standard input methods and graphical user interfaces are used to interact with users and devices, IoT intelligence is only concerned with device interaction (Zhang, 2022).

9.8.3 Dynamic Nature

Data collection from its environment, which is accomplished through dynamic changes in the environment surrounding the devices, is the most crucial aspect of IoT. The likelihood of these devices being connected or disconnected fluctuate rapidly. The number of devices, temperature, location, and speed of the devices, as well as their number, alter dynamically with a person, time, and place (Zhang, 2022).

9.8.4 Security

IoT devices may be compromised by cyberattacks. Concerns regarding privacy and transparency are prevalent in the Internet of Things. It is essential to secure endpoints, networks, and the data that flows between them when developing a security paradigm (Shen et al., 2022).

9.8.5 Sensing

Without sensors that produce data, can interact with it, and detect or measure changes in the environment, the Internet of Things cannot be imagined.

Even though sensory data is the analogue input to the physical world, it can give a thorough comprehension of the complicated reality (Shen et al., 2022).

9.8.6 Heterogeneity

IoT devices can communicate with the platforms of other devices via a variety of networks and are built on a variety of hardware platforms and networks. Direct network connectivity between heterogeneous networks must be supported by IoT architecture. Scalability, modularity, extensibility, and interoperability are all necessary for IoT heterogeneous networks (Shen et al., 2022).

9.8.7 Enormous Scale

The number of connected devices will far outnumber the number of devices communicating with one another. The management and interpretation of these devices for application purposes is more crucial. Gartner (2015) estimates that 6.4 billion connected devices will be used worldwide in 2016, up 30% from 2015. This demonstrates the enormous scope of the Internet of Things. Every day, 5.5 million new things will be connected, according to Gartner. By 2020, the report projects that there will be 20.8 billion connected devices (Shen et al., 2022).

9.9 ADVANTAGES OF IOT

There are many ways that businesses can benefit from the Internet of Things. While some benefits are applicable to all industries, others are industry-specific. Keeping an eye on all business procedures, enhancing the experience of customers (CX), reducing expenses and time, boosting the productivity of employees, adapting and integrating business models, making better decisions for business, IoT encourages businesses to reevaluate their business strategies and provides them with the tools to do so, as well as increasing revenue generation. Where sensors and other IoT devices are utilized, manufacturing, transportation, and utility organizations benefit most from IoT. However, it has also been used in infrastructure, agriculture, and home automation, encouraging digital transformation in some businesses. The IoT can make farming easier. To assist in automating farming practices, sensors, among other things, can collect data on temperature, humidity, rainfall, soil content, and rainfall. The Internet of Things has an impact on every sector, including manufacturing, finance, healthcare, and retail (Carri et al., 2021).

9.10 KEY FEATURES OF IOE

1. Data processing that is decentralized data is processed in a decentralized manner in an IoE environment, with several distributed nodes playing important roles.
2. Data input and output IoE refers to a networked environment, so devices can input external data and exchange it with other network elements as necessary.
3. Networking with other technologies IoE interacts with other technologies such as AI, machine learning, IoT, big data, cloud, fog, and edge computing. Furthermore, advancements in IoE are linked to the technologies that businesses use for digital transformation processes (Carri et al., 2021).

9.11 CONCLUSION

IoT technology is used in almost every industry today, including smart cities, healthcare, and agriculture. The IoT is used in applications like regular patient health monitoring and drug traceability, among others, in the healthcare industry. Block chain is a distributed technology that can be used to enhance system security. However, integrating IoT with Block can address a variety of security concerns associated to the Internet of Things. An overview of issues, security enhancements, applicability, features, and cryptocurrency types can be found in this chapter. From its fundamental theories to its current forms, the chapter traces the development of blockchain. It provides diverse business applications with usable case studies and successful implementations in cloud/edge computing, smart cities, and IoT. This chapter attempted to enumerate various possible ways in which IoT technology and crypto currency can be integrated into the healthcare sector to improve overall performance and strengthen the current infrastructure. For the next generation of internet applications, block chain will be just as crucial as the public cloud, micro service architectures, and developments are for the current generation. Make sure to factor in the impact of blockchain in all of your current and future application architecture plans.

REFERENCES

Adenugba, Favour, Sanjay Misra, Rytis Maskeliūnas, Robertas Damaševičius, and Egidijus Kazanavičius. "Smart irrigation system for environmental sustainability in Africa: An Internet of Everything (IoE) approach." *Mathematical Biosciences and Engineering* 16, no. 5 (2019): 5490–5503.

Aiello, G., A. Camillo, M. Del Coco, E. Giangreco, M. Pinnella, S. Pino, and D. Storelli. "A context agnostic air quality service to exploit data in the IoE era." In *2019 4th International Conference on Smart and Sustainable Technologies (SpliTech)*, pp. 1–8. IEEE, 2019.

Alkhabbas, Fahed, Romina Spalazzese, and Paul Davidsson. "Characterizing internet of things systems through taxonomies: A systematic mapping study." *Internet of Things* 7 (2019): 100084.

Alsuwaidan, L. Data management model for internet of everything. In *Lecture Notes in Computer Science (including subseries Lecture Notes in Artificial Intelligence and Lecture Notes in Bioinformatics)*; Springer: Berlin/Heidelberg, Germany, 2019; pp. 331–341.

Anand, Sruthy, and Maneesha Vinodini Ramesh. "Multi-layer architecture and routing for internet of everything (IoE) in smart cities." In *2021 sixth international conference on wireless communications, signal processing and networking (WiSPNET)*, pp. 411–416. IEEE, 2021.

Badr, Mohamed, Mohamed M. Aboudina, Faisal A. Hussien, and Ahmed N. Mohieldin. "Simultaneous multi-source integrated energy harvesting system for IoE applications." In *2019 IEEE 62nd International Midwest Symposium on Circuits and Systems (MWSCAS)*, pp. 271–274. IEEE, 2019.

Carri, Andrea, Alessandro Valletta, Edoardo Cavalca, Roberto Savi, and Andrea Segalini. "Advantages of IoT-based geotechnical monitoring systems integrating automatic procedures for data acquisition and elaboration." *Sensors* 21, no. 6 (2021): 2249.

Deepa, Natarajan, Quoc-Viet Pham, Dinh C. Nguyen, Sweta Bhattacharya, B. Prabadevi, Thippa Reddy Gadekallu, Praveen Kumar Reddy Maddikunta, Fang Fang, and Pubudu N. Pathirana. "A survey on blockchain for big data: approaches, opportunities, and future directions." *Future Generation Computer Systems* (2022).

Di Martino, Beniamino, Kuan-Ching Li, Laurence Tianruo Yang, and Antonio Esposito. "Trends and strategic researches in internet of everything." *Internet of Everything: Algorithms, Methodologies, Technologies and Perspectives* (2018): 1–12.

Farias da Costa, Viviane Cunha, Luiz Oliveira, and Jano de Souza. "Internet of everything (IoE) taxonomies: A survey and a novel knowledge-based taxonomy." *Sensors* 21, no. 2 (2021): 568.

Khan, Latif U., Ibrar Yaqoob, Nguyen H. Tran, SM Ahsan Kazmi, Tri Nguyen Dang, and Choong Seon Hong. "Edge-computing-enabled smart cities: A comprehensive survey." *IEEE Internet of Things Journal* 7, no. 10 (2020): 10200–10232.

Lazaroiu, Cristian, and Mariacristina Roscia. "Smart district through IoT and blockchain." In *2017 IEEE 6th international conference on renewable energy research and applications (ICRERA)*, pp. 454–461. IEEE, 2017.

Mitridati, Lesia, Jalal Kazempour, and Pierre Pinson. "Design and game-theoretic analysis of community-based market mechanisms in heat and electricity systems." *Omega* 99 (2021): 102177.

Mohanty, Saraju P., Venkata P. Yanambaka, Elias Kougianos, and Deepak Puthal. "PUFchain: A hardware-assisted blockchain for sustainable simultaneous device and data security in the internet of everything (IoE)." *IEEE Consumer Electronics Magazine* 9, no. 2 (2020): 8–16.

Muhammed, Thaha, Rashid Mehmood, Aiiad Albeshri, and Iyad Katib. "UbeHealth: A personalized ubiquitous cloud and edge-enabled networked healthcare system for smart cities." *IEEE Access* 6 (2018): 32258–32285.

Nakamoto, Satoshi. "Bitcoin: A peer-to-peer electronic cash system." *Decentralized Business Review* (2008): 21260.

Novo, Oscar. "Blockchain meets IoT: An architecture for scalable access management in IoT." *IEEE Internet of Things Journal* 5, no. 2 (2018): 1184–1195.

Nozari, Hamed, Agnieszka Szmelter-Jarosz, and Javid Ghahremani-Nahr. "The ideas of sustainable and green marketing based on the Internet of Everything—The case of the dairy industry." *Future Internet* 13, no. 10 (2021): 266.

Ravindra, S. "The role of blockchain in cybersecurity." *available at:* www. infosecurity-magazine. com/next-gen-infosec/blockchain-cybersecurity/ *(accessed 22 May 2020)* (2018).

Ripple, "Ripple Net", https://ripple.com (Last accessed September 2022).

Ryoo, Jungwoo, Soyoung Kim, Junsung Cho, Hyoungshick Kim, Simon Tjoa, and Christopher DeRobertis. "IoE security threats and you." In *2017 International Conference on Software Security and Assurance (ICSSA)*, pp. 13–19. IEEE, 2017.

Shahinzadeh, Hossein, Jalal Moradi, Gevork B. Gharehpetian, Hamed Nafisi, and Mehrdad Abedi. "Internet of Energy (IoE) in smart power systems." In *2019 5th Conference on Knowledge Based Engineering and Innovation (KBEI)*, pp. 627–636. IEEE, 2019.

Shen, Xinyi, Guolong Shi, Yongxing Zhang, and Shizhuang Weng. "Wireless volatile organic compound detection for restricted internet of things environments based on cataluminescence sensors." *Chemosensors* 10, no. 5 (2022): 179.

Shilpa, A., V. Muneeswaran, D. Devi Kala Rathinam, Grace A. Santhiya, and J. Sherin. "Exploring the benefits of sensors in internet of everything (IoE)." In *2019 5th International Conference on Advanced Computing & Communication Systems (ICACCS)*, pp. 510–514. IEEE, 2019.

Sirma, Muharrem, Adnan Kavak, and Burak Inner. "Cloud based IoE connectivity engines for the next generation networks: challenges and architectural overview." In *2019 1st International Informatics and Software Engineering Conference (UBMYK)*, pp. 1–6. IEEE, 2019.

Snyder, Tom, and Greg Byrd. "The internet of everything." *Computer* 50, no. 06 (2017): 8–9.

Sunyaev, Ali. *Internet computing: Principles of Distributed systems and emerging internet-based technologies*. Springer Nature, 2020.

Tariq, Usman, Atef Ibrahim, Tariq Ahmad, Yassine Bouteraa, and Ahmed Elmogy. "Blockchain in internet-of-things: A necessity framework for security, reliability, transparency, immutability and liability." *IET Communications* 13, no. 19 (2019): 3187–3192.

Tian, Feng. "A supply chain traceability system for food safety based on HACCP, blockchain & Internet of things." In *2017 International conference on service systems and service management*, pp. 1–6. IEEE, 2017.

Tse, Daniel, Bowen Zhang, Yuchen Yang, Chenli Cheng, and Haoran Mu. "Blockchain application in food supply information security." In *2017 IEEE international conference on industrial engineering and engineering management (IEEM)*, pp. 1357–1361. IEEE, 2017.

Vaya, Dipesh, and Teena Hadpawat. "Internet of Everything (IoE): A new era of IoT." In *ICCCE 2019: Proceedings of the 2nd International Conference on Communications and Cyber Physical Engineering*, pp. 1–6. Springer Singapore, 2020.

Wang, Jingzhong, Mengru Li, Yunhua He, Hong Li, Ke Xiao, and Chao Wang. "A blockchain based privacy-preserving incentive mechanism in crowdsensing applications." *Ieee Access* 6 (2018): 17545–17556.

Wang, Kun, Yun Shao, Lei Shu, Chunsheng Zhu, and Yan Zhang. "Mobile big data fault-tolerant processing for ehealth networks." *IEEE Network* 30, no. 1 (2016): 36–42.

Yli-Huumo, Jesse, Deokyoon Ko, Sujin Choi, Sooyong Park, and Kari Smolander. "Where is current research on blockchain technology?—a systematic review." *PloS One* 11, no. 10 (2016): e0163477.

Yu, Jihong, Pengfei Zhang, Lin Chen, Jiangchuan Liu, Rongrong Zhang, Kehao Wang, and Jianping An. "Stabilizing frame slotted aloha-based IoT systems: A geometric ergodicity perspective." *IEEE Journal on Selected Areas in Communications* 39, no. 3 (2020): 714–725.

Zhan, Jinsong, Shaofeng Dong, and Wei Hu. "IoE-supported smart logistics network communication with optimization and security." *Sustainable Energy Technologies and Assessments* 52 (2022): 102052.

Zhang, Xiaofeng. "Development and construction of Internet of Things training practice platform for employment skills assessment." *Journal of Sensors* 2022 (2022).

Zheng, Zibin, Shaoan Xie, Hong-Ning Dai, Xiangping Chen, and Huaimin Wang. "Blockchain challenges and opportunities: A survey." *International Journal of Web and Grid Services* 14, no. 4 (2018): 352–375.

10 An Anti-Counterfeiting Architecture Ensuring Authenticity and Traceability for Expensive Medicines Using QR Codes

Sk. Yeasin Kabir Joy, Md. Nafees Imtiaz Ahsan, Nuzhat Tabassum Progga, Aloke Kumar Saha, Mohammad Shahriar Rahman, and Abdullah Al Omar

CONTENTS

DOI: 10.1201/9781003366010-13

10.1 INTRODUCTION

Counterfeiting is one of the most concerning problems right now in the world. Counterfeiting means the process of imitating any authentic products illegally and trying to project them as originals. It not only causes financial losses for legitimate brands but also deceives consumers by representing fake products as authentic ones. Also, such unfair activities damage economic growth in a very bad way. A recent study by the International Chamber of Commerce (ICC) shows that by 2022 the projected job losses will be between 4.2 and 5.4 million because of counterfeiting [1]. Especially, in the medical sector, counterfeit medicines cause disastrous losses for the consumers. The World Health Organization (WHO) estimates that one out of every ten medical products in developing nations is low in quality or fraudulent [2].

The usability of medicines in our daily lives has become quite a regular phenomenon. Furthermore, the dependency on these medicines is not decreasing, rather it is increasing rapidly. For this reason, adversaries often target this sector to make counterfeit medicines. Moreover, some medicines are crucial for costly treatment which are highly expensive (e.g. Chemotherapy Medicines, Insulin, Vascular Rings). If these types of medicines get counterfeited, it will be threatening to our lives as well as people will face massive financial losses [3]. Hence, restricting the production of counterfeit medicines is very important. Recently, IoE is being introduced to fight against counterfeiting in the healthcare sector. For networked interactions to be as valuable and practical as possible, the Internet of Everything (IoE) integrates the four pillars of people, processes, data, and things [4].

Ensuring authenticity and tracking traceability are the two most significant ways to restrict the counterfeiting of a product. Here, authenticity means getting medicines from a legitimate manufacturer. Keeping the supply chain unaffected by any unwanted party is quite important, especially in the pharmaceutical industry. Traceability means the ability to track specific stages in a supply chain of a product, production location, and identification of the history of its distribution process [5]. Through traceability information, both the patients and the manufacturers come to know that their products are not being counterfeited [6]. But keeping the traceability of a product in the whole supply chain is not that easy, rather it is a very difficult task [7].

The remaining chapters are structured as follows: in section 10.2 related works about anti-counterfeiting are discussed. The necessary preliminaries are discussed in section 10.3. Section 10.4 illustrates the proposed architecture. Section 10.5 reflects upon the security analysis. The implementation details are described in section 10.6. Section 10.7 describes the result analysis. Lastly, section 10.7 concludes the chapter.

10.2 RELATED WORKS

Multiple studies on traceability and anti-counterfeiting have been conducted using a variety of technologies. M. Bala Krishna et al. [8] proposed a product authentication scheme using QR codes. In their paper, they use QR codes for every product and the QR code can be scanned only once for checking the authenticity. Their proposed method did not offer any traceability information about products.

In another paper [9], Shundao Xie et al. proposed an anti-counterfeiting architecture for a traceability system using a modified two-level QR code. The paper proposed a two-level QR code consisting of a copy detection pattern that cannot be copied and re-printed. Using a web-based architecture, the original QR code will provide traceability information like details of production, packaging, and warehouse. For checking the authenticity of the product, the customer needs to take a QR code's picture. After that, the customer has to upload the image to the database. The authentication process is based on assessing the correlation between the two patterns, the original one with the printed pattern. The response time is 25–30 seconds which may not be feasible for some customers.

For implementing the traceability of a medical product, most of the paper used blockchain technology [10], [11]. Blockchain is a distributed decentralized network. Here, information is gathered as blocks which form a chain. The blockchain offers decentralization, transparency, traceability, and immutability which ensures secured transactions among peers [12]. Randhir Kumar et al. [11] have introduced a method using a private blockchain, digital signature, and QR code for ensuring drug safety and authenticity. Customers can get the details of drug information from legitimate manufacturers through the blockchain. But how to get specific drug details is not mentioned properly in their proposal.

Mueen Uddin [10] suggested a novel system that uses smart contracts, it can track and trace drug supply chain systems that use the hyperledger fabric blockchain platform. He offered a blockchain-enabled traceability platform that displays the processes between various pharmaceutical supply chain stakeholders. He also implemented some smart contracts to store and share the transaction information of different pharmaceutical supply chain stakeholders. Moreover, users will need to identify themselves using digital certificates. However, the proposed private fabric solution faces some significant challenges like scalability limitations, especially for the healthcare industry.

The authenticity of a product can be ensured by Physical Unclonable Functions (PUF). PUF provides hardware security by acting as a biometric fingerprint. It is impossible to clone PUF because it depends on various parameters while manufacturing [13]. Using the idea of PUF, Riikka Arppe-Tabbara et al. [14] implemented a method of product authentication where all products are integrated with unique PUF tags during the production. Validation and unique identification of the product is possible by the end-user throughout the supply chain. They also claimed and showed that PUF tags can be added to QR codes. They used three methods including spray coating, knife coating, and stamping. Their system relies on ink containing specific microparticles.

Another technology to fight against counterfeiting is the introduction of Radio-Frequency Identification (RFID) tags. It is frequently used as an anti-counterfeiting method because of its low expense. Mainly, RFID tags have three components, a tag, a reader, and a database [15]. However, to monitor the scheme flawlessly, it requires a continual internet connection between the supply chain's partners and the end customers.

Hoda Maleki et al. [16] seemed to have solved the consistent online connection problem of RFID by proposing a new scheme where they integrated a Non-Volatile Memory (NVM) in the RFID tags so that the persistent online connection will not be needed. The authors showed that all the necessary data can be stored in the NVM keeping it confidential and secured. The authors also showed in their framework that the verification will only be needed at the endpoints after the delivery of the products to customers. Though they have managed to get rid of the persistent online connection by integrating NVM, that may increase the cost of the whole system.

Recently, some researchers have used Near-field communication (NFC) to build anti-counterfeiting frameworks [17], [18]. NFC is a technology that is used over short distances for exchanging data without any contact. Naif Alzahrani et al. [17] proposed an architecture where they used NFC tags and blockchain to implement an anti-counterfeit system. They have proposed a new block validation protocol in their system that has two steps. Firstly, the blockchain initiators select the validation leader randomly and secondly, the validation leaders select log (n) validators and ask them to validate the block. However, the validation selection is not truly random which means validation selection can be predicted. This may hamper the blockchain network's security.

Using (NFC), Mohammad Wazid et al. [18] introduced a novel medication anti-counterfeiting technique for verifying the authenticity of the medication's dose forms which is convenient for IoT environments. However, there is no explicit description of time consumption for the users.

10.3 PRELIMINARIES

The preliminaries that are necessary for our work are briefly mentioned in this section.

10.3.1 Encryption

Encryption is the process of converting plain text into ciphertext by using mathematical procedures. Elliptic Curve Cryptography (ECC) is used for the encryption and decryption of carton ID and product ID. ECC is asymmetric encryption (public-key encryption) scheme that means different keys will be used for encryption and decryption. The encryption and decryption will take place on the web app which is connected to the database. In Gayoso Martínez et al. [19] elaborated on the advantages of using ECC. ECC provides the same amount of security with less key length compared to RSA. The Elliptic Curve Discrete Logarithm Problem, on which ECC is built, determines how secure it is (ECDLP). Curve25519 high-speed elliptic curve cryptography with a key size of 256 bits is used here. A total of four keys (2 public keys and two private keys) are needed. Mathematical notation for the cryptographic operations is shown below.

TABLE 10.1
Overview of the related works

Research	Authenticity	Traceability Information	Unforgeable QR	Digital Signature	Time Overhead
M. Bala Krishna et al.	*Y*	*N*	*N*	*N*	*N*
Shundao Xie et al.	*Y*	*Y*	*Y*	*N*	*Y*
Randhir Kumar et al.	*Y*	*Y*	*N*	*Y*	*Y*
Mueen Ud-din	*Y*	*Y*	*X*	*Y*	*Y*
Riikka Arppe-Tabbara et al.	*Y*	*N*	*Y*	*N*	*N*
This Chapter	*Y*	*Y*	*N*	*Y*	*N*

Y=yes, N=No, X= not used

$$\text{PrivetKey}_{\text{sign}} = \text{PRIV}_{\text{s}}$$
$$\text{PublicKey}_{\text{verif y}} = \text{PUB}_{\text{v}}$$
$$\text{PublicKey}_{\text{encryption}} = \text{PUB}_{\text{enc}}$$
$$\text{PrivateKey}_{\text{decryption}} = \text{PRIV}_{\text{dec}}$$

Here, Encryption is a function which takes a public key and a message and later encrypts that message. Decryption is another function taking a private key and a cyphertext and later it decrypts that cyphertext.

A private key and a message are taken by the Sign function which digitally signs the message. Verify is a function which takes a public key and a signed message. After that, it verifies the signed message.

$$\text{Cyphertext} = \text{Encryption}\ (\text{PUB}_{\text{enc}}, \text{Sign}\ (\text{PRIV}_{\text{s}}, \text{Message}))$$
$$\text{Message} = \text{Verify}\ (\text{PUB}_{\text{v}}, \text{Decryption}\ (\text{PRIV}_{\text{dec}}, \text{Cyphertext}))$$

10.3.2 QR Codes

A 2-dimensional barcode that contains data is called a Quick Response code (QR code). The machine-readable QR code is formed of a grid of black and white squares. It is typically scanned with a traditional scanner or a smartphone scanner app. QR code is used to decode data at a high speed [20]. There are a total of 40 versions of QR codes. The higher the version gets, the more data it can store. QR code uses the Red-Solomon algorithm which enables it to scan the code and extract the data even if the QR code is damaged [20]. There are four levels of error correction and those are given below in Table 10.2.

TABLE 10.2
QR code error correction level

Level L (Low)	*up to 7% damage*
Level M (Medium)	*up to 15% damage*
Level Q (Quartile)	*up to 25% damage*
Level H (High)	*up to 30% damage*

Error correction level M is used in implementation that can restore the data if it is damaged up to 15%. QR code version 5 is also used with error correction level M that can store up to 122 alphanumeric characters. Focardi et al. [21] showed that 300*300-pixel version 5 QR code has a hundred percent success rate of scanning frequency outcome.

10.3.3 Website and Database

Python Django Framework is used to implement the web app for implementing the architecture. The manufacturer will use the web app for creating the QR codes and store the carton and product details. SQLite is used as the database. Built-in security functions are used to secure the database.

10.3.4 Blockchain

The term "blockchain" refers to a distributed, decentralised ledger [22], which is used to store transactional data after ensuring its validation. Formally, blockchain is described as a public ledger that stores all previous transactional records [23]. It is a decentralized network independent of any single authority. Blockchain has given rise to new types of peer-to-peer communication platforms that are both secure and trustworthy [24]. A consensus mechanism is a method by which a majority of network validators come to an agreement on the status of a ledger. It's a collection of rules and procedures that allow several participating nodes to maintain a consistent set of facts [25].

10.3.4.1 Blockchain Functionalities

Transactions are saved in blocks, which are permanent time-stamped units. A cryptographic hash created from the previous block's contents links each block to the one before it. Each block carries a timestamp, the preceding block's hash value ("parent"), and a nonce, which is a random number for hash verification [26]. Because of the cryptographic hash, it's difficult to alter data in one block without affecting every other block in the chain at the same time. This means that any attempt to edit or delete data will cause the cryptographic chain to break, signaling the problem to all nodes in the network. The ledger is not automatically updated with new transactions. Instead, the consensus mechanism makes sure that these transactions are maintained in a block for a specific amount of time before being published to the ledger. After that, changes to the data in the blockchain are no longer possible.

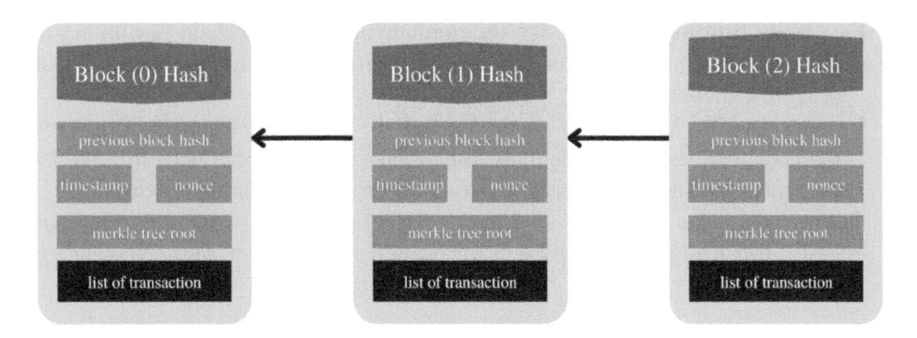

FIGURE 10.1 Block structure.

Decentralization, immutability, transparency, and auditability are all properties of blockchain that make transactions more secure and tamper-proof [12].

Decentralization: In contrast to centralized systems, the blockchain network allows transactions to be made between any two peers without the need for third party authentication. By employing a variety of consensus techniques, blockchain can help to alleviate trust concerns.

Immutability: Immutability refers to the inability to change. An immutable object in computer science is one whose state cannot be changed after it is created. Blockchain technology's immutability is one of its most important advantages. Any entity can't change, replace, or fabricate data saved on the network since transactions are immutable. The immutability of data allows for a high level of data integrity.

Transparency: Because of the decentralized nature of blockchain, all transactions can be examined in real-time by utilizing a personal node or blockchain explorers. As new blocks are confirmed and added, each node's copy of the chain is updated. This feature ensures the transparency of the blockchain.

Auditability: All transactions in a blockchain network are tracked by a digital distributed ledger, which also verifies each one with a digital timestamp. As a result, it is possible to reach any node in the network to audit and trace previous data. According to [27] followings are the major issues of blockchain.

Lack of privacy: The network's transaction history is stored by every node individually. This might be advantageous in a secure setting and the quality of some applications, but it poses a problem for the use of situations when privacy is needed.

High costs: The blockchain's fundamental processing, which replicates all transaction history across all nodes, and involves intensive computing. This feature offers security benefits, but it might be a sticking point for larger networks. Security model: For transaction verification and execution, blockchain use public-key encryption. This technique, while extremely secure, required the usage of both public and private keys. If a party misplaces or accidentally discloses their private key, the system has no safeguards to ensure enhanced security.

Flexibility limitations: Although the immutability of blockchain guarantees the integrity of transactions, for use cases requiring transaction updates, it could provide a challenge.

Latency: In a blockchain, every node stores all the transaction history which secures the network's security credentials. But whenever new blocks need to be added, the process of storing transaction history becomes computationally expensive.

Governance: The decentralized nature of blockchain has benefits for some use cases. But it might be a problem for general management and governance by organizations.

10.3.4.2 Smart Contract

Smart contracts use blockchain technology to ensure that contracts are executed correctly [28]. Smart contracts are self-executed programs that run when specific conditions are fulfilled. For being assured of the outcome immediately from all parties, smart contracts are used where no intermediaries are involved. Another advantage of smart contract is upon fulfilling required criteria, it can automate any workflow.

After validation of the fulfilled conditions, the computer network operates the necessary operations. For example, activities such as payment procedure, vehicle registration or buying tickets are included. Blockchain updates as soon as the transaction occurs. Only those who have the official access granted by authority can see the results. Moreover, these transactions are unalterable.

Smart Contract's life cycle consists of four steps [28].

Creation of smart contracts: Different programming languages are used for writing Smart Contracts. Before writing the contract, an agreement is needed among stakeholders which involves rights, functions, and limitations on the contract.

Deployment of smart contracts: The smart contracts are deployed after validation. Contracts are deployed on different applications which are built on blockchains. If anything needs to be changed, the whole structure requires a new formation of the contract. After the contract deployment, all parties can invoke the smart contracts through the blochchains.

Execution of smart contracts: When the smart contract is deployed and specific conditions are fulfilled, the functions of the smart contract will be executed automatically. As smart contracts are basically computer programs, they consist of logical statements. After predefined criteria are met, the associated functions will be executed, that means a transaction will occur.

Completion of smart contracts: After a smart contract is executed, the state of the blockchain changes accordingly and the new state will be updated to each copy of the blockchain that is stored by the participant. As a result, the new transaction as well as the previous transaction history will be saved.

10.3.4.3 Types of Blockchain

According to [29], blockchain can be categorized into four groups based on how accessible the data is.

Public Blockchain: This type of blockchain is open to all. Anyone has the accessibility of reading or initiating transaction.

Private Blockchain: This type of blockchain is not accessible by all. Only the parties who have authorized access can read or initiate a transaction in private blockchain. Most of the time, this type of blockchain is used within organizations or similar group of organizations.

Consortium Blockchain: Multiple companies establish a consortium and are authorized to submit transactions and read transactional data in this type of Blockchain.

Hybrid Blockchain: As the name suggests, two or more of the abovementioned blockchains can be blended together. Different features or characteristics co-exist in this type of blockchain. A Hybrid Blockchain allows a Blockchain platform to be configured in many ways. For example, sometimes a public blockchain can have some features from a private blockchain.

Also, based on the need for authorization, blockchain is categorized as permissionless blockchain, permissioned blockchain and hybrid blockchain [29].

Permissionless Blockchain: This sort of Blockchain does not require prior permission to participate. Users can join this type of blockchain from anywhere by using their computer after getting verified.

Permissioned Blockchain: Unlike permissionless blockchain, anyone cannot be a part of the blockchain. Before joining, users must go through certain verification process. After getting verified, users will be eligible to join the blockchain.

Hybrid Blockchain: When a node participates in both permissionless and permissioned blockchains to promote inter-blockchain communication, this type of blockchain is also known as Hybrid Blockchain.

10.3.5 Hyperledger Fabric

Hyperledger Fabric is a private (permissioned) blockchain. Two of the main features of this blockchain ensure the consistency of data and provide permission control. It is used to create distributed ledger solutions that offer large-scale flexibility, high scalability, improved security, and so on. Moreover, the modularity and adaptability of this design make it a great choice for industrial applications [30]. Although the consensus and transaction validation are done by a small group of nodes, thus decreasing the pressure on consensus algorithm, it cannot solve the bottleneck problem of the blockchain entirely [31].

10.3.5.1 Hyperledger Fabric Architecture

The primary components of the hyperledger fabric will be covered in this part.

Membership service provider: The membership service provider (MSP) is responsible for maintaining the node's identities of the system and issuing node credentials for authentication and authorization [32]. Digital certificates are used for this. The nodes utilize these certificates to ensure that only the authenticated nodes are allowed to join the blockchain network.

Certificate Authorities: Certificate authorities (CA) offer different functions to various blockchain nodes. Registration, issuance, and management of digital certificates are the key functions executed by CA [33]. For the issuance and management of digital certificates, CA depends on the Public Key Infrastructure (PKI). This dependency ensures all network nodes have access to a public and private key combination. For making a transaction on the blockchain network, these nodes require the public and private key-pairs.

Nodes: There is a difference between the permissionless blockchain and permissioned blockchain regarding rights to join, access and validating transactions. In the permissionless blockchain, all nodes have the same joining, accessing and validating rights. Meanwhile, as hyperledger fabric is a permissioned blockchain, it does not offer the same rights to the nodes as permissionless blockchain.

Peers: The component peers are crucial to the blockchain network's operation. Peers represent a company or business. Each of them can have one or more peers representing them in the blockchain network. They take on different roles depending on the duty that was allocated to them during the network setup. A peer may be a member of more than one organization or channel. It is in charge of maintaining the ledgers and chaincodes.

Channels: A channel in hyperledger fabric provides a private network for initiating and managing private transaction. This private transaction occurs between multiple network participants. Before the transaction takes place on a channel, the participating network nodes require going through an authentication process to get the permission for submitting the transactions. A unique identity is provided to every peer who joined a channel. This identity is given by MSP and later used for authentication purposes.

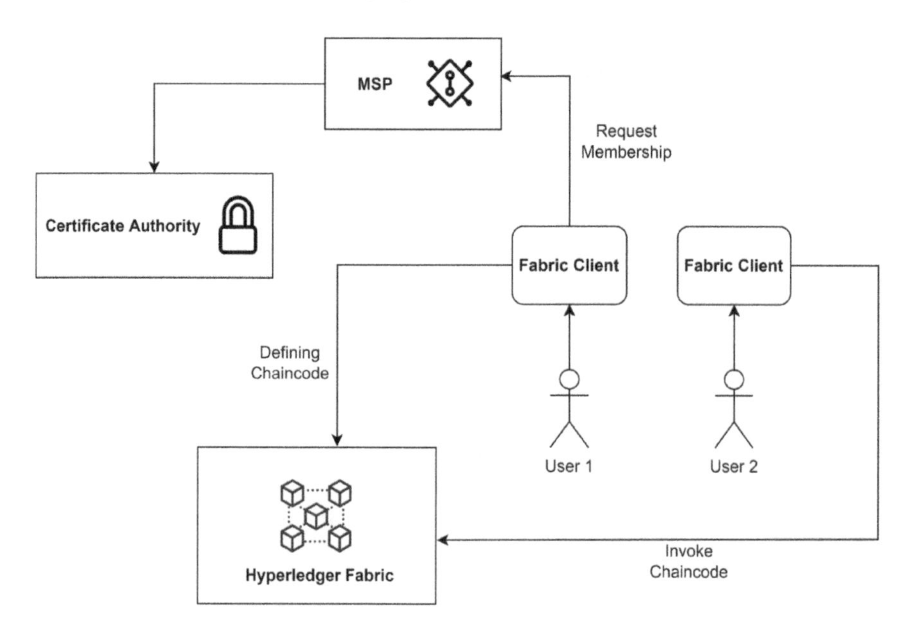

FIGURE 10.2 Hyperledger fabric architecture.

Smart Contracts/Chaincode: In hyperledger fabric, smart contracts are programs written in Go, JavaScript/TypeScript, or Java that include business logic to be performed as blockchain transactions. Nodes with membership define chaincodes into the hyperledger fabric using the fabric client application. Application user can invoke the chaincode using the client application.

World State: The world state and blockchain are two different parts of the hyperledger fabric ledger. World state is a database for storing key-pair values. It stores cache of the current values. Blockchain stores all the transaction history that are linked together in blocks to form a chain [34]. The world state, unlike the blockchain, is mutable. It contains the most recent values for all keys that have been updated in the transaction. Even though a smart contract reads and saves the world state data, the world state does not change throughout the smart contract execution. A smart contract always has access to the previous block's world state, and the outcome of execution is a read/write set, which includes a list of keys to read and a list of keys and values to change.

10.3.6 Security Definition

- **Privacy:** Privacy means each user is protected from others' interference.
- **Traceability:** Traceability is the ability to check a product's details throughout the whole supply chain.
- **Integrity:** Integrity means only authorized changes will be applicable to carton information but product information will be unchanged throughout its life cycle.
- **Confidentiality:** Confidentiality means information being preserved from unauthorized parties.
- **Non-repudiation:** The state of ensuring the validity of a product ID and carton ID.

10.4 PROPOSED METHOD

In the proposed system, both traceability and authenticity of a product in a supply chain have been implemented. Here, it is assumed that the products are expensive drugs. In the framework, it is considered that the supply chain consists of the manufacturer, distributor, delivery person, pharmacist and consumer. Figure 10.3 represents the pictorial representation of the proposed architecture. The start point of the supply chain is the manufacturer and the endpoint is the consumer. The framework can be divided into two parts. The first one is the manufacturer to the pharmacist and the second one is the pharmacist to the consumer.

10.4.1 Working Procedure

1. The manufacturer will produce the products and attach a unique QR code to each product. Also, they will attach a different QR code to each carton which consists of a set of products. The product QR code will contain the website URL concatenated with encrypted product ID and the carton QR

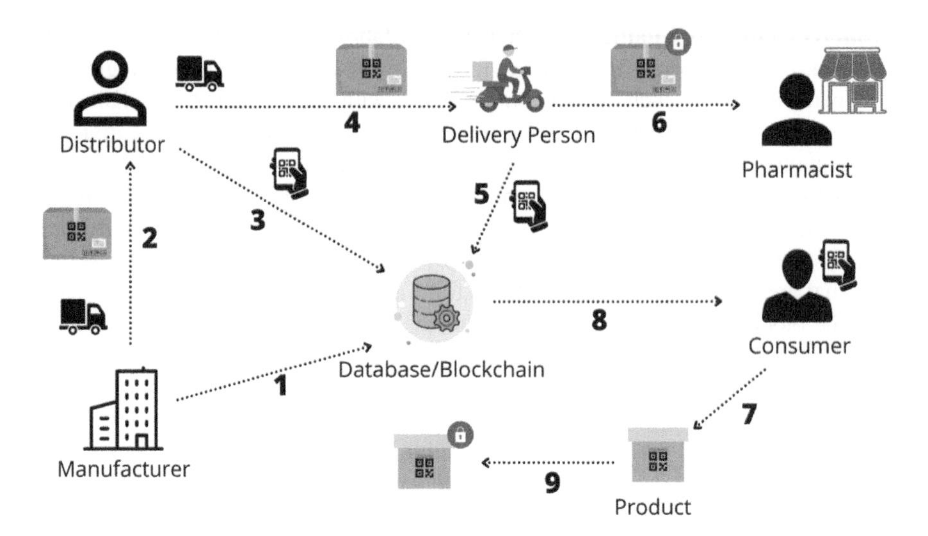

FIGURE 10.3 Proposed anti-counterfeit architecture.

code contains encrypted carton ID. The QR code generation process is implemented by a web app and the carton ID and corresponding product ID will be stored in the database. These Product IDs and Carton IDs will be mapped with each other for necessary steps afterward. For example, suppose a carton ID is AA001 then the product ID will be AA001001. The traceability of the products in the supply chain is ensured by the process of mapping carton ID and product ID.

Furthermore, for ensuring additional security padding has been added to the carton ID and product ID. Padding is the method to add some extra data in the front, middle, or end of the original data before encryption. The padding helps to make it secure from the adversary by not revealing any kind of information about the patterns between carton ID and product ID.

2. The manufacturer will deliver the carton to the distributor.
3. After receiving the carton, the distributor will scan the carton QR code by any mobile scanner app and the distributor's information will be added to the database. It is assumed that the manufacturer company already would have registered the distributor as their legitimate distributor. Through this process, the detailed information of the distributor will be added to the database.
4. Then the distributor will deliver it to the pharmacist through a delivery person.
5. The delivery person will scan the QR code with any mobile scanner app. After the scanning, the delivery person will add the pharmacist's information to the database. Also, the delivery person's information will be added to the database automatically after scanning. Both of the information will be added against the carton ID. Here, it is assumed that the delivery person is also a registered user of the system. After this step, the carton QR code will be blocked from scanning and adding information further.

6. After successfully adding the information, the delivery person will confirm the delivery to the pharmacist.
7. While purchasing a product, the authenticity of the product can be checked by scanning the product QR code by any mobile scanner app. It can be done by a customer or pharmacist himself/herself.
8. The scanning will let the customer see detailed information about the product and the manufacturer, distributor and pharmacist regarding it. Through this process, the traceability of the product can be checked.
9. After confirming the purchase, the QR code scanning must be done and once the QR is scanned, it will be blocked.

If the QR code of a product is scanned but any of the information of the supply chain is not found, then it will mean that the product broke the supply chain and it might be a counterfeit product. Also, if someone wants to scan the QR code more than once, it will send a message saying that the QR code is not valid. As a result, this feature will help to restrict the QR code copy-paste and that ensures product authenticity.

Here, we have mentioned the database as a traditional database. But when we have implemented the architecture, both traditional database and blockchain are implemented.

In the blockchain implementation, hyperledger fabric is used. The blockchain was used to store the data only the authentication process is done using a traditional database.

10.4.2 Data Transition Overview

Here, the data transition overview of the proposed methodology is described. Table 10.3 listed out the notations that are used in this section.

Figure 10.4 shows the data transition overview of our architecture. It shows how the data flow among different units and users in our architecture.

System Users: Manufacturer, Distributor, Delivery Person, Consumer.

Login Unit: In this unit, the Manufacturer, Distributor, and Delivery Person will log in with their necessary credentials.

Mobile Scanner App: Distributors, Delivery persons and Consumers will use any mobile scanner app.

Web App: Our developed web app will be used by all the users. It will be used to interact with the database/blockchain.

Database: All necessary information will be stored in the database.

Blockchain: Carton and product information will be stored in the blockchain.

Steps in the System:

Step 1: MAN login using ID and PWD.

Step 2: MAN gets access to the web app.

Step 3: MAN sends the PD and CD to the DB/BC (through the web app) and creates CQR and PQR.

Step 4, 5: DB/BC sends CNFRM to the MAN (through the web app).

TABLE 10.3
Terminology table

Notations	Description
ID	*Username*
PWD	*Password*
PD	*Product Details*
CD	*Carton Details*
CNFRM	*Confirmation*
DI	*Distributor Information*
PI	*Pharmacist Information*
DPI	*Delivery Person Information*
PID	*Product ID*
PDT	*Product Traceability Information*
CID	*Carton ID*
SCN	*Scan*
MAN	*Manufacturer*
DIS	*Distributor*
DEL	*Delivery Person*
PQR	*Product QR Codes*
CQR	*Carton QR Codes*
DB	*Database*
BC	*Blockchain*

Step 6, 7, 8: DIS SCN the CQR and DI is added against the CID after logging in with ID and PWD. CID is tracked by SCN the carton QR code.
Step 9, 10: DB/BC sends CNFRM to the DIS (through the web app).
Step 11, 12, 13: DEL SCN the CQR and PI, DPI is added against the CID after logging in with the ID and PWD. CID is tracked by SCN the CQR.
Step 14, 15: DB/BC sends CNFRM to the DEL (through the web app).
Step 16, 17: Consumer requests a specific PD by SCN the PQR. The PID is sent to the DB/BC.
Step 18, 19: DB/BC sends PD to the consumer (through the web app).

10.5 SECURITY ANALYSIS

This section explains the security analysis of the proposed method.

- **Privacy:** User accounts will be password protected. User credentials will be stored in a database. This information will be non-accessible to other users. This ensures the privacy of the users. It is assumed that manufacturers will not reveal user credentials under any circumstance.

  User Credentials = USRCR
  USRCR = Secured

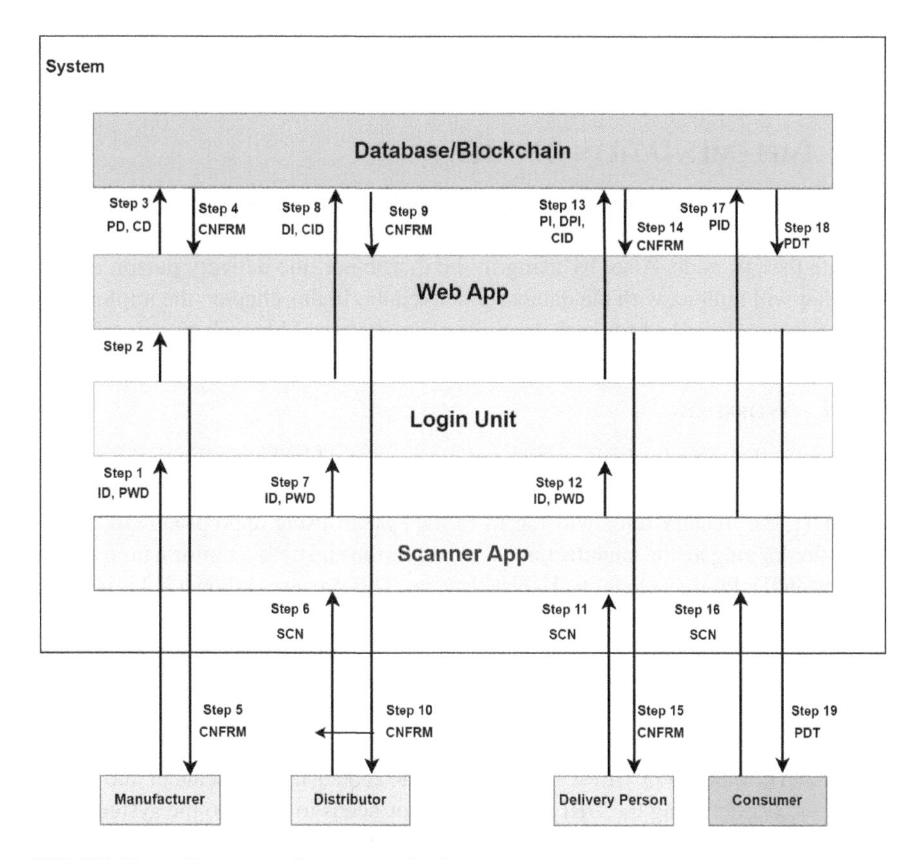

FIGURE 10.4 Overview of data transition in our system.

- **Traceability:** By scanning the QR code from the product, traceability information will be available to the consumer. Moreover, in blockchain implementation, blockchain itself provides a traceability feature. The ledger keeps all the previous records of product and carton information.
- **Integrity:** Registration unit and user credentials will provide data integrity for both centralized database and blockchain.
- **Confidentiality:** Registration will be completed by the manufacturer so different types of users will have specific access permissions. As a result, the user credentials will be prevented from unauthorized access. Also, encrypted IDs and padding will provide confidentiality.

UnauthorizedAccess = U A
UnauthorizedAccessDenied = UAD
USRCR/= U A
UAD(U SRCR)

- **Non-repudiation:** We have ensured non-repudiation by using digital signature. Cyphertext = Encryption(PU B_{enc}, Sign($PRIV_{s,}$ Message))

Here, digital signature is added before encryption, so it is adding an extra layer of security that ensures non-repudiation.

10.6 IMPLEMENTATION DETAILS

Two prototypes (using a centralized database and blockchain) according to our architecture are developed. A web app is developed by which the manufacturers will generate the QR code. Also, by using it, the distributor, the delivery person and the consumer will interact with the database/blockchain. In this chapter, the implementation details are described for both centralized database and blockchain

10.6.1 Workflow

Workflow is the same for both centralized database and blockchain implementations.

Step 1: The manufacturer will log in to the system using appropriate credentials. After logging in, the manufacturer will add carton and product information such as carton ID, Product Quantity, Expiry date etc. This process is shown in Figure 10.5.

Product IDs are generated automatically against Carton IDs because they are mapped with each other. Different QR codes for products are also generated automatically in the same process. In Figure 10.6, QR codes are generated for a specific carton.

Step 2: The distributor will scan the carton QR code using any scanner and will go to the system using the URL. The distributor needs to log in to the system using appropriate credentials. After that, the distributor information will be automatically added against the carton ID. The process is shown in Figure 10.7.

Step 3: The delivery person will scan the carton QR code with any scanner and will go to the system using the URL. The delivery person needs to log in to the system using appropriate credentials. After that, the delivery person will add pharmacy information. Both the pharmacy's information and the delivery person's information will be added against the carton ID. The process is shown in Figure 10.8.

Step 4: After confirming the purchase, the customer or pharmacist can check the QR code of the product by scanning with any scanner. By doing this, the customer/pharmacist can see the product details. By this process, authentication of

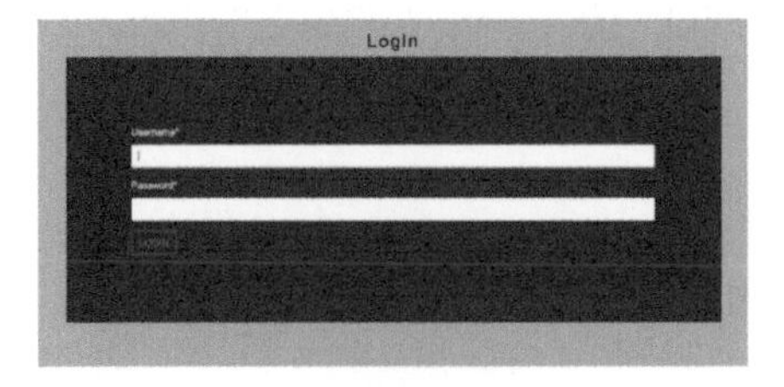

Login Page

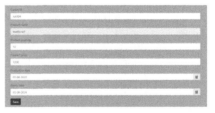

Carton and Product Adding

FIGURE 10.5 Adding information by manufacturer.

FIGURE 10.6 Adding information by manufacturer.

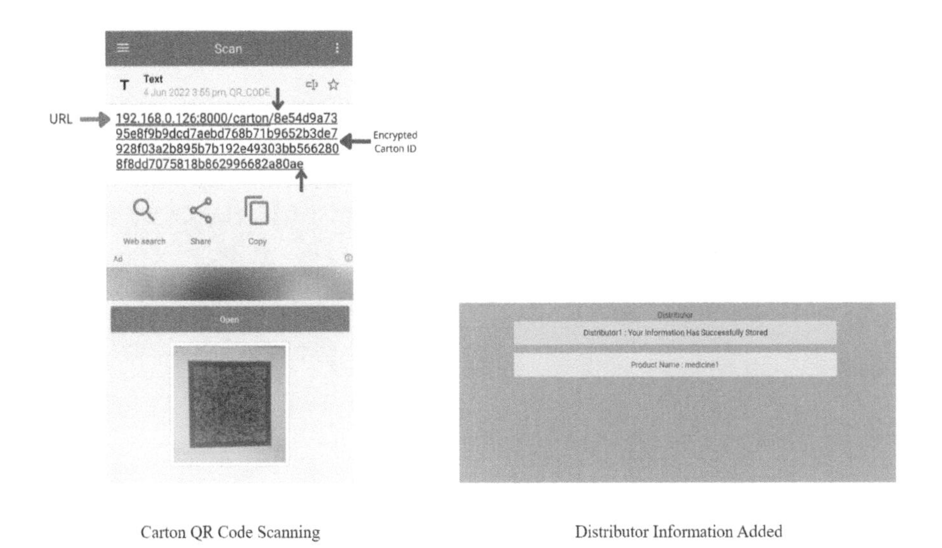

FIGURE 10.7 Distributor scanning and adding information.

the product and all the traceability information can be checked. The whole process is shown in Figure 10.9.

If the QR code shows 'already scanned before' or any information is missing then the customer can file a complaint by clicking 'click here' button. This will take him/her to another page for submitting the complaint. By adding complain details, the customer can submit the complaint to the manufacturer. This process is shown in Figure 10.10.

10.6.2 Blockchain Implementation

In the blockchain implementation, only the carton and product information is stored in the blockchain rather than a centralized database. The authentication and complain

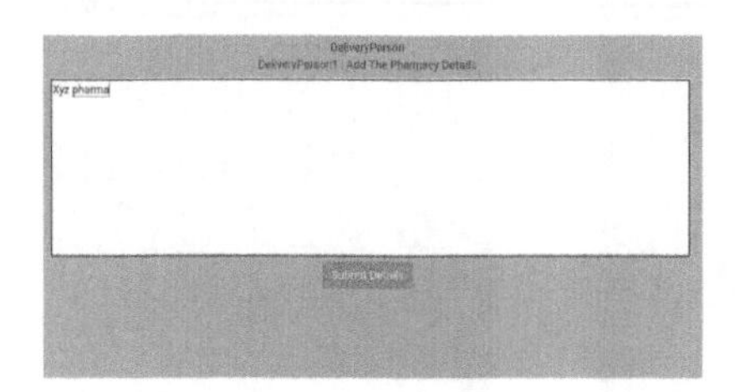

Pharmacy Information Added by Delivery Person

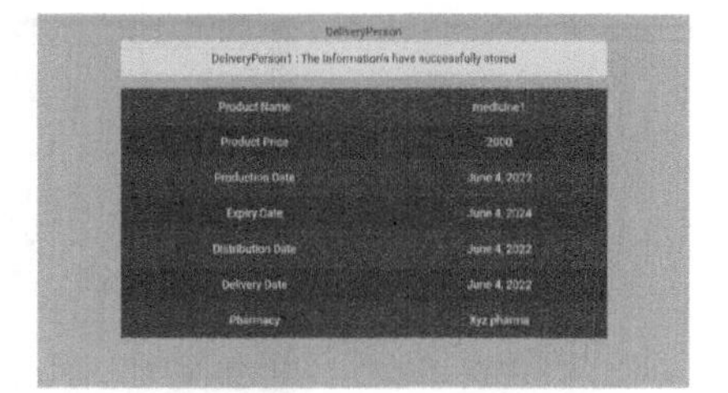

Pharmacy and Delivery Person Information Added

FIGURE 10.8 Product QR scanning and checking information.

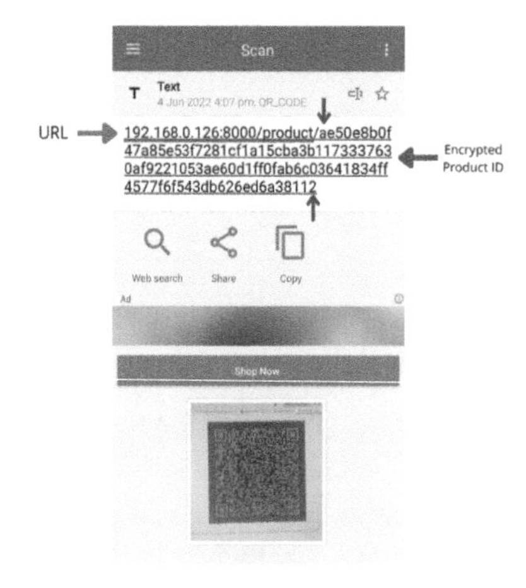

Product QR Code Scan

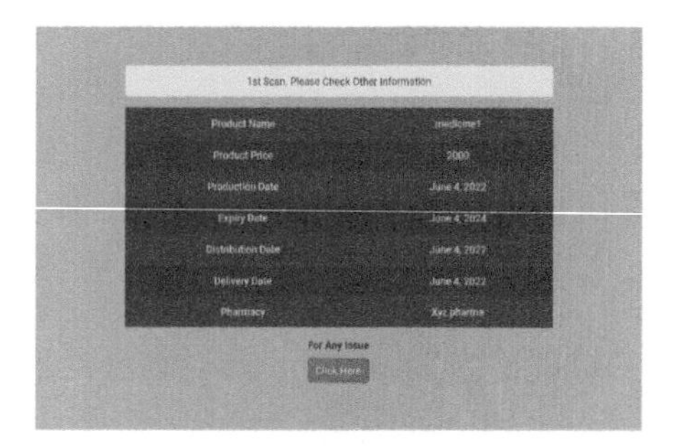

Product Details

FIGURE 10.9 Product QR scanning and checking information.

submission process is handled in centralized database. For the blockchain simulation, hyperledger fabric version 2.2 test-net is used which includes two organizations consisting of one peer each. Also, for world state we have used CouchDB. CouchDB is the hyperledger fabric's standard peer state database, which stores ledger data in JSON format. It allows for efficient queries across enormous datasets. In Figure 10.11, it can be seen that blockchain data are stored in CouchDB.

10.7 RESULT ANALYSIS

In this section, both centralized database implementation and blockchain implementation are analyzed and compared.

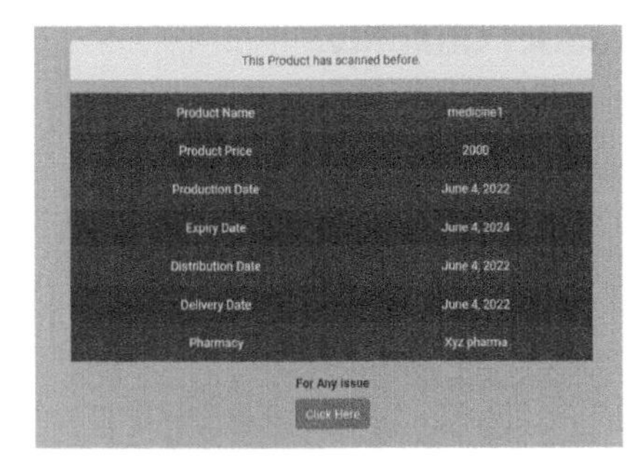

QR Code Scanned before

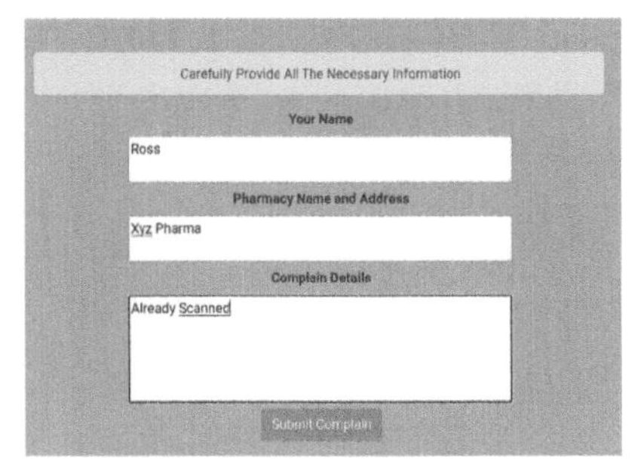

Submit Complain

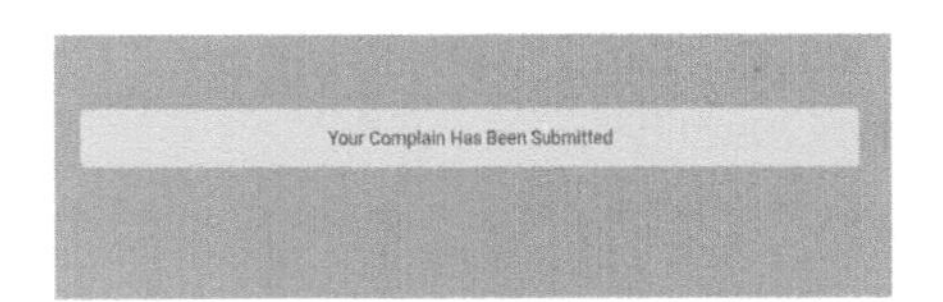

Complain Submitted

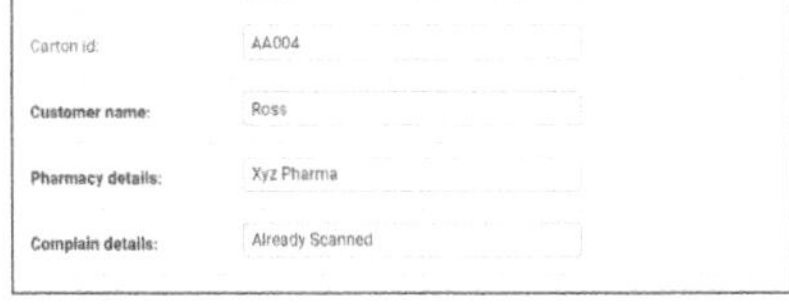

Complain Received at Manufacturer's End

FIGURE 10.10 QR code scanning and complain submission.

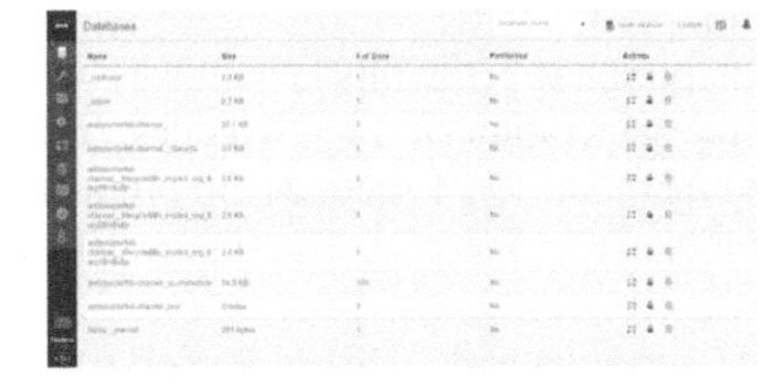

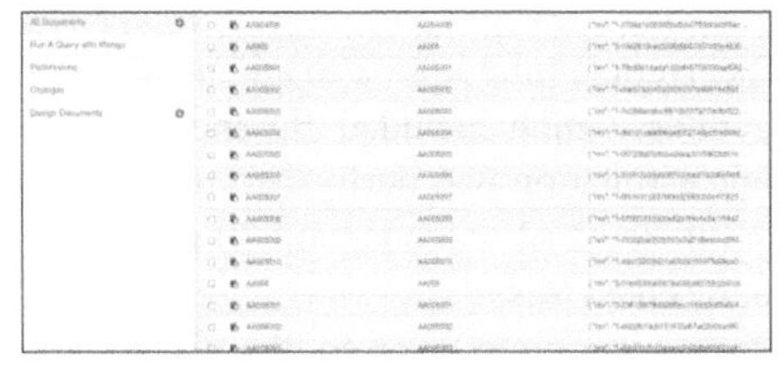

FIGURE 10.11 CouchDB storing blockchain information.

10.7.1 Centralized Database Analysis

When manufacturers add details of a carton, a specific number of products will be added automatically. Each product ID will be mapped with a corresponding carton ID. For adding a carton and ten products, the average time taken was 2.76 seconds which includes encryption of IDs and a total of 11 QR codes generation. We have done this time estimation for 25 products that take 5.96 seconds, 50 products that take 11.70 seconds and 70 products that take 14.51 seconds. This was reflected in Figure 10.12.

The average time taken after scanning for adding/accessing information by the distributor (0.070 seconds), Delivery Person (0.100 seconds) was analyzed. Also, the average time taken to view the product information by consumer (0.102 seconds).

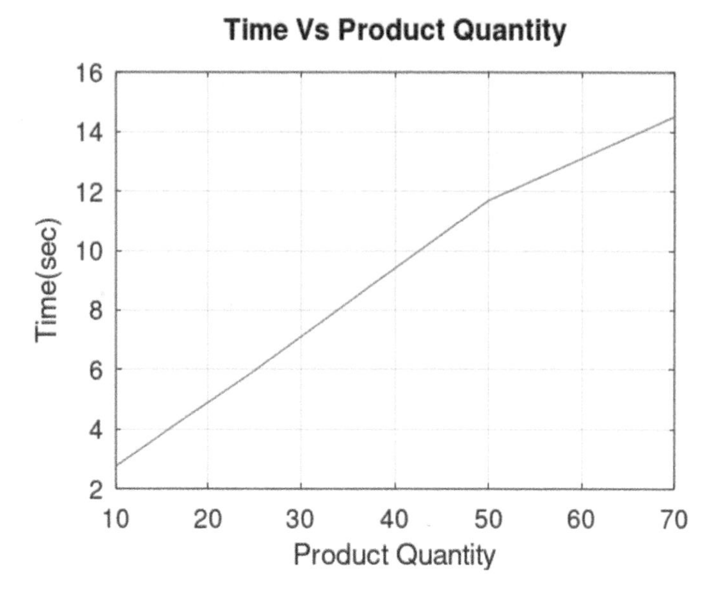

FIGURE 10.12 Time vs product quantity (CD).

This information is shown in Figure 10.13. Note that the web application is hosted on localhost. System configuration Intel core i5, 8 GB ram, Windows 10.

10.7.2 Blockchain Analysis

The average time it took to add a carton and ten products to the blockchain was 5.07 seconds, which included ID encryption and the generation of 11 QR codes. This time estimation was performed for 25 products that took 19.71 seconds, 50 products that took 30.12 seconds, and 70 products that took 41.39 seconds. The Figure 10.14 represented this.

The average time taken by the distributor (0.41 seconds) and Delivery Person (0.64 seconds) after scanning for adding the information was evaluated. Consumer's average time spent seeing product information (0.36 seconds) was also examined. Figure 10.15 depicts this information.

10.7.3 Centralized Database vs Blockchain

Both centralized database and blockchain have different advantages and disadvantages. Here in this section, centralized database and blockchain are compared. When comparing private blockchain with databases, the first thing to notice is how authority is handled. A private blockchain is partially decentralized because of the fact that a permissioned group handles transaction validation. Databases are always centralized.

Blockchain in general offers traceability. In our proposed method whenever new information of a carton or product is added blockchain will not replace the previous information. So the manufacturer can track back and obtain the full life

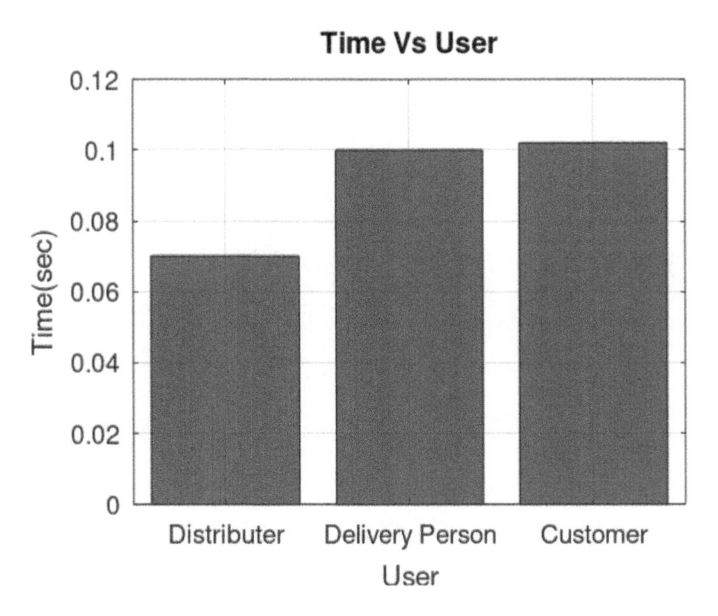

FIGURE 10.13 Time vs user (CD).

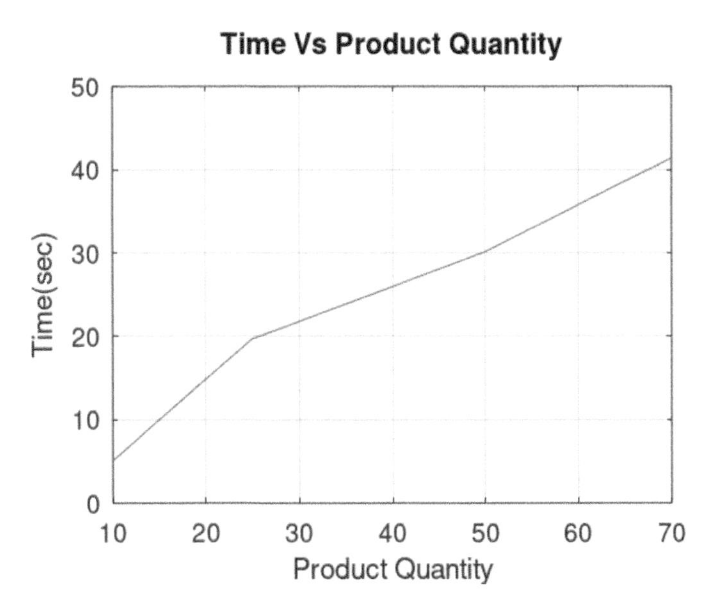

FIGURE 10.14 Time vs product quantity (BC).

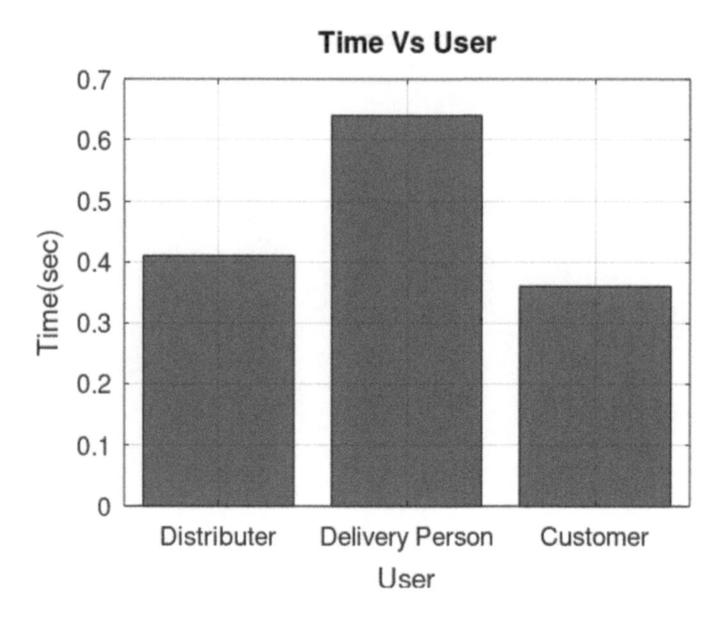

FIGURE 10.15 Time vs user (BC).

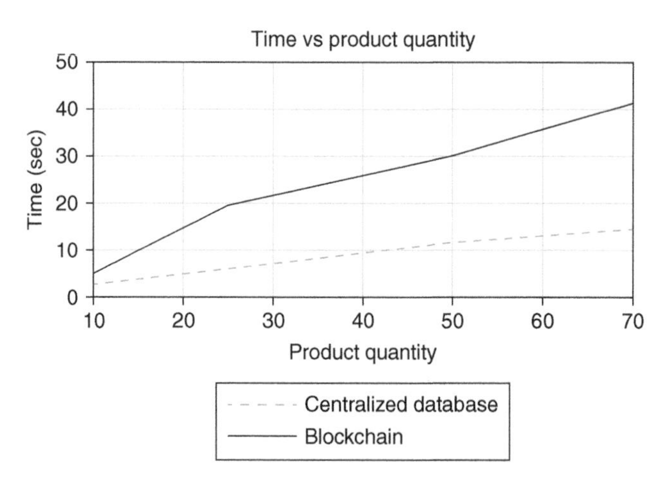

FIGURE 10.16 Time vs product quantity comparison.

cycle of a carton and the products inside it. This feature will provide traceability for manufacturers also. Whereas in the traditional databases the new information replaces the old information which blocks the traceability feature.

When it comes to performance, database is extremely fast in terms of data transactions. In comparison to public blockchain, private blockchain is fast but not as fast as centralized database [35]. In our blockchain implementation, the time taken for adding product and individual time needed for distributor, delivery person, and

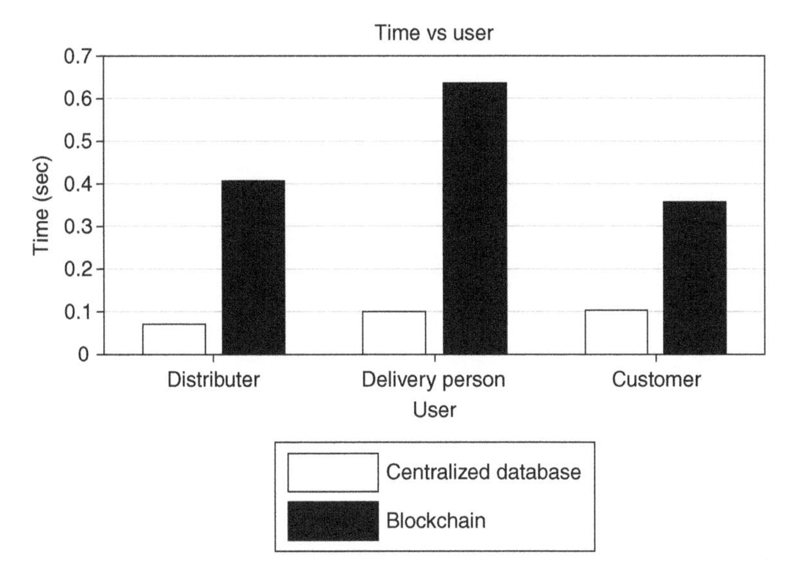

FIGURE 10.17 Time vs user comparison.

consumer for adding and receiving data is higher than database implementation. This is shown in Figures 10.16 and Figure 10.17. From the consumer perspective time taken for accessing the product information is feasible (0.012 seconds in a centralized database and 0.36 seconds in the blockchain).

Blockchain has data redundancy. Every peer in our hyperledger fabric blockchain has a copy of the ledger. On the other hand, in a centralized database no data redundancy occurs. Using a centralized/traditional database is subjected to single point of failure [36]. Blockchain is more secured in terms of no single point of failure as well as using cryptographic function for its underlying implementation. Also, the immutability and transparency of blockchain help to eliminate human errors [3].

10.7.4 Cryptographic Functions

The plaintext size is linearly increasing with the ciphertext size. The encryption is included with digital signature. The plaintext size against the ciphertext size is plotted in Figure 10.18.

Figure 10.19 shows the encryption time plot against the size of the plaintext. Encryption process also includes the digital signature. When the data size is 500 KB, the average encryption time is 0.0021 seconds, according to the graph. Average encryption time is 0.0043 seconds when the data size is 1000 KB. Average encryption time for 1500 KB is 0.0063 seconds, for 2000 KB the average encryption time is 0.0085 seconds. So, the average encryption time is not linear for different size of data-set.

Figure 10.20 shows the decryption time plot against the size of the plaintext. Decryption process includes the verification of digital signature. When the data size

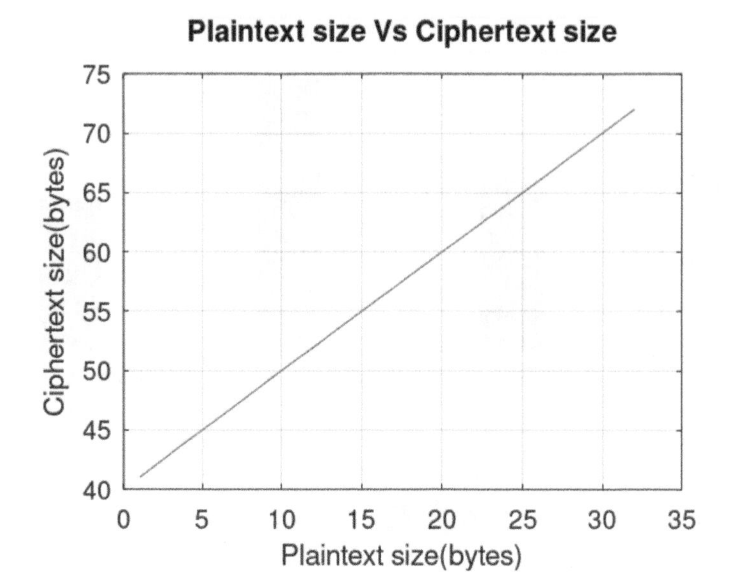

FIGURE 10.18 Plaintext size vs ciphertext size.

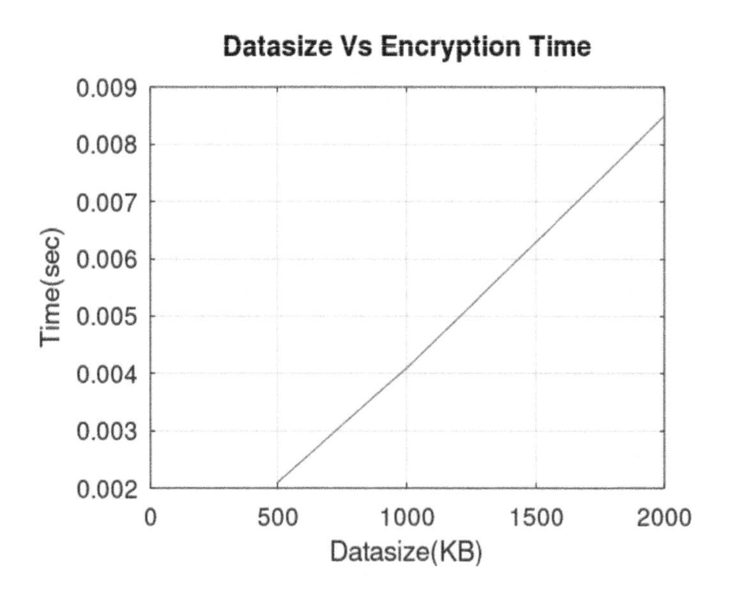

FIGURE 10.19 Datasize vs encryption time.

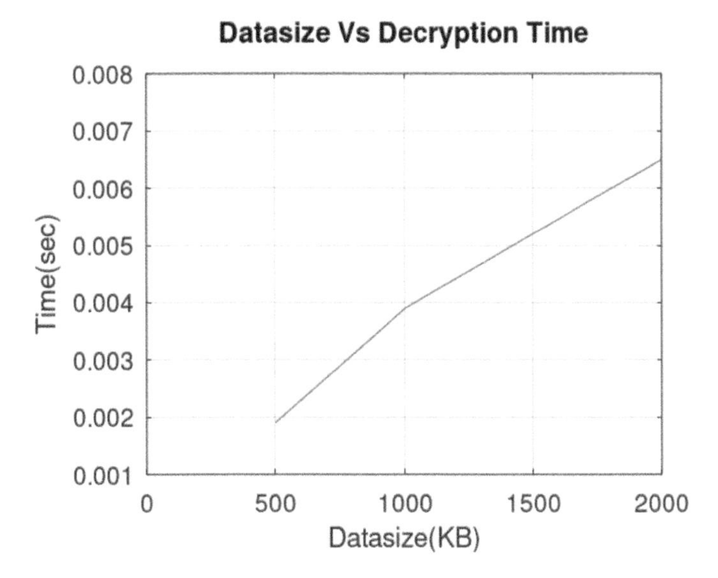

FIGURE 10.20 Datasize vs decryption time.

is 500 KB, the average decryption time is 0.0019 seconds, according to the graph. Average decryption time is 0.0039 seconds when the data size is 1000 KB. Average decryption time for 1500 KB is 0.0052 seconds, for 2000 KB the average decryption time is 0.0065 seconds. So, the average decryption time is also not linear for different size.

The encryption scheme used (curve25519 high-speed elliptic curve cryptography) for the implementation of the proposed architecture is fast. Figures 10.19 and 10.20 shows encryption takes a little bit more time than decryption but the overall time taken for encryption and decryption is extremely fast. So for both the implementations, encryption and decryption of the carton ID and product ID are efficiently implemented.

10.8 CONCLUSION

This chapter presents an anti-counterfeiting architecture for expensive medicines. The proposed architecture provides traceability in the supply chain and the authenticity of every product. This architecture has the potential to successfully put an end to the counterfeiting issue in the pharmaceutical industry. With this framework, both consumers and manufacturers will be benefited. For expensive drugs, our architecture provides an easy-to-implement framework at a minimal cost. This architecture is implemented using both traditional database and hyperledger fabric separately. The traditional database implementation is faster than the blockchain implementation in terms of adding and accessing data. But adding blockchain makes the architecture more secure by eliminating the single point of failure and providing immutability. Adding QR codes in product packaging and maintaining the involved parties (distributor and delivery person) is still a concern. In the future, this architecture can be improved in terms of involving all the parties in the blockchain with proper access control and eliminate the usage of the database for authentication.

REFERENCES

[1] Economics, Frontier. "The economic impacts of counterfeiting and piracy: Report Prepared for BASCAP and INTA." (2017).

[2] WHO. "1 in 10 medical products in developing countries is substandard or falsified." (2017).

[3] Pandey, Prateek and Ratnesh Litoriya. "Securing e-health networks from counterfeit medicine penetration using blockchain." *Wireless Personal Communications* 117 (2021): 7–25.

[4] Miraz, Mahdi H., Maaruf Ali, Peter S. Excell, and Rich Picking. "A review on Internet of Things (IoT), Internet of everything (IoE) and Internet of nano things (IoNT)." *2015 Internet Technologies and Applications (ITA)* (2015): 219–224.

[5] Wimmers, Heidi. "Why is drug traceability important in a hospital?." In *COST*, p. 74. 2015.

[6] Huang, Yan, Jing Wu, and Chengnian Long. "Drugledger: A practical blockchain system for drug traceability and regulation." In *2018 IEEE international conference on internet of things (iThings) and IEEE green computing and communications (GreenCom) and IEEE cyber, physical and social computing (CPSCom) and IEEE Smart Data (SmartData)*, pp. 1137–1144. IEEE, 2018.

[7] Griebel, Lena, Hans-Ulrich Prokosch, Felix Köpcke, Dennis Toddenroth, Jan Christoph, Ines Leb, Igor Engel, and Martin Sedlmayr. "A scoping review of cloud computing in healthcare." *BMC medical informatics and decision making* 15, no. 1 (2015): 1–16.

[8] Bala Krishna, M. and Arpit Dugar. "Product authentication using QR codes: A mobile application to combat counterfeiting." *Wireless Personal Communications* 90 (2016): 381–398.

[9] Xie, Shundao and Hong-Zhou Tan. "An anti-counterfeiting architecture for traceability system based on modified two-level quick response codes." *Electronics* 10, no. 3 (2021): 320.

[10] Uddin, Mueen. "Blockchain Medledger: Hyperledger fabric enabled drug traceability system for counterfeit drugs in pharmaceutical industry." *International Journal of Pharmaceutics* 597 (2021): 120235.

[11] Kumar, Randhir, and Rakesh Tripathi. "Traceability of counterfeit medicine supply chain through Blockchain." In *2019 11th international conference on communication systems & networks (COMSNETS)*, pp. 568–570. IEEE, 2019.

[12] Monrat, Ahmed Afif, Olov Schelén, and Karl Andersson. "A survey of blockchain from the perspectives of applications, challenges, and opportunities." *IEEE Access* 7 (2019): 117134–117151.

[13] Gao, Yansong, Damith C. Ranasinghe, Said F. Al-Sarawi, and Derek Abbott. "Secure goods supply chain and key exchange with virtual proof of reality." *Cryptology ePrint Archive* (2015).

[14] Arppe-Tabbara, Riikka, Mohammad Tabbara, and Thomas Just Sørensen. "Versatile and validated optical authentication system based on physical unclonable functions." *ACS applied materials & interfaces* 11, no. 6 (2019): 6475–6482.

[15] Khalil, Ghaith, Robin Doss, and Morshed Chowdhury. "A comparison survey study on RFID based anti-counterfeiting systems." *Journal of Sensor and Actuator Networks* 8, no. 3 (2019): 37.

[16] Maleki, Hoda, Reza Rahaeimehr, Chenglu Jin, and Marten Van Dijk. "New clone-detection approach for RFID-based supply chains." In *2017 IEEE International Symposium on Hardware Oriented Security and Trust (HOST)*, pp. 122–127. IEEE, 2017.

[17] Alzahrani, Naif, and Nirupama Bulusu. "Block-supply chain: A new anti-counterfeiting supply chain using NFC and blockchain." In *Proceedings of the 1st Workshop on Cryptocurrencies and Blockchains for Distributed Systems*, pp. 30–35, 2018.

[18] Wazid, Mohammad, Ashok Kumar Das, Muhammad Khurram Khan, Abdulatif Al-Dhawailie Al-Ghaiheb, Neeraj Kumar, and Athanasios V. Vasilakos. "Secure authentication scheme for medicine anti-counterfeiting system in IoT environment." *IEEE Internet of Things Journal* 4, no. 5 (2017): 1634–1646.

[19] Gayoso Martínez, Víctor, Luis Hernández Encinas, and Carmen Sánchez Ávila. "A survey of the elliptic curve integrated encryption scheme." *Journal of Computer Science and Engineering* 2 no. 2 (2010): 7–13.

[20] Tiwari, Sumit. "An introduction to QR code technology." In *2016 international conference on information technology (ICIT)*, pp. 39–44. IEEE, 2016.

[21] Focardi, Riccardo, Flaminia L. Luccio, and Heider AM Wahsheh. "Usable security for QR code." *Journal of Information Security and Applications* 48 (2019): 102369.

[22] Berdik, David, Safa Otoum, Nikolas Schmidt, Dylan Porter, and Yaser Jararweh. "A survey on blockchain for information systems management and security." *Information Processing & Management* 58, no. 1 (2021): 102397.

[23] Zheng, Zibin, Shaoan Xie, Hong-Ning Dai, Xiangping Chen, and Huaimin Wang. "Blockchain challenges and opportunities: A survey." *International journal of web and grid services* 14, no. 4 (2018): 352–375.

[24] Li, Zhi, Ali Vatankhah Barenji, and George Q. Huang. "Toward a blockchain cloud manufacturing system as a peer to peer distributed network platform." *Robotics and computer-integrated manufacturing* 54 (2018): 133–144.

[25] Swanson, Tim. "Consensus-as-a-service: A brief report on the emergence of permissioned, distributed ledger systems." *Report, available online* 28 (2015).

[26] Nofer, Michael, Peter Gomber, Oliver Hinz, and Dirk Schiereck. "Blockchain." *Business & Information Systems Engineering* 59 (2017): 183–187.

[27] Hughes, Laurie, Yogesh K. Dwivedi, Santosh K. Misra, Nripendra P. Rana, Vishnupriya Raghavan, and Viswanadh Akella. "Blockchain research, practice and policy: Applications, benefits, limitations, emerging research themes and research agenda." *International Journal of Information Management* 49 (2019): 114–129.

[28] Zheng, Zibin, Shaoan Xie, Hong-Ning Dai, Weili Chen, Xiangping Chen, Jian Weng, and Muhammad Imran. "An overview on smart contracts: Challenges, advances and platforms." *Future Generation Computer Systems* 105 (2020): 475–491.

[29] Shrivas, Mahendra Kumar and Thomas Yeboah. "The disruptive blockchain: Types, platforms and applications." *Texila International Journal of Academic Research* 2019 (2019): 17–39.

[30] Aggarwal, Shubhani and Neeraj Kumar. "Hyperledger." In *Advances in computers*, vol. 121, pp. 323–343. Elsevier, 2021.

[31] Gorenflo, Christian, Stephen Lee, Lukasz Golab, and Srinivasan Keshav. "FastFabric: Scaling hyperledger fabric to 20 000 transactions per second." *International Journal of Network Management* 30, no. 5 (2020): e2099.

[32] Androulaki, Elli, Artem Barger, Vita Bortnikov, Christian Cachin, Konstantinos Christidis, Angelo De Caro, David Enyeart et al. "Hyperledger fabric: A distributed operating system for permissioned blockchains." In *Proceedings of the thirteenth EuroSys conference*, pp. 1–15, 2018.

[33] Lu, Ning, Yongxin Zhang, Wenbo Shi, Saru Kumari, and Kim-Kwang Raymond Choo. "A secure and scalable data integrity auditing scheme based on hyperledger fabric." *Computers & Security* 92 (2020): 101741.

[34] Foschini, Luca, Andrea Gavagna, Giuseppe Martuscelli, and Rebecca Montanari. "Hyperledger fabric blockchain: Chaincode performance analysis." In *ICC 2020–2020 IEEE International Conference on Communications (ICC)*, pp. 1–6. IEEE, 2020.

[35] Wüst, Karl and Arthur Gervais. "Do you need a blockchain?." In *2018 crypto valley conference on blockchain technology (CVCBT)*, pp. 45–54. IEEE, 2018.

[36] Kuo, Tsung-Ting, and Lucila Ohno-Machado. "Modelchain: Decentralized privacy-preserving healthcare predictive modeling framework on private blockchain networks." *arXiv preprint arXiv:1802.01746* (2018).

[37] Dinh, Tien Tuan Anh, Ji Wang, Gang Chen, Rui Liu, Beng Chin Ooi, and Kian-Lee Tan. "Blockbench: A framework for analyzing private blockchains." In *Proceedings of the 2017 ACM international conference on management of data*, pp. 1085–1100, 2017.

11 Reducing Counterfeit Medicine through Blockchain

Asha K and Anil George K

CONTENTS

11.1 INTRODUCTION

The World Health Organization is currently facing numerous problems with forged drugs. In recent months, laws have discovered a significant increase in pharmaceutical companies operating around the clock and dispensing bogus medications in numerous regions of India. There have been reports about the critical Pfizer dose's use and manufacture all across the world. These drugs are linked to the treatment of COVID-19 infection. Falsified medications have the potential to endanger people's lives. This chapter provides a systematic study of reducing feign mechanisms for drugs using blockchain technology. A ledger that can be used in absolute, clear, and redistributed

DOI: 10.1201/9781003366010-14

fashion is blockchain. Ledger contains records of transactions, and blockchain can be visualized as a collection of blocks where each block copies data from a preceding module. Each block of the chain has a timestamp assigned to it based on the contents that were changed. Interfering with data that is stored in the blockchain and includes determining the reliability of work and methods for retrieving results is difficult. This suggested medicine design will be referred to as Hyperledger Fabric under any hospitals with the most cutting-edge technology. It is part of the company logistics integrity management (Sukhwani, 2018). The suggested approach is proof of a workable experiment that uses decentralized blockchain technology to retain data on discrete drug records. It gives hospitals the right to acquire patient medical records, handle them, and initiate their inherent use for the entire individual medication life cycle in a safe and responsible manner through detailed logs. It also allows hospitals to include other parties under one roof.

11.1.1 Drivers in the Pharmaceutical Industry

The World Health Organization (WHO) has elucidated forged medicine as "one which is an unscrupulously miscategorized with respect to specification" and its quality (Burns, 2006; Adjei and Ohene, 2015). Fabricating creates problems and poses different hazards to rightful pharmaceutical products. This should be brought to the attention of the public and civic awareness to mitigate the earning loss of the sanctioned manufacturing organizations (WHO, 2016). The report published by the International Chamber of Commerce Geneva stated that the yearly sales of forged wares in the world amount to U.S. $ 650 billion (Malik et al., 2013). In real-time, numerous methods are followed to identify forged drugs in the medical supply chain. None of the methods has the capability to authorize product authenticity and what the manufacturer claims. To reduce the flow of forged medicine in the drug business logistics chain, the proposed blockchain method outperforms the series of transaction ledger and tracking of the logistics at the individual drug level (Haq, 2018).

In a survey conducted in 2020 by Indian Pharmaceutical Association, the various drivers of market growth in the pharma industry are reported as proprietary drugs, off-patent drugs, healthcare reform, skyrocketing drug prices, changing drug status, the merging of healthcare and technology, novelties in medical treatment and chronic diseases.

The series of connected data structure track tasks implemented on allocated and peer-to-peer networks. Every other module forms a chain linked similarly like a permanent and irreversible history when used in a real-time audit trial by many to judge the results of the records by reviewing data itself. The security in the system comes from its use of cryptographic functions. Different blockchain platforms have different cryptographic algorithms to make the data or ledger tamper-proof. These consensus mechanisms ensure that for a transaction to be recorded in blockchain it has to satisfy the defined restrictions, and each of these valid transactions is cryptographically signed and stored in the system which makes the system tamper-proof and secure. Thus, the blockchain keeps itself secure and virtually unbreakable.

11.1.2 Design Issues in the Pharmaceutical Business Logistics Chain

A pharmaceutical business logistics chain has multiple stakeholders such as stock suppliers, mass producers, merchandisers, autonomous bodies, druggists, hospitals, and patients. The intricacy of the wares and transaction flow in the pharmaceutical supply chain needs an actual traceability system to determine the contemporary and all foregoing product ownerships. Digitizing the shipment process provides substantial benefits for governing autonomous bodies to overlook and ensures product safety. Blockchain-based drug traceability offers a proven solution to create a distributed shared data platform for an immutable, reliable, accountable, and explicit system in the pharmaceutical supply chain (Swan, 2015). Blockchain is jointly combined by all network participants and amending an existing ledger is not rationally possible. The usage of cryptographic algorithms is the main key to this. Immutability is implemented using a hash value, a digital fingerprint of data. Every block has a reference to a previous block's hash value and thus gives a strict order to the blockchain. Following hashes from the current block ends with block 0—called the genesis block. On particular blockchains, it is a first-created block, and the type of data structure supports provenance. Since all blockchain transactions are timestamped and immutable, fraudulent drug traffickers can be easily identified. Blockchains come in dual formats: public and private. The medical care companies must enlist largely in private blockchain networks if they want to guarantee the veracity and authenticity of their products (Pilkington, 2016). The private blockchain is hosted by an intermediate entity and the distinct manufacturer or supplier has access to the so-called drug blockchain show beyond doubt on its authenticity. Once the drug/medicine goes into transit from the maker to the vendor, operational data are certified in the blockchain. In the olden days, many blockchain systems were built on a permissionless network, where the privacy of the user data was at stake. In the permissionless blockchain, just by creating its own address any user can breach the network.

In contrast to the permissionless blockchain network, the proposed solution takes care of all aspects when it comes to managing the coherence and privacy of data connected to medical history, drug management, and other associated reports. This in turn satisfies the intent of healthcare-based solutions to keep medical data secure and transparent. The dispensation rate is limited and many challenges still exist in the current blockchain-based healthcare system. Moreover, this increasing complexity spirals up the overhead costs and henceforth affects the availability and the purchasing power of consumers. By switching to safe automated methods, the development of blockchain technology resulted in a paradigm shift in the way the traditional drug supply chain in healthcare operated. Hyperledger Fabric and smart contracts enable the above entities under healthcare to use the software platform without a third-party provider.

11.2 BLOCKCHAIN TECHNOLOGY IN THE PHARMACEUTICAL INDUSTRY

The vital focus in today's real-time environment and the pharmaceutical industry is to confirm protected transactions along with the supply chain reliability. The stakeholders suffer losses due to theft while the drugs are in transit and this is because of the absence of a robust tracking and tracing system in place (Huang et al., 2018). In the marketplace drug governing authorities have exposed the inferior and forged products. As a result, many conglomerates in the pharmaceutical industry resorted to deploy blockchain technology to connect and restructure tracking thereby promising the patient safety (Huang et al., 2018). Forged medications were always a security hazard addressed through serialization (Alshahrani, 2021). Blockchain codes and quick start guides are arranged from the mass producer to the druggists and end users to keep up the immaculate standards (Makarov and Pisarenko, 2019).

11.2.1 Counterfeit Prevention

Consumers typically choose serialized and sequentially packaged products with safety characteristics, which sets them apart from counterfeit drugs. When blockchains are used in conjunction with established assurance procedures to identify counterfeit medications, safety is improved and lives are saved (Adsul and Kosbatwar, 2020).

Several lines, including the Anti-Counterfeit Medicine System, exploiting Ethereum blockchain had implemented and evaluated the program for small (Kar et al., 2019). The signature is then confirmed by approving peers, these are secured and sent to the ordering services (Kumar et al., 2019).

11.2.2 Merchandise Issuance Channel

The existence of other intermediaries in the drug industry provides a place for misconduct that weakens supply chain operations. Blockchain has been instrumental in preventing the flow of substandard pharmaceuticals (Hulea et al., 2018). Degraded pharmaceuticals are sequestered, and the entry into the pharmaceutical business logistics management is thoroughly verified. Blockchain information is devoid of any illegal access that jeopardize security firewalls (Dwivedi et al., 2020). The Internet of Things (IoT) in the drug business delivery system bolsters competency (Botcha and Chakravarthy, 2019).

11.2.3 Tracking and Tracing

The transit of shipment needs to be tracked from dispatch to destination and the delay affects business operations. The wares in transit are delivered as per turnaround time to the desired terminals, with good and secure tracking, enabling continuous pharmaceutical business and patient management. A protected international registry was created to facilitate the seamless distribution of drugs and the technology provides cosmic opportunities for large and small pharmaceutical firms (Garankina et al. 2018).

11.2.4 Safety and Security

Due to the immense value and need of the hour requirements, ad hoc measures are necessary to protect drugs. The blockchain technology has its cryptographic features and forms that has a pivotal part in the safety and security of pharmaceutical industry

(Sinclair et al., 2019). The forged pharmaceuticals find it difficult in front of bolstered security against theft (Plotnikov and Kuznetsova, 2018). This mitigates unauthorized drug modification, blocking illegal pharmaceutical stakeholders who modify drugs which challenge quality. Business logistics management that are in charge of tight safety and standard procedures generate an alert when quality is compromised.

11.3 CHALLENGES IN CURRENT BLOCKCHAIN

Leading pharmaceutical industries evaluate new technologies and tools such as blockchain for drug business logistics management. The blockchain has the upper hand of not requiring any single entity to access all the information and while the stakeholders verify the authenticity of the record, organizations don't need to share the proprietary information. Figure 11.1 shows the challenges with data having a linear flow of material and data across manufacturing with end-to-end traceability (Jamil et al., 2019). The slow transfer of the electronic information back to the requesting entity may be outdated by the time the information is received.

The isolated information may not be communicated across organizations due to issues of trust and due to the certainty issues in sharing data. The primary benefit of blockchain is that it allows organizations to confirm the veracity of the record, does not require a single central owner of all information, and only requires that the record be substantiated and shared transparently across all parties involved in the logistics chain, all of which alleviate any worries about the security and trustworthiness of data and systems.

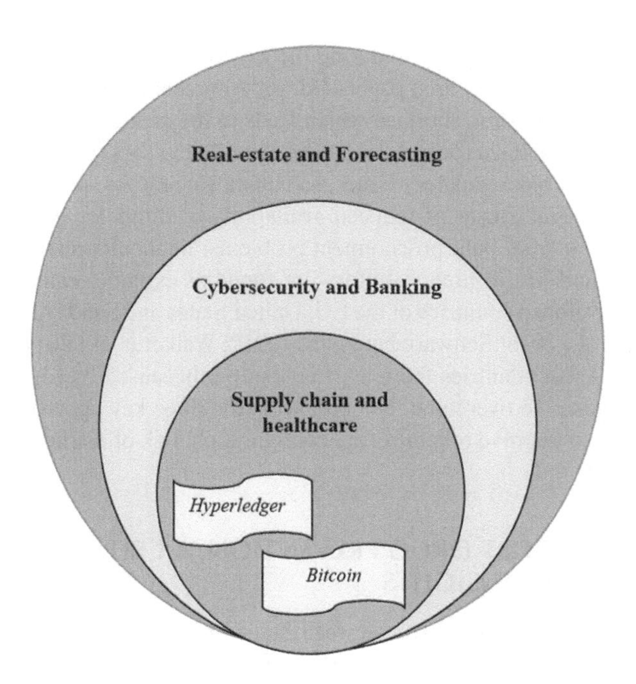

FIGURE 11.1 Challenges in the area of current blockchain.

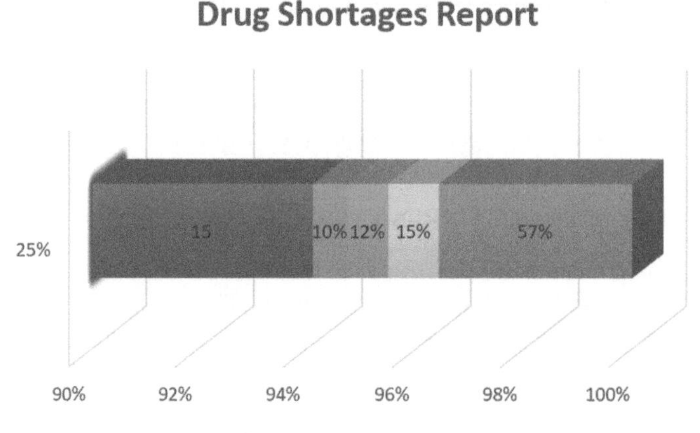

FIGURE 11.2 Drug shortage report during Covid pandemic.

11.3.1 Agility in Mitigating Drug Shortage

The prevalence of drug shortages worldwide has been affecting different economies. In developed economies, the reduced availability of drugs has been massive in conjunction to other countries. The causes include bulk procurement problems, business compromise agreements, stock grist unavailability, and autonomous body issues. There is a lack of standardized definition of drug shortage globally. There are a multitude of reasons for the drug dearth during the Covid pandemic, based on the type of drug reported in several research papers (Mazer-Amirshahi et al., 2014). The reports have shown an overall drug shortage which leads to the creation of forged medicines widespread. Drug shortage root causes can be classified as supply and demand, manufacturing, raw material, regulatory issues, and unknown reasons (refer to Figure 11.2).

The multifactorial causes of reduced availability of drugs are due to myriads of reasons extending from bulk procurement problems, financial pressures, stock grist unavailability, and just in time inventory are found as essential causes of medicine shortages in developed countries of the EU, United States and Saudi Arabia (Hedman, 2016; Awad et al., 2016; Schwartzberg et al., 2017; Walker et al., 2017; Jenzer et al., 2019). But in Asian countries there are frequent perils caused by logistics management interruptions. To overcome the bottleneck, the three key approaches that companies may use to improve reliability are configuration, risk of disruptions, and speed of recovery.

11.4 COUNTERFEIT DRUG PREVENTION: A CRITICAL REVIEW AND SOME INSIGHTS

The statistical study findings estimate that the commercial sector industry amounts to a 200 billion dollar financial loss yearly (Malik et al., 2013). Numerous solutions have been proposed to cover the issues in drug forging, but it still remains because of the poor drug management structure especially at the stage of tracking the process during manufacturing and logistics.

Haq (2018) proposed a case and used blockchain in healthcare. Kumar and Tripathi (2019) highlighted a conceptual system to add reliability, visibility, and traceability using blockchain. Pandey and Litoriya (2021) implemented the system with blockchain. A framework suggestion by Singh et al. (2020) was to use blockchain and fast response codes when dealing with drug safety fraud. Jamil et al. (2019) projected a full medicare system to suppress drug imitations in India. Abbas et al., (2020) used an ameliorated health system based on IoT and blockchain to follow all stages of drug business logistics chain. Kumar et al., (2019) proposed a good novel medicare logistics using permissioned blockchain Web application. Botcha and Chakravarthy (2019) introduced a novel health system based on blockchain. Toscano (2011) specified the state of under-development needs to be catapulted when they maximize the network. Sahoo et al. (2019) mentioned the quality of forged materials is often life threatening, leading to a surge in death rates, especially in the pharmaceutical industry.

Kumiawan et al. (2020) said that forged drugs may contain undue amounts of spurious elements which within the body affects its activity, for example the process of absorption by the body. Kumar et al., (2019) presented details of forged drugs and their after effects on various domains using blockchain technology. Khizar et al. (2020) introduced system consisting of blockchain-based drug logistics management and machine learning-based drug suggestions for customers. Tseng et al. (2018) explained regarding the forging of the Gcoin-named medicine, and its lifespan is introduced, including every stage from production to post-market. Bryatov (2019) addressed a problem in forged drugs. Jamil et al. (2019) stated on a Forged Drug, where a continuous digital Blockchain model for drug delivery was created for the system's transparency and security. Pashkov and Soloviov (2019) addressed forged drug and logistics costs. Raj et al. (2019) specified proof of ownership in establishing anti-counterfeiting in the supply chain of pharmaceuticals. Clark and Burstal (2018) presented a study on how blockchain can assist in preventing drug forging.

11.5 DESIGN OF DRUG SUPPLYCHAIN FRAMEWORK FOR OPTIMIZATION

With the system being implemented in a blockchain it gets all the benefits such as powerful cryptographic methods to secure the network and transactions (Jamil et al., 2019). With the help of the suggested model, it is possible to determine where each event or action first entered the process in question and trace its provenance from there.

- Fraud reduction through increased audit trail transparency.
- The lack of an ethical third party is not required for reaching an agreement or validating particular claims regarding the transfer of medications.

The network serves as the backbone of the proposed architecture. All the cryptographic algorithms, endorsement policies to endorse transactions, are implemented in the network.

Once the network is deployed, the required business logic is developed in Hyperledger Composer. The chain code enforces restrictions in the network for a transaction to be valid. The developed chain code is installed in the deployed network. The features of the chain code can be accessed via its APIs, which makes developing applications really easy and further users are created in the network. The users of the network can communicate with the chain code via the exposed APIs through which transfer of medicines takes place.

11.5.1 Development Environment

The pharmaceutical supply chain plays a significant role in healthcare because it keeps track of all pertinent information about medications while in transit (Yue et al., 2016). The current system uses the track and trace method, which makes it easier to trace products, thus making it possible to reduce the volume of forged medicine (Dalianis et al., 2015). The suggested system is modular because it may be divided into smaller submodules and connected by a straightforward structure (Arsene, 2019). The technique of extracting existing software components from the previous context for which they were originally built and using them again in contexts similar is known as Modular Composition (Mettler, 2016). Modular continuity states that the structure of individual modules is impacted by modest changes in the problem specification. If an abnormal state arises in a module at run time and its effect is contained to that module or only spreads to a small number of nearby modules, the proposed system satisfies the modularity protection criterion (Mettler, 2016). Hyperledger is a type of blockchain technology that employs smart contracts to control transactions and contains a ledger. With the help of hyper ledger fabric, one might create a unique blockchain network for their particular use case. A participant who is a part of one channel cannot view the transaction taking place on another channel (Jamil et al., 2019). Each channel in a network would have a unique shared ledger that sets it apart from other channels. The network's participants all possess copies of one or more shared ledgers, depending on the channels (Bocek et al., 2017). These shared ledgers contain both transaction log information and a world state.

The ledger's current state is described by the world state component (Allison, 2016). The transaction log component oversees all the transactions that contributed to world state's current value (Abeyratne and Monfared, 2016). Because of this, fabric is incredibly adaptable when it comes to integrating with existing systems that use conventional database systems like SQL, Mongo DB, CouchDB, and so on.

11.5.2 Building Component of Business Network

A blockchain-based framework and the design specifications for medication traceability have been discussed in this section. The two blockchain platforms on which the proposed architecture is built are discussed with detailed study to identify the major differences (Singh and Singh, 2016). It has been demonstrated that Hyperledger Fabric provides greater transaction limits and thereby detects the process speed capability gauging up to thousands of counts per second (Guo and Liang, 2016). Hyperledger

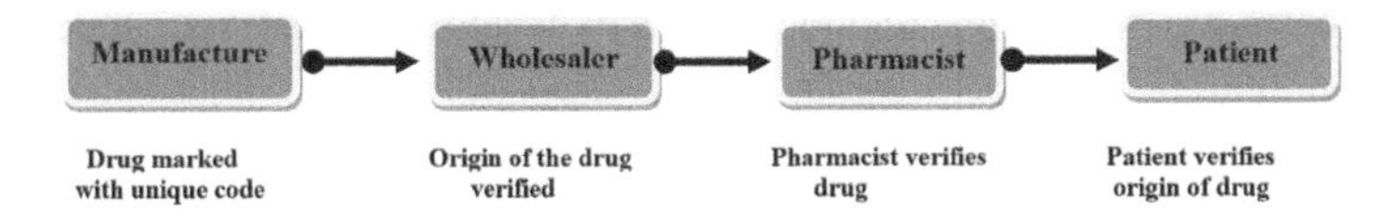

FIGURE 11.3 Dataflow using blockchain-based medicine supply chain.

Fabric is a flawless choice for complicated business logistics management systems that involve several concrete and solid logical processes and participants (Dhillon, 2017). The adoption obstacle for the technology is lower while developing smart contracts using general-purpose programming languages than when using specialized programming languages. The dataflow of the blockchain-based pharmaceutical logistics chain is demonstrated in Figure 11.3.

An enterprise-level business logistics chain denotes a pattern which is provided by the Hyperledger Fabric drug traceability framework as suggested in this chapter, bringing forward various players in the pharmaceutical logistics chain and the discussion of their equivalent notions (Fan et al., 2018).

Channels allow variant contributors in the same segment to clearly differentiate the business logic and data firewall protection norms (Sousa et al., 2018). A permissioned private blockchain network is built in the proposed Hyperledger Fabric framework, where all participating organizations and end users are found and added by a membership service provider component (MSP). The Certificate Authority (CA) provided by Hyperledger Fabric is the pluggable MSP component (Ekblaw et al., 2016). MSP has by-laws in which various contributors are managed, checked for authenticity, and access enabled to blockchain resources to connect to an ethical environment between unknown players (Griggs et al., 2018). This permits easy activity tracing and preserves the privacy and confidentiality of all parties involved. The MSP is an extensive unique design that, by decentralizing identity management, gives a new and improved form for removing ambiguity, fatigue caused by lack of supply and its source, and the intrusion of firefighting between participating contributors in the pharmaceutical logistics chain (Androulaki et al., 2018). At the heart of the Hyperledger Fabric framework is the mechanism and an element of network to host ledgers and smart contracts and ordering service (OS) (Gatteschi et al., 2018). The OS receives authorized transactions from client applications, arranges them into blocks with cryptographic signatures, then broadcasts these blocks to the committed participants in the blockchain network for permitted policy validation.

Each chain code developed both in go and node js and serves as the business logic of the network. Each transaction is initiated by a peer in the network that has to satisfy the restrictions defined and noted as a valid transaction in the blockchain (Cheung and Choi, 2011). Thus, developers implement logic in which the network should function in the chaincode. Hyperledger Composer is a framework that makes the developing and testing chaincode a simple task. Ideally, chaincodes can be developed and tested directly in fabric network, but this involves tedious procedures. In composer, a developer can code and test chaincode directly without having the chaincode deployed into a fabric network (Li et al., 2018). This also makes the testing process much faster.

The composer has both online and local IDEs to develop and test chaincodes.

a) Modeling language
This is an object-oriented language that allows defining the model of network with such resources as users (participants), assets, concepts, and so on. Like any other language, modeling language also has its data types which includes String, Double, Integer, Long, Date Time, and Boolean. In composer, Model File of .cto format holds all the definitions done by the modeling language. Apart from the resources of the network, each transaction and the events that are defined in chaincode should be declared in the model file.

b) Script File
The file which is implemented in node js defines the business logic of the network through which restrictions are imposed on each transaction so only valid transactions are committed to the blockchain. For every transaction or event declared in the model file, definitions are implemented in the script file. It uses the concept of promises in node js to execute in a synchronous manner.

c) Access Control Language
For the restricted access to users, a resources and transactions code provides an access control language which is used to access the control file. The particular requirements and limitations by which users should access different network resources are specified using ACL language developer codes. Developers can implement special privileges to admin users and so on through an ACL file.

11.5.3 Smart Contract Modeling of Drug Business Logistics Management

The medication supply chain process prevents unauthorized and dishonest participants from handling medicines and from being present in the network (Cheung and Choi, 2011). The main component in the code is the medicine log which is staged through the blockchain technology and stores for each medicine – id, a description, the identity of the creator, and the complete history of owners up to the current one, including the time at which changes of ownership occurred.

Figure 11.4 depicts the complete transaction processing stages in the proposed framework.

The different phases in transaction processing are detailed as follows:

- A registered supplier or manufacturer submits a transaction proposal.
- A chain code function is invoked with certain arguments to read, write, and update the ledger.
- The proposal is submitted to the entities for approval. The client's cryptographic credential and transaction payload comprising the name of the chaincode function to be executed with input arguments are the primary contents.
- The transaction proposal is put into effect by endorsing peers and the endorsement policy.

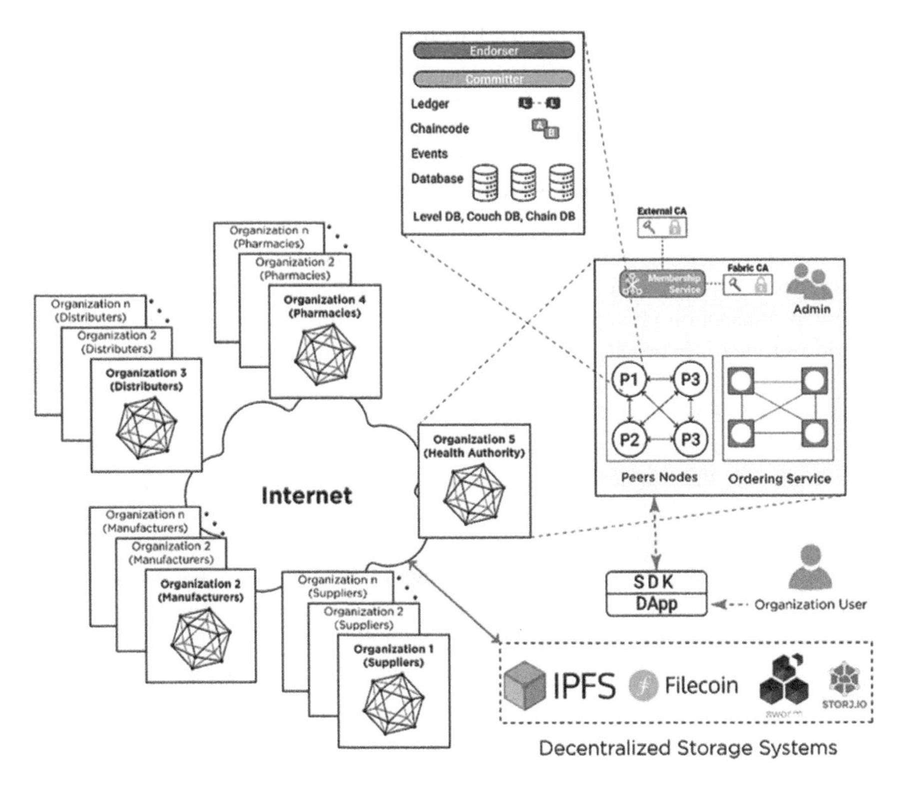

FIGURE 11.4 Framework followed for medicine traceability.

- The results will be encrypted with cryptographic signatures with read-set and write-set, no changes are made to the document at this point.
- When the required responses are received, it starts to analyze the similarity of read-write sets.
- Transaction proposal and responses are within the ordering service. Read-write sets, endorsing peer signatures, and channel identifier are included here.
- The decentralized ordering service uses a consent protocol to institute the implementation phase of all the submitted transactions in each channel.
- The final execution phase in the Hyperledger Fabric network has the OS broadcasting the new blocks to the leading participants.
- The leading participants then distribute the blocks to others between the organization using the gossip protocol.
- The chain code methods are then used to verify the endorsement validity by the participants.
- The most valid process updates are published and invalid ones are retained.
- The client app submits the final result and this will be highlighted by individual peer on the network.

According to the architecture specified with Hyperledger Fabric, a distributed network is created with a docker. A docker allows the fabric to work platform independent. Initially, fabric images from the docker hub together with the necessary binaries which do the complex cryptographic functions like producing a public key and a private key for each user are downloaded. Once the environment is set than necessary docker files are created. Fabric network should consist of at least one order peer, ca-server, optional database peer, and multiple client peers. The necessary endorsement policies should be defined in the network, this makes sure when a client peer or user initiates a transaction, one of the endorsing peer's checks if the generated transaction is authenticated or not. If the transaction is through by using ordering service it is broadcasted into the entire network. Here the ordering service makes sure that transactions are broadcast in the correct order so as to maintain the integrity of the network. Once the fabric network is deployed, that is, when the docker containers are up and running, then it's time to install the chaincode in the network. From composer the business network archive (.bna) file which includes chaincode can be exported. The .bna file can be uploaded to deployed fabric network with composer command line tools. Once the chaincode is successfully deployed in each peer of the network, a rest server is initiated through a composer-reset-server command line tool which exposes APIs to communicate with the chaincode. Multiple users and the assets can be created in the network through these exposed APIs. For each user created there will be corresponding user cards which contain the credentials for that user to authenticate to the network.

11.5.4 Planning and Scheduling of Drug Supply Chain Management

Project scheduling and planning are two components of the drug business logistics management process. Planning focuses on tasks necessary to complete the project successfully. The goal of scheduling is to establish the time frame needed to accomplish the project. The following are the few techniques applied after information is collected from the planning activities.

Medicine Creation: The function uploads the drug details to the blockchain using the drug id and description as inputs. Given that id is created when a medication is consumed, this process aids in maintaining the integrity of the drug throughout its entire lifecycle. The participant address and participant ID that are initially created are likewise set for other properties like creator and owner. When using the Medicine Creation function, it is first checked to see if a medicine already exists with the same ID; if so, the duplicate medicine is not created.

Medicine Transfer: The method transfers ownership to the provided address after receiving the medication ID and address as input. Before setting the owner to the new owner if the person calling the function is the owner of the medication, the method first checks to see if it already exists.

Medicine Deletion: This function removes the associated medication from the blockchain using the supplied medicine id. It first verifies that the person who calls themself the originator of the drug, and if so, it deletes the entry for the medicine from the blockchain. The Medicine Deletion function removes the associated medicine from the asset register because medicine is an asset. Sometimes this is necessary because the medication is no longer effective or appropriate for the instance being looked into. No participant has the ability to delete actual medications, but they can make a transaction indicating that a specific medication is no longer necessary for a given situation.

Medicine Display: The function accepts a medicine id as input and returns the blockchain data about the drug. This function merely checks to see if a particular medication already exists.

In order to find the optimal solution that satisfies all the requirements, a feasibility study is carried out. It is supported with a report outlining all the consequences.

The transaction-oriented smart chain is shown in Figure 11.5. An open-source software platform known as docker alongside an operating system can create, deploy, and manage virtualized application containers. The main sponsor of the open-source application is Docker Inc.; the business that created docker in the first place has accomplished many administrative, operational, and development functions.
The Hyperledger Composer command line application and composer can be used to perform multiple administrative, operational, and development tasks.

The Hyperledger Fabric SDK for Node.js provides a powerful API for interacting with a Hyperledger Fabric v2.0 blockchain. For use with the SDK, the Node.js JavaScript runtime is intended. With the SDK, JavaScript runtime is designed for use.

For implementation purposes, a custom fabric network was implemented and deployed using docker. A screenshot of deployed docker containers and the response time for requests during implementation are shown in Figures 11.6 and 11.7.

Figure 11.8 shows the average throughput of proposed design and it is clearly visible that the framework can process up to 300 transactions per second in the same network. The line represents the existed design implementation and the ash color line indicates the average throughput of proposed design.

11.6 DIGITAL MEDICINE AND BLOCKCHAIN: NEW RESEARCH AVENUES

The drug regulatory authorities look ahead for quality and oversee the illegal drug production and trafficking of counterfeit medicine which is adequately sanctioned. Regulating bodies have a more important and complex role in blockchain-based solutions since it is challenging for them to establish the lawful parameters and framework for blockchain technology. When a new transaction is invoked in the network, authorities will find it hard to demarcate the jurisdiction and licit obligations of the participants. Blockchain technology is therefore conflicting with current laws and rules governing the pharmaceutical business logistics management. Potential investors will refrain from such networks since this could lead to un-competitiveness, especially when competitors exist in the same business logistics chain (Androulaki et al., 2018). Existing drug traceability technologies, such as serialization, bar codes, RFID tags, and e-pedigree, as well as blockchain-based systems, do not fully interoperate due to the absence of standardized solutions (Zhang and Linx, 2018). In addition, under the Hyperledger umbrella, many blockchain platforms are addressing problems with interoperability, maximum scalability, and adaptability for facilitating internal and external communication between business units. A recent cybersecurity

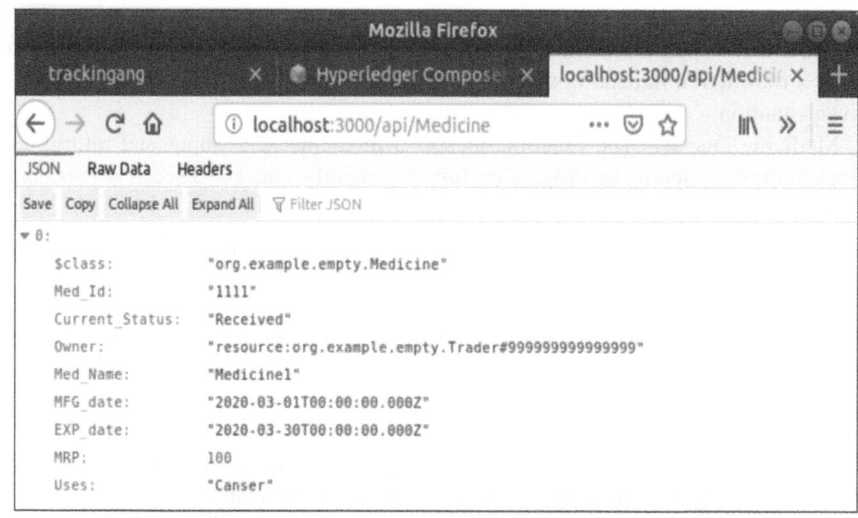

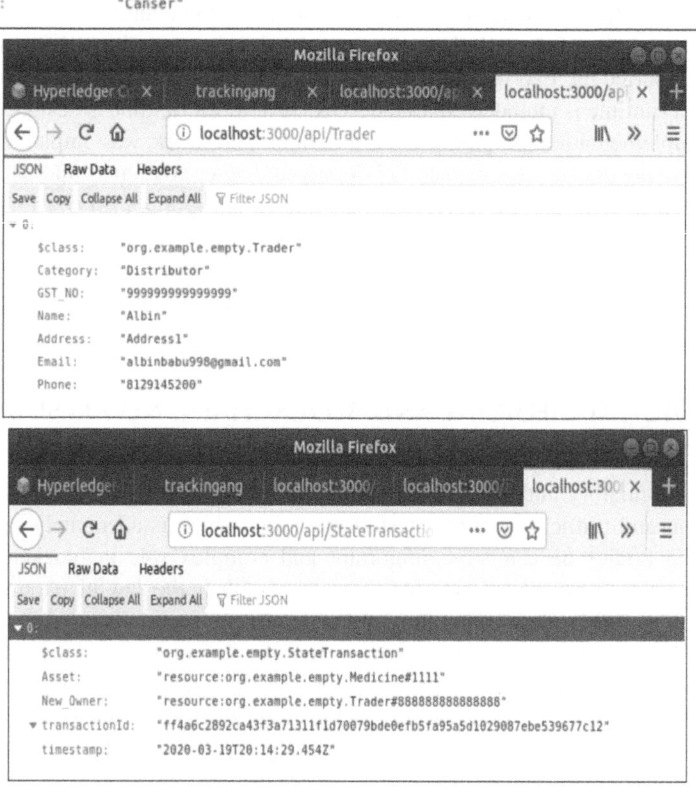

FIGURE 11.5 Smart chain dashboard created for assets, participant and transaction.

```
robin@albin-ubuntu: ~/fabric-dev-servers
File Edit View Search Terminal Help
robin@albin-ubuntu:~/fabric-dev-servers$ docker ps
CONTAINER ID        IMAGE
                    COMMAND                  CREATED             STATUS              PORTS
                        NAMES
371a2389ed1a        dev-peer0.org1.example.com-tracking-0.0.1-13d6e108c464e68023fcd19703e3fe8df94550591175314
4b140cc2d031e03e7   "/bin/sh -c 'cd /usr…"   6 minutes ago       Up 6 minutes
                        dev-peer0.org1.example.com-tracking-0.0.1
f9931369d4c8        hyperledger/fabric-peer:1.2.1
                    "peer node start"        9 minutes ago       Up 9 minutes        0.0.0.0:7051->7051/tcp,
0.0.0.0:7053->7053/tcp   peer0.org1.example.com
7218070bb1cf        hyperledger/fabric-couchdb:0.4.10
                    "tini -- /docker-ent…"   9 minutes ago       Up 9 minutes        4369/tcp, 9100/tcp, 0.0.
0.0:5984->5984/tcp      couchdb
a18d9c37f7e9        hyperledger/fabric-ca:1.2.1
                    "sh -c 'fabric-ca-se…"   9 minutes ago       Up 9 minutes        0.0.0.0:7054->7054/tcp
                        ca.org1.example.com
c2046072c0bd        hyperledger/fabric-orderer:1.2.1
                    "orderer"                9 minutes ago       Up 9 minutes        0.0.0.0:7050->7050/tcp
                        orderer.example.com
robin@albin-ubuntu:~/fabric-dev-servers$
```

FIGURE 11.6 Deployed fabric network using docker containers.

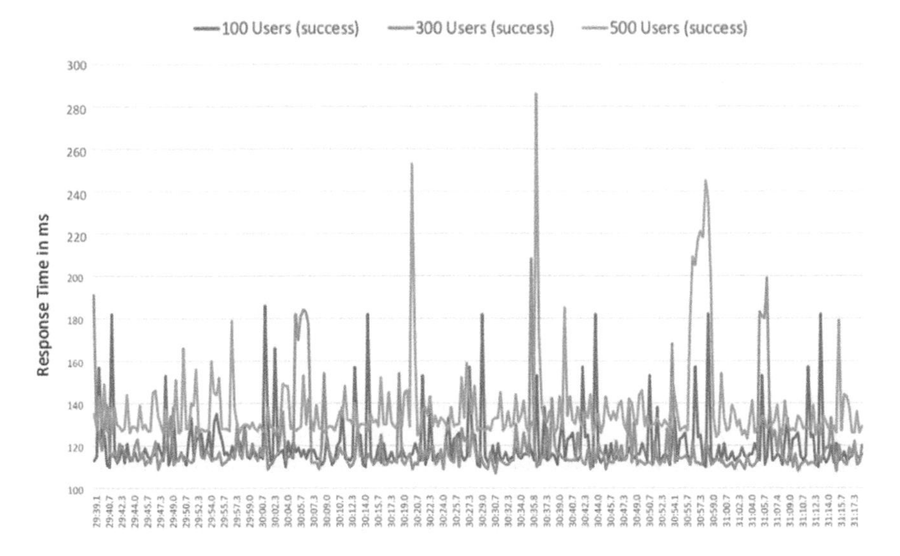

FIGURE 11.7 Response time for requests.

assessment identifies a number of security threats, including rogue actors and impersonating attacks, that could affect the blockchain network and reveal weaknesses (Dagher et al., 2018).

11.6.1 Open Challenges

Since most of the solutions are still in development, creating the ideal blockchain application is challenging. Costs associated with implementation and energy are among the biggest issues that businesses, particularly those in the pharmaceutical

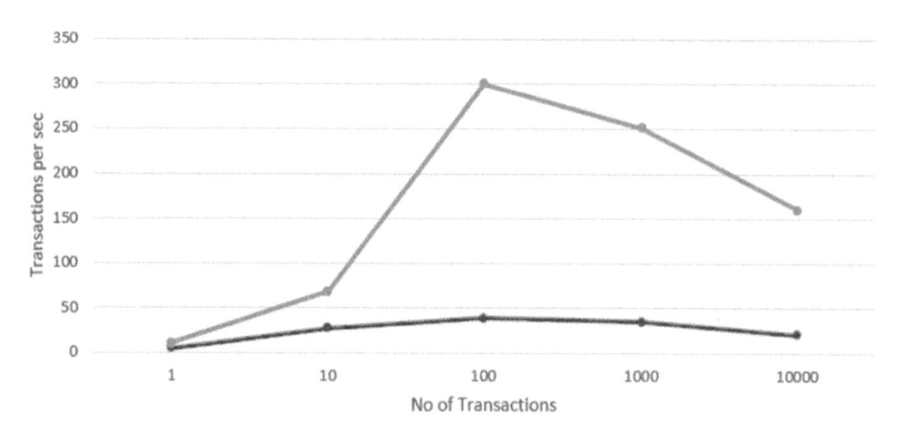

FIGURE 11.8 Comparison between proposed and existing design.

business logistics chain, must contend with (Peterson and Krug, 2015). Abiding platforms and legacy software systems execute transactions inefficiently and centrally, which results in high implementation and maintenance costs (Cohn et al., 2019). While the evolution of processes and systems are taking place, the present-day blockchain deployment have intrinsic deficiencies and defects (Mettler, 2016).

In the absence of procedures, malware, implementation exploiting, and phishing scams trigger difficulties that need to be addressed going forward (Metcalf, 2019).

The energy community sector has been captivated by the disruptive potential of blockchain and other distributed ledger technologies (Li et al., 2019). Blockchain in energy is moving into a new phase as the initial buzz fades: the variety of parties directly involved in pilot projects and trials is growing, and business models are being examined and tested for their viability as part of a large-scale commercial adoption process (Swan, 2015).

In the financial sector, the supply chain and passenger trip chain management have the potential to undergo significant transformation (Hulseapple, 2015). The larger transport and mobility sector is home to hundreds of proof-of-concept projects, many of which have been running effectively for some time. These platforms include several that have drawn a sizeable portion of major international players in the logistics industry to collaborate and participate.

Outside of the government sector, blockchain technology has uses in the private sector as well. Over 200 blockchain initiatives are currently active in global government organizations (Bagozzi, 2017). This led to the most recent advancements in blockchain-enabled government technology, as well as how businesses and governments are experimenting and innovating to better people's lives and revolutionize the way and how they conduct business (Akunyili, 2004). Divergent viewpoints on the degree to which governments should investigate GovTech projects based on blockchain were also raised.

It can be used in food logistics management to set up shipping and logistics management, keep forged items away, and authenticate products while verifying the food source (Swan, 2015; Pilkington, 2016).

Small and medium-sized businesses and governments will both benefit from the adoption of digital technologies like blockchain since the former could conserve resources while the latter could more quickly and readily gather credible data (Blackstone et al., 2014). If all parties recognize the advantages of going digital, adoption might proceed much more quickly.

11.7 CONCLUSION

The blockchain is a potent technology and the framework can guarantee both medicine security and manufacturer legitimacy. The current medical supply chain relies on third-party conviction, which is not particularly good for the safety of drugs. A suggested solution uses a digital signature to guard against impersonating attacks and replay. Blockchain may be one of greatest options for maintaining and tracing the medical supply chain because it is designed to enforce integrity, transparency, authenticity, security, and auditability. Future work will focus on creating smart contracts with a more adaptable framework for storing medicine, deploying the system's overall components, creating a user interface by increasing the number of transactions events in the system, and integrating a database that will allow it to store digital data.

REFERENCES

Abbas, K., Afaq, M., Ahmed Khan, T., and Song, W. C. (2020) "A blockchain and machine learning-based drug supply chain management and recommendation system for smart pharmaceutical industry." Electronics. Vol. 9, no. 5: 852.

Abeyratne, S. and Monfared, R. (2016) "Blockchain ready manufacturing supply chain using distributed ledger." *Int. J. Res. Eng. Technology*. Vol. 5: 1–10.

Adjei, H. K., and Ohene, P. (2015) "Counterfeit drugs: the relentless war in Africa." *Pharm Pharmacol Int J*, Vol. 2, *no.* 2, 00016.

Adsul, K. B., and Kosbatwar, S. P. (2020) "A novel approach for traceability & detection of counterfeit medicines through blockchain." *International Journal of Current Engineering and Technology*.

Akunyili, D. (2004) "Fake and counterfeit drugs in the health sector: The role of medical doctors." Ann. Ib. Postgrad. Med. Vol. 2: 19–23.

Allison, I. (2016) "Skuchain: Here's how blockchain will save global trade a trillion dollars." *International Business Times, 8.*

Alshahrani, A. (2021) "A frame analysis of the language used by eight US media to describe the role of China and Chinese in spreading Covid-19 during late January to early June 2020." *Journal of Language and Linguistic Studies*. Vol. 17: 1129–1140.

Androulaki, E., Barger, A., Bortnikov, V. et al. (2018) "Hyperledger fabric: A distributed operating system for permissioned blockchains." In: Proceedings of the thirteenth EuroSys conference, pp. 1–15. New York: Association for Computing Machinery.

Arsene, C. (2019) "Hyperledger project explores fighting counterfeit drugs with blockchain." Accessed: July 5, 2020.

Awad, H., Al-Zu'bi, Z. B. M., and Abdallah, A. B. (2016) "A quantitative analysis of the causes of drug shortages in Jordan: a supply chain perspective." *International Business Research*. Vol. 9, no. 6: 53–63.

Bagozzi, D. (2017) "CL 1 in 10 medical products in developing countries is substandard or falsified."

Blackstone, E. A., Fuhr, J. P., Jr., and Pociask, S. (2014) "The health and economic effects of counterfeit drugs." *Am. Health Drug Benefits*. Vol. 7: 216–224.

Bocek, T., Rodrigues, B. B., Strasser, T., and Stiller, B. (2017) "Blockchains everywhere-a use-case of blockchains in the pharma supply-chain." In Proceedings of the 2017 IFIP/IEEE Symposium on Integrated Network and Service Management (IM), Lisbon, Portugal, May, pp. 772–777.

Botcha, K. M. and Chakravarthy, V. V. (2019) "Enhancing traceability in pharmaceutical supply chain using Internet of Things (IoT) and blockchain." In 2019 IEEE International Conference on Intelligent Systems and Green Technology (ICISGT) (pp. 45–453). IEEE.

Bryatov, S. R. and Borodinov, A. A. (2019, May). "Blockchain technology in the pharmaceutical supply chain: Researching a business model based on Hyperledger Fabric." In *Proceedings of the International Conference on Information Technology and Nanotechnology (ITNT), Samara, Russia* (Vol. 10, pp. 134–140).

Burns, W. (2006) "WHO launches taskforce to fight counterfeit drugs." *Bulletin of the World Health Organization, 84*, 689–690.

Cheung H. H. and Choi, S. H. (2011) "Implementation issues in RFID-based anti-counterfeiting systems." *Comput. Ind.* Vol. 62, no. 7: 708–718.

Clark, B. and Burstall, R. (2018) "Blockchain, IP and the pharma industry—how distributed ledger technologies can help secure the pharma supply chain." *Journal of Intellectual Property Law & Practice*. Vol. 13, no. 7: 531–533.

Cohn, J. M., Finn, P. G., Nair, S. P., Panikkar, S. B., and Pureswaran, V. S. (2019) *U.S. Patent No. 10,257,270*. Washington, DC: U.S. Patent and Trademark Office.

Dagher, G. G., Mohler, J., Milojkovic, M. et al. (2018) "Ancile: Privacy-preserving framework for access control and interoperability of electronic health records using blockchain technology." *Sustain Cities Soc.* Vol. 39: 283–297.

Dalianis, H., Henriksson, A., Kvist, M., Velupillai, S., and Weegar, R. (2015) "HEALTH BANK: A workbench for data science applications in healthcare." In Proceedings of the CAiSE'15, 27th International Conference on Advanced Information Systems Engineering, Stockholm, Sweden, 8–12 June; pp. 1–18.

Dhillon, V., Metcalf, D., and Hooper, M. (2017) "The hyperledger project." In *Blockchain Enabled Applications,* Springer: Berlin/Heidelberg, Germany, pp. 139–149.

Dwivedi, S. K., Amin, R., and Vollala, S. (2020) "Blockchain based secured information sharing protocol in supply chain management system with key distribution mechanism." *Journal of Information Security and Applications, 54*, 102554.

Ekblaw, A., Azaria, A., Halamka, J. D., and Lippman, A. (2016) "A case study for blockchain in healthcare: 'MedRec' prototype for electronic health records and medical research data." In Proceedings of IEEE Open & Big Data Conference. Vol. 13: 13.

Fan, K., Wang, S., Ren, Y., Li, H., and Yang, Y. (2018) "Medblock: Efficient and secure medical data sharing via blockchain." *J. Med. Syst.* Vol. 42, no. 8: 136.

Garankina, R. Y., E. R. Zakharochkina, I. F. Samoshchenkova, N. Y. Lebedeva, and A. V. Lebedev (2018) "Blockchain technology and its use in the area of circulation of pharmaceuticals." *Journal of Pharmaceutical Sciences and Research.* Vol. 10, no. 11: 2715–2717.

Gatteschi, V., Lamberti, F., Demartini, C., Pranteda, C., and Santamaría, V. (2018) "Blockchain and smart contracts for insurance: Is the technology mature enough?" *Future internet.* Vol. 10, no. 2: 20.

Griggs, K. N., Ossipova, O., Kohlios, C. P., Baccarini, A. N., Howson, E. A., and Hayajneh, T. (2018) "Healthcare blockchain system using smart contracts for secure automated remote patient monitoring." *J. Med. Syst.* Vol. 42, no. 7: 130.

Guo, Y. and Liang, C. (2016) "Blockchain application and outlook in the banking industry." *Financ. Innov*. Vol. 2: 24.

Haq, I. and Esuka, O. M. (2018) "Blockchain technology in pharmaceutical industry to prevent counterfeit drugs." *International Journal of Computer Applications*. Vol. 180, no. 25: 8–12.

Hedman, L. (2016) "Global approaches to addressing shortages of essential medicines in health systems." *WHO Drug Inf.* Vol. 30: 180.

Huang, Y., Wu, J., and Long, C. (2018, July) Drugledger: A practical blockchain system for drug traceability and regulation. In *2018 IEEE international conference on internet of things (iThings) and IEEE green computing and communications (GreenCom) and IEEE cyber, physical and social computing (CPSCom) and IEEE Smart Data (SmartData)*, Espoo, Finland (pp. 1137–1144). IEEE.

Hulea, Mihai, Ovidiu Rosu, Radu Miron, and Adina Aştilean. "Pharmaceutical cold chain management: Platform based on a distributed ledger." In *2018 IEEE International conference on automation, quality and testing, robotics (AQTR)*, Romanis, pp. 1–6. IEEE, 2018.

Hulseapple, C. (2015) "Block verify uses blockchains to end counterfeiting and 'make world more honest'." Accessed: June 5, 2020.

Jamil, F., Hang, L., Kim, K., and Kim, D. (2019) "A novel medical blockchain model for drug supply chain integrity management in a smart hospital." *Electronics*. Vol. 8, no. 5: 505.

Jenzer, H., Sadeghi, L., Maag, P., Scheidegger-Balmer, F., Uhlmann, K., and Groesser, S. (2019) "The European medicines shortages research network and its mission to strategically debug disrupted pharmaceutical supply chains." In *Pharmaceutical Supply Chains-Medicines Shortages*. Springer Cham: New York City, 1–22.

Kar, S. K., Yasir Arafat, S. M., Kabir, R., Sharma, P., and Saxena, S. K. (2020) "Coping with mental health challenges during COVID-19." *Coronavirus Disease 2019 (COVID-19) Epidemiology, Pathogenesis, Diagnosis, and Therapeutics*, Springer Singapore, 199–213.

Khizar Abbas, Muhammad Afaq, Talha Ahmed Khan, and Wang-Cheol Song, (2020) "A blockchain and machine learning-based drug supply chain management and recommendation system for smart pharmaceutical industry." *MDPI Electronics*. Vol. 9: 852.

Kumar, A., Choudhary, D., Raju, M. S., Chaudhary, D. K., and Sagar, R. K. (2019) "Combating counterfeit drugs: A quantitative analysis on cracking down the fake drug industry by using Blockchain technology." In 2019 9th International Conference on Cloud Computing, Data Science & Engineering (Confluence) (pp. 174–178). IEEE.

Kumar, R. and Tripathi, R. (2019) "Traceability of counterfeit medicine supply chain through Blockchain." In 2019 11th International Conference on Communication Systems & Networks (COMSNETS) (pp. 568–570). IEEE.

Kumiawan, H., Kim, J., and Ju, H. (2020) "Utilization of the blockchain network in the public community health center medicine supply chain." In 21st Asia-Pacific Network Operations and Management Symposium (APNOMS) (pp. 235–238). IEEE.

Li, P., Nelson, S. D., Malin, B. A., and Chen, Y. (2019) "DMMS: A decentralized blockchain ledger for the management of medication histories." *Blockchain in Healthcare Today, 2*.

Li, Y., Marier-Bienvenue, T., Perron-Brault, A., Wang, X., and Paré, G. (2018) "Blockchain technology in business organizations: A scoping review." In Proceedings of the 51st Hawaii International Conference on System Sciences, Hilton Waikoloa Village, Hawaii.

Makarov, A. M. and Pisarenko, E. A. (2019, December) "Blockchain technology in the production and supply of pharmaceutical products." In *International Scientific and Practical Conference on Digital Economy (ISCDE 2019)* (pp. 646–650). Atlantis.

Malik, M., Hassali, M. a. A., Shafie, A. A., and Hussain, A. (2013) "Why hospital pharmacists have failed to manage antimalarial drugs stock-outs in Pakistan? A qualitative insight." Malar. Res. Treat. Vol. 2013: 342843.

Mazer-Amirshahi, M., Pourmand, A., Singer, S., Pines, J. M., and van den Anker, J. (2014) Critical drug shortages: implications for emergency medicine. *Academic Emergency Medicine*, Vol. 21, no. 6, 704–711.

Metcalf, D. (2019) Blockchain in Healthcare: Innovations that Empower Patients, Connect Professionals and Improve Care. Taylor & Francis: Abingdon, UK.

Mettler, M. (2016) "Blockchain technology in healthcare: The revolution starts here." In Proceedings of the 2016 IEEE 18th International Conference on e-Health Networking, Applications and Services, Munich, Germany, 14–16 September, pp. 1–3.

Pandey, P. and Litoriya, R. (2021) "Securing e-health networks from counterfeit medicine penetration using blockchain." *Wireless Personal Communications*, Vol. 117: 7–25.

Pashkov, V. and Soloviov, O. (2019) "Legal implementation of blockchain technology in pharmacy." In *SHS Web of Conferences* (Vol. 68, p. 01027). EDP Sciences.

Peterson, J. and Krug, J. (2015) "Augur: A decentralized, open-source platform for prediction markets." *arXiv preprint arXiv:1501.01042*, 507.

Pilkington, M. (2016) "Blockchain technology: principles and applications." In *Research Handbook on Digital Transformations* (pp. 225–253). Edward Elgar Publishing.

Plotnikov, V., and Kuznetsova, V. (2018) "The prospects for the use of digital technology 'blockchain' in the pharmaceutical market." In *MATEC web of conferences* (Vol. 193, p. 02029). EDP Sciences.

Raj, R., Rai, N., and Agarwal, S. (2019, October). "Anticounterfeiting in pharmaceutical supply chain by establishing proof of ownership." In *TENCON 2019-2019 IEEE Region 10 Conference (TENCON)* (pp. 1572–1577). IEEE.

Sahoo, M., Singhar, S. S., Nayak, B., and Mohanta, B. K. (2019) "A blockchain based framework secured by ecdsa to curb drug counterfeiting." In 2019 10th International Conference on Computing, Communication and Networking Technologies (ICCCNT) (pp. 1–6). IEEE.

Schwartzberg, E., Ainbinder, D., Vishkauzan, A., and Gamzu, R. (2017) "Drug shortages in Israel: Regulatory perspectives, challenges and solutions." Isr. J. Health Pol. Res. Vol. 6: 1–8.

Sinclair, D., Shahriar, H., and Zhang, C. (2019, January). Security requirement prototyping with hyperledger composer for drug supply chain: a blockchain application. In *Proceedings of the 3rd International Conference on Cryptography, Security and Privacy* (pp. 158–163).

Singh, R., Dwivedi, A. D., and Srivastava, G. (2020) "Internet of things based blockchain for temperature monitoring and counterfeit pharmaceutical prevention." Sensors. Vol. 20, no. 14: 3951.

Singh, S. and Singh, N. (2016) "Blockchain: Future of financial and cyber security." In Proceedings of the 2016 2nd International Conference on Contemporary Computing and Informatics (IC3I), Noida, India, 14–17 December; pp. 463–467.

Sousa, J., Bessani, A., and Vukolic, M. (2018) "A byzantine fault-tolerant ordering service for the hyperledger fabric blockchain platform." In 48th annual IEEE/IFIP international conference on dependable systems and networks (DSN), Luxembourg, 25–28 June, pp. 51–58. New York: IEEE.

Sukhwani, H. (2018) Performance Modeling & Analysis of Hyperledger Fabric (Permissioned Blockchain Network), Duke University, NC.

Swan, M. (2015) Blockchain: Blueprint for a New Economy. O'Reilly Media, Inc.: Sebastopol, CA, USA.

Toscano, Paul. (2011) "The dangerous world of counterfeit prescription drugs." *Retrieved October* 23 (2011): 2017.

Tseng, J.-H., Liao, Y.-C., Chong, B., and Liao, S.-w. (2018) "Governance on the drug supply chain via gcoin blockchain." *Int. J. Environ. Res. Public Health.* Vol. 15: 1055.

Walker, J., Chaar, B. B., Vera, N., Pillai, A. S., Lim, J. S., Bero, L. et al. (2017) "Medicine shortages in Fiji: A qualitative exploration of stakeholders' views." PloS one. Vol. 12: e0178429.

WHO (2016) Substandard, spurious, falsely labelled, falsified and counterfeit (SSFFC) medical products. *Fact sheet. Updated January.*

Yue, X., Wang, H., Jin, D., Li, M., and Jiang, W. (2016) "Healthcare data gateways: Found healthcare intelligence on blockchain with novel privacy risk control." *J. Med Systems.* Vol. 40: 218.

Zhang, A. and Lin, X. (2018) "Towards secure and privacy-preserving data sharing in e-health systems via consortium blockchain." *Journal of Medical System.* Vol. 42, no. 8: 140.

12 Home Automation Using Block Chain-enabled Cyber Physical System

Shaik Qadeer, Qazi Basheer, and Mohammad Sanaullah Qaseem

CONTENTS

12.1 INTRODUCTION

The following are some of the benefits of smart automation: To begin with, it will speed up routine daily tasks like using geysers, lights, and other home appliances. While at work, users need not be concerned about whether or not the lights, geyser, or air conditioner have been turned OFF and from everywhere on the earth, users can access our refuge away from home [1].

Second, in terms of security, consumers can keep their homes safe while they are gone. There are systems that continuously monitor the home and alert the owners when an incident occurs. The notice might be as basic as an SMS or a voice call, and it can be used to notify and take action. These devices can also be powered by batteries. So there's no need to be concerned even if the power is switched off [2].

Third, users don't have to physically go through each room to see if any light or fan is on and turn it off; alternatively, they can switch the lights in a hall or throughout the house ON and OFF. Users will be able to control equipment using their tablets or cellphones [3].

Safety is the next issue. Every home experiences circumstances where parents must leave and kids are left alone at home. In this case, automation helps to safeguard children. From anywhere on the planet, surveillance cameras can be installed and monitored via smartphones. There are also gadgets that make it impossible to unlock the main entrance from the outside after locking it from the inside. Installing sensors

DOI: 10.1201/9781003366010-15

outside will enable automatic lighting at night [4]. The cyber physical system is the fundamental idea behind Industry 4.0's creation of smart factories. In addition to smart factories, many CPS-based applications have been established, including smart cities, smart housing, smart buildings, and smart grid, Internet of vehicles, and healthcare [5-10]. CPS comes in a variety of architectures, the most prevalent of which is the 5C protocol architecture [11-20]. In this chapter, a simple implementation of smart house automation is proposed using the 5C protocol architecture of CPS.

This system's hardware architecture comprises of a Node MCU and a cellphone. The Internet is used to communicate wirelessly between the Node MCU and the cellphone. The android operating system includes a built-in voice recognition function called Google Assistant, which was leveraged to create a cellphone app that can manage home appliances through the use of vocal commands. The user's voice command is converted into text by this program, which then transmits the text to the Adafruit libraries, which are linked to the Micro Controller via the IFTTT service. IFTTT stands for "IF THIS THEN THAT" and is a webpage that enables you to create basic conditional statements called applets. The user merely needs to pronounce the appliance name followed by a four-digit password with a voice-controlled home automation system into the cellphone microphone and command it to turn the appliances OFF or ON. This enables users to easily and conveniently control their home appliances. Users can add more home appliances to the system via this simple application, which has an interface with user-friendly capabilities. This technology can be used to control electrical or electronic appliances and devices in any structure. The biggest advantage is that its array can be extended because we are using the Internet instead of Bluetooth, which has a restricted limit. However, this method is not cost-effective. Another advantage is the ease with which control messages may be entered using Google Assistant to pass control signals. Because many present home automation systems are reliant on wired connections, home automation is entirely wireless.

Because it improves security and anonymity, the block chain concept has recently attracted a lot of interest in distributed technologies like the Internet of Things. To be used successfully in CPS applications, it must have a few essential characteristics, such as the capacity to execute orders automatically (using a smart contract, for example), the kind of network access (permissioned or permissionless), and the type of network (public or private). A blockchain is a distributed ledger that participants of a peer-to-peer (P2P) network can mimic and share. Recently, the blockchain concept has gained popularity in distributed technologies like the Internet of Things (IoT) due to its improved security and privacy, increased system fault tolerance, provision of faster settlement and reconciliation with a scalable network, and helps in saving cost and time by removing intermediaries. By addressing interoperability, data integrity, security, privacy, and resilience, a block chain enabled CPS intends to address the difficulties associated with the practical implementation of the 5C-CPS structure.

12.2 BACKGROUND

Let's look at similar work first before discussing it, such as: Ransing and Rajput [21] used the Zigbee protocol because of its low power requirements for temperature monitoring. The smart home system employed by Bluetooth and system-integrated sensors was the subject of Kumar and Lee's [22] research. They employed web services with a RESTful architecture as the interoperable layer. To send and receive commands, they have created an android application. A Zigbee-based smart automation system that Young-Guk Ha [23] has created has been tested for security and warning purposes. They have magnetic sensors mounted to a door or window for home security. Sensors and modern or "smart" nursing homes are needed, according to Stankovic and Kiran [24]. However, in this chapter, a simple implementation of smart house automation is proposed using the 5C protocol architecture of CPS.

12.3 THE PROPOSED SYSTEM DESIGN

12.3.1 The Architecture of the Proposed System

The architecture of the proposed system for home automation using the 5C protocol of CPS is shown in Figure 12.1. It consists of the ESP32 Microcontroller, IFTTT, Adafruit IO, Google Assistant, and Channel Relay Module. A brief discussion of components is given here:

a) ESP 32: This controller contains the code for the final control element. It has a Wi-Fi module built in, which aids in online connectivity.

a) IFTTT: You can make applets, which are collections of elementary conditional statements, using the free web programme IFTTT. An applet may be activated by changes to other web services like Gmail, Facebook, Telegram, Instagram, or Pinterest.
b) Adafruit IO: This particular cloud service company specializes in IoT cloud solutions. It works with a variety of firmware, like Raspberry Pi, ESP32, and Arduino.
c) Google Assistance: this is virtual assistant software based on artificial intelligence that lets users maintain all of the apps on their device. Most of the applications on users' personal devices may be controlled and instructed using voice commands.
d) It is connected to the microcontroller, which performs the control operations and serves as the Final Control Element of this process, allowing the household appliances to be turned ON and OFF.

The final control circuit of the proposed system for one connection of a home appliance is shown in Figure12.2.

12.3.2 Mapping of Proposed System with 5C Protocol

In CPS systems, the 5C protocol can be used for implementation. It is as follows: Connection, Conversion, Cyber, Cognition, and Configuration as shown in Figure 12.3. The Google Assistant is used at the connection level in our system, where the input is delivered to the system as a voice command via Google Assistant. At the

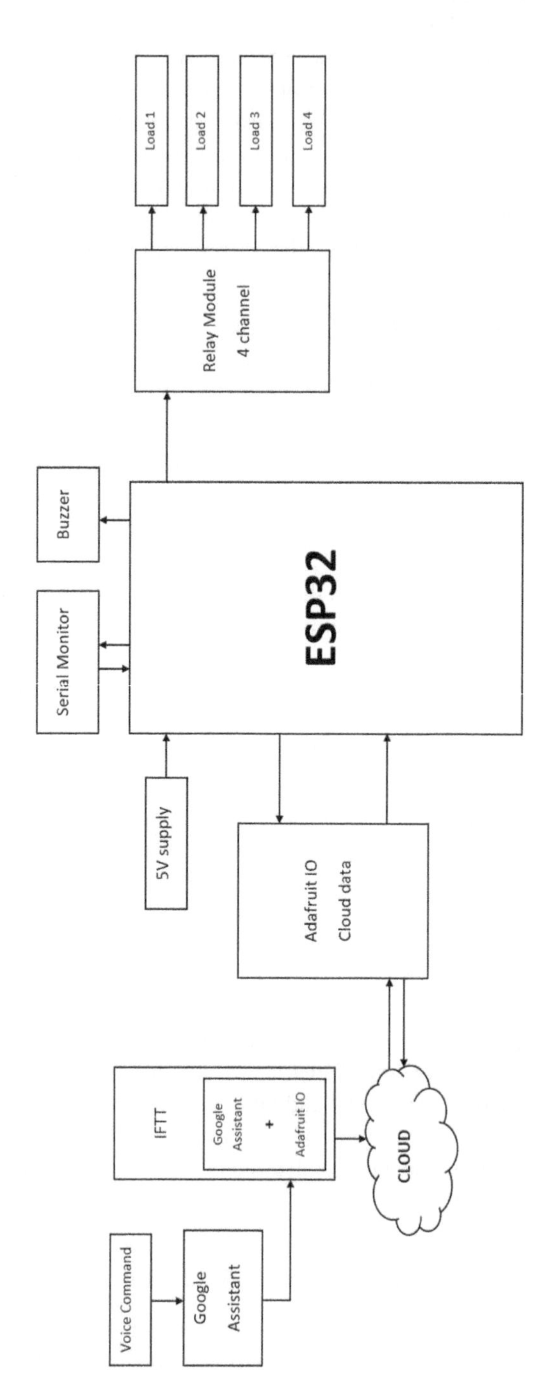

FIGURE 12.1 Proposed home automation system architecture.

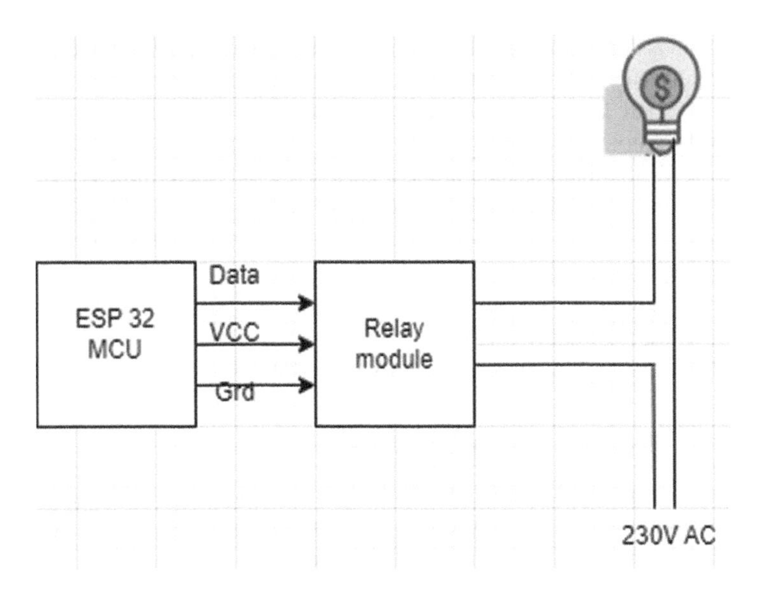

FIGURE 12.2 Final control circuit of the proposed system.

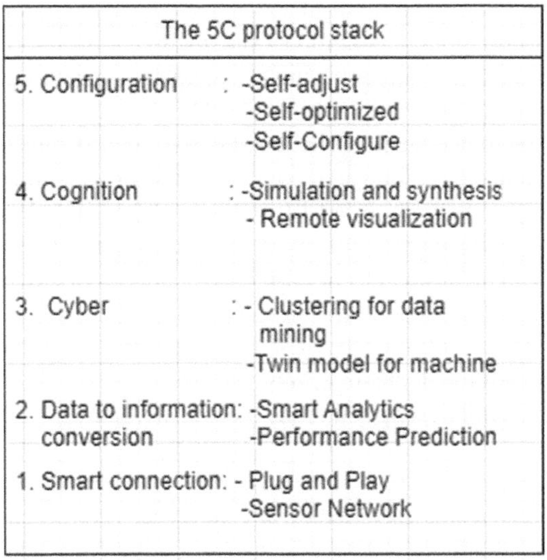

The 5C protocol stack	
5. Configuration :	-Self-adjust -Self-optimized -Self-Configure
4. Cognition :	-Simulation and synthesis - Remote visualization
3. Cyber :	- Clustering for data mining -Twin model for machine
2. Data to information: conversion	-Smart Analytics -Performance Prediction
1. Smart connection:	- Plug and Play -Sensor Network

FIGURE 12.3 The 5C protocol stack.

conversion level, the input is translated into a suitable form to perform the operation. The data provided by Google Assistant is converted to the Adafruit command using the IFTTT server. The Cyber level of control serves as a central information hub, storing the translated data from IFTTT in the cloud as well as any information related to the output. The converted commands are saved in the MQTT cloud server in our system, which is subsequently collected by the smart final control element (ESP32) to deliver a control signal to the smart actuator. Operators are given an appropriate presentation of analytic information for decision-making at the cognition level of control. At this level, remote collaboration and decision-making are both possible. The priority of jobs for the maintenance process may be quickly determined because comparative data and individual machine states are readily available. The configuration level transmits data from the cyber world to the real world. This level is responsible for supervising how machines configure themselves, adapt themselves, and optimize themselves. It functions as a resilience control system (RCS) that applies controls to the supervised appliances based on choices taken at the cognition level. The ESP-32 microcontroller is used in our system to create a smart actuator.

12.3.3 The Proposed System Operation

Home Automation with Voice Control is based on a system that is integrated with a cell phone-based application to offer the aged and handicapped the ability to control home utilities completely from their phone using voice commands and a four-digit password. For non-technical people, the gadget is made to be easy to travel with, install, configure, operate, and maintain. The concept of Google Assistant-controlled Home Automation is to use voice commands to control home gadgets (command followed by security key). The Adafruit account is connected to the IFTTT webpage, which generates if-else types of conditional statements. Adafruit is a free, cloud-based IoT web server for building virtual machines. Voice control and security mechanisms for Google Assistant are now available on the IFTTT website. The user can instruct the Google Assistant to operate household equipment such as lights, fans, and motors, which are accompanied by a four-digit password. The orders, along with a four-digit password, are decrypted (through IFTTT and the Adafruit IO cloud) and brought to the CPU, which forwards them to the relays connected to it. The gadget connected to the specific relay can be turned ON or OFF based on the user's request to the Google Assistant.

The key needs for each tier of the 5C-CPS are outlined in Table 12.1, along with any potential effects of block chain technology [25].

The following is a full description of the blockchain-enabled CPS architecture (see Figure 12.4).

Management Net: In this case, data from the cyber level is employed in a data-driven decision support system to enhance productivity, resilience, and ultimately industrial sustainability.

Cyber Net: In order to achieve integrity, fault tolerance, and resilience, the cyber net is in charge of managing cyber physical and cyber-cyber interactions and converting data into useful information.

Connection Net: The key features at this level are advanced connectivity, data management, integrity, and security. Interoperability is the most crucial component of this global connectivity and integration. The blockchain uses sophisticated cryptographic methods and a global consensus mechanism to overcome issues with security and privacy [26].

TABLE 12.1
Requirements and characteristics of CPS-5C architecture and its corresponding blockchain impacts

	Mapping Features		
Architecture of CPS-5C	**Characteristics**	**Requirements**	**Impact on Block chain (A-F band)**
Connection	1. Smart Sensors and Actuators 2. Physical layer interface	1.Interoperability 2.1. Connectivity (BW, availability etc.) 2.2. Privacy and security	1.D-F 2.1 D-F 2.2. A-G
Conversion	1. Deep Learning` 2. Analytical tools of AI 3. Edge Computing	1. Fast Computation 2. Design Complexity handling 3. Robust, Reliable and Adaptive	1. D-F 2.F 3. A-E
Cyber	1. Big data	1. Trust worthy of Data Science	1. A-C
Cognition	1. Integrated simulation and Synthesis	1. Real-Time data access	1. E and F
Configuration	Decision-making needs to be Intelligent	Self- optimize /adjust / configure	1.B-F

12.3.4 The Implementation Details of the Proposed System

The Adafruit IO must first be configured. The second stage is connecting the ESP32 (Final control element) and the final step is connecting to Google Assistant through IFTTT. On the webpage Adafruit IO, users can build virtual switches that can be turned ON or OFF using voice control with the Google Assistant.

a) Adafruit IO: Initially, go to www.Adafruit.io and create an account, then go to Adafruit and build a dashboard. This dashboard is a user interface that allows you to control things from a distance. For the button, enter 0 as the (OFF) text and 1 as the (ON) text, then click "create" on your dashboard, this will create a toggle button that can be used to control items remotely. The dashboard is now available for IoT applications such as home automation.
b) IFTTT: This web page is used to translate voice input into a command and a security key that Adafruit IO can understand. The steps to make this website are as follows: Create an IFTTT account using the same e-mail address as Adafruit. After you've created an account, go to My Applets and then New Applet. According to the program, any term can be entered. For each phrase, use a separate applet (like for ON and OFF). The user will now be prompted to choose the option for linking Google Assistant to Adafruit on a new page that

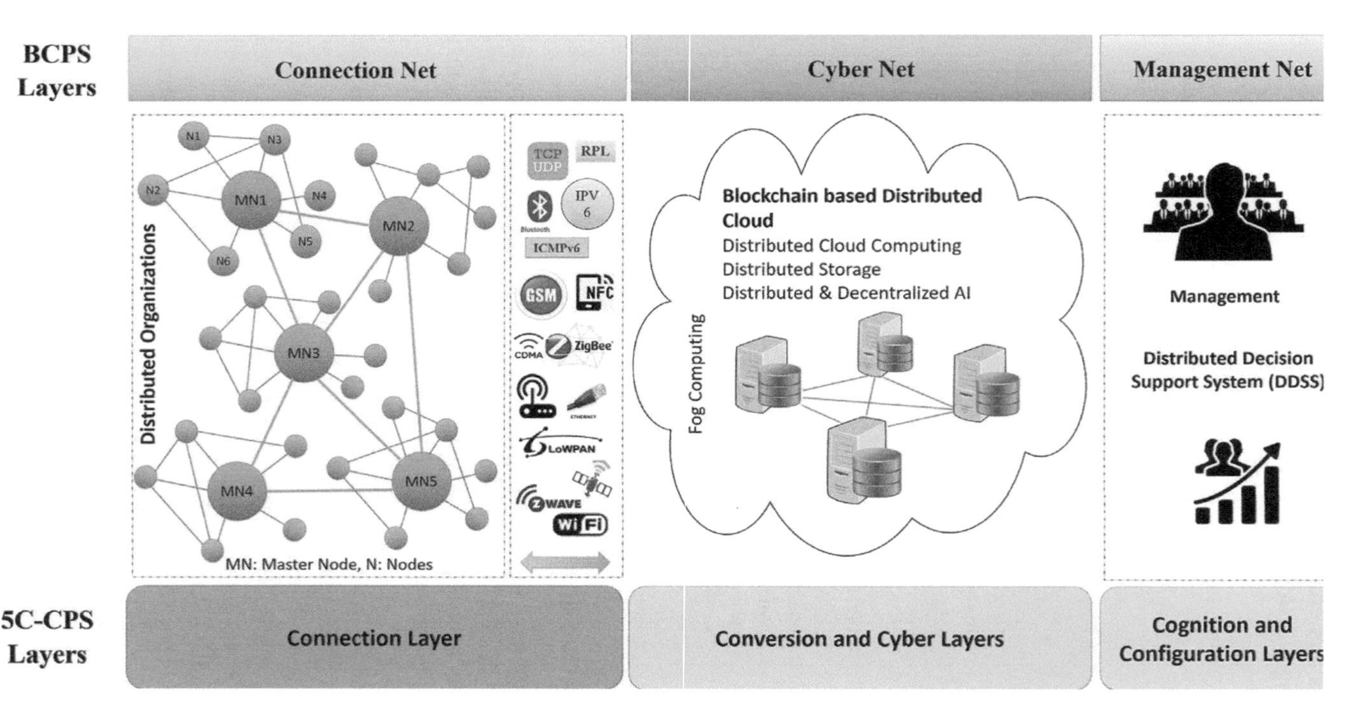

FIGURE 12.4 The proposed three-layer Block chain enabled CPS architecture.

TABLE 12.2
Comparison of proposed work with similar work

Parameters for Comparison	[23]	Proposed system
Aim	Home automation using Wi-Fi	Home automation suing blockchain enabled CPS
Protocol used	No protocol needed as no cloud is used	IFFT protocol is used as it is cloud-based
Security	Not applicable	A 4 bit security password is added for Adding security feature

will now open. Then, after selecting Adafruit, choose actions such as different ways a user can feed the sentences.

c) ESP 32: The circuit for this is depicted in Figure 12.2, and the code inside this controller is presented in Table 12.3 to make it a smart actuator. It reads the control signal from the Adafruit IO cloud server (which has Wi-Fi) and sends it to the appropriate loads.

A brief comparison between similar work [27] and the proposed work is given in Table 12.2. The similarity between both works is that they perform home automation using IoT. The differences are listed in the table; notable things are that the proposed system is cloud-based and blockchain-enabled, and the resultant implementation is CPS type, whereas the work shown [27] is a conventional one.

12.4 CONCLUSION

In this chapter, a case study (home automation) of CPS implementation is given. This can be easily extended to other industrial applications of Industry 4.0. All 5C operations are included along with the features of blockchain-enabled CPS. An effective block chain mapping and implementation can protect CPS systems from vulnerable security risks while also providing a new direction for CPS application research. The economically higher nature of the block chain with higher bandwidth overhead remains to be the most challenging scope when applied towards CPS.

12.5 CHALLENGES AND FUTURE ASPECTS

In this segment of the article, we are going to examine the problems that the proposed work has not included the implementation of cognition feature of 5C protocol and the anticipated future work to surmount these challenges. This feature will add the use of machine learning trained model to be used as security factor.

TABLE 12.3
Final control elements firmware code

Header files Local and global variables declarations	"WiFi.h"
	"Adafruit_MQTT.h"
	"Adafruit_MQTT_Clie"
	// Configration of Pins
	//Let there are 4 appliances
	Buzzer= 26
	Relay1= 32
	Relay2= 13
	Relay3= 14
	Relay4= 27
	// Assign Access Point for WiFi
	WLAN_SSID= "Sohel"
	WLAN_PASS= "ABC@1234"
	// Assign Setup for Adafruit.io
	Assignt1: "AIO_SERVER= "io.adafruit.com""
	Assignt2: "AIO_SERVERPORT= 1883 "
	// for SSL use 8883
	Assignt3: "AIO_USERNAME= "Sohelpasham""
	Assignt4: AIO_KEY= "aio_avbx594BieV8Pb6NWk4dzRvFTxli"
	//This is Global state (we need not to change this)
	//Using MQTT protocol connect to Adafruit cloud server
	Adafruit_MQTT_Client mqtt(&client, AIO_SERVER, AIO_SERVERPORT, AIO_USERNAME, AIO_KEY);
	// Path for MQTT:
	<username>/feeds/<feedname>
	// ON-OFF statements.
	Statement1:"Adafruit_MQTT_Subscribe Fan" = "Adafruit_MQTT_Subscribe(&mqtt, AIO_USERNAME"/feeds/dc-fan")";
	Statement2:"Adafruit_MQTT_Subscribe Extra" = "Adafruit_MQTT_Subscribe(&mqtt, AIO_USERNAME" "/feeds/extra")";
	Statement3:"Adafruit_MQTT_Subscribe Charger" = "Adafruit_MQTT_Subscribe(&mqtt, AIO_USERNAME" ""/feeds/charger")";
	Statement4:"Adafruit_MQTT_Subscribe Light "= "Adafruit_MQTT_Subscribe(&mqtt, AIO_USERNAME " "/feeds/bulb")";

TABLE 12.3 (Continued)
Final control elements firmware code

Code which executed just once	void setup() {
	Command to set up baud rate "Serial.begin(110052)";
	delay(20);
	//Configuration of ports
	Buuzzer: "pinMode(buzzer,OUTPUT)";
	Relay4: "pinMode(Relay4, OUTPUT)";
	Relay3: " pinMode(Relay3, OUTPUT)";
	Realy2: "pinMode(Relay2, OUTPUT)";
	Realy1: "pinMode(Relay1, OUTPUT)";
	
	//To send initially ON command
	Realy4: "digitalWrite(Relay4,HIGH)";
	Realy3: " digitalWrite(Relay3,HIGH)";
	Realy2: "digitalWrite(Relay2,HIGH)";
	Relay1: " digitalWrite(Relay1,HIGH)";
	//to Send command on monitor
	" Serial.println(F("Adafruit MQTT demo"))";
	// Now established connection to WiFi.
	Command1: "Serial.println()";
	Command2: "Serial.print("Connecting to ")";
	Command3: "Serial.println(WLAN_SSID)";
	Command4: "WiFi.begin(WLAN_SSID, WLAN_PASS)";
	Command5: "while (WiFi.status()!= WL_CONNECTED) {
	delay(600);
	Serial.print(".");
	}"
	Command6: "Serial.println()";
	Command7: "Serial.println(WiFi connected)";
	Command8: "Serial.println(IP address:)";
	Command9: "Serial.println(WiFi.localIP())";
	// Now for onoff feed MQTT subscription.
	Command1:"mqtt.subscribe(&Fan)";
	Command2: "mqtt.subscribe(&Light)";
	Command3: "mqtt.subscribe(&Extra)";
	Command4: "mqtt.subscribe(&Charger)";
	}

(Continued)

TABLE 12.3 (Continued)
Final control elements firmware code

Main Code

```
void loop() {
// Connect to MQTT server
Command1: " MQTT_connect()";
// Do delay
Command2: "Adafruit_MQTT_Subscribe *subscription";
   while ((subscription = mqtt.readSubscription(15000))) {

   if (subscription == &Fan) {
   Serial.print(F("Got: ")); Serial.println((char *)Fan.lastread);
   int Light2_State = atoi((char *)Fan.lastread);
   digitalWrite(Relay3, Light2_State);
   }
if (subscription == &Light) {
   Serial.print(F("Got: "));
   Serial.println((char *)Light.lastread);
   int Light1_State = atoi((char *)Light.lastread);
   digitalWrite(Relay1, Light1_State);
   }
if (subscription == &Extra) {
   Serial.print(F("Got: "));
   Serial.println((char *)Extra.lastread);
   int Fan2_State = atoi((char *)Extra.lastread);
   digitalWrite(Relay4, Fan2_State);
   }
if (subscription == &Charger) {
   Serial.print(F("Got: "));
   Serial.println((char *)Charger.lastread);
   int Fan1_State = atoi((char *)Charger.lastread);
   digitalWrite(Relay2, Fan1_State);
   }
}"
// check for connection using polling
}
// MQTT connection function
void MQTT_connect() {
   int8_t ret;}
```

REFERENCES

[1] S. -Y. Chien, M. Lewis, K. Sycara, A. Kumru, and J. -S. Liu, "Influence of Culture, Transparency, Trust, and Degree of Automation on Automation Use," in IEEE Transactions on Human-Machine Systems, vol. 50, no. 3, pp. 205–214, June 2020, doi: 10.1109/THMS.2019.2931755.

[2] D. Wang, "An Enterprise Data Pathway to Industry 4.0," in IEEE Engineering Management Review, vol. 46, no. 3, pp. 46–48, 1 third quarter, Sept. 2018, doi: 10.1109/EMR.2018.2866157.

[3] C. J. Turner, J. Oyekan, L. Stergioulas and D. Griffin, "Utilizing Industry 4.0 on the Construction Site: Challenges and Opportunities," in IEEE Transactions on Industrial Informatics, vol. 17, no. 2, pp. 746–756, Feb. 2021, doi: 10.1109/TII.2020.3002197.

[4] J. J. Fuertes, M. Á. Prada, J. R. Rodríguez-Ossorio, R. González-Herbón, D. Pérez, and M. Domínguez, "Environment for Education on Industry 4.0," in IEEE Access, vol. 9, pp. 144395–144405, 2021, doi: 10.1109/ACCESS.2021.3120517.

[5] M. H. Cintuglu, O. A. Mohammed, K. Akkaya and A. S. Uluagac, "A Survey on Smart Grid Cyber-Physical System Testbeds," in IEEE Communications Surveys & Tutorials, vol. 19, no. 1, pp. 446–464, Firstquarter 2017, doi: 10.1109/COMST.2016.2627399.

[6] S. Xin, Q. Guo, H. Sun, B. Zhang, J. Wang, and C. Chen, "Cyber-Physical Modeling and Cyber-Contingency Assessment of Hierarchical Control Systems," in IEEE Transactions on Smart Grid, vol. 6, no. 5, pp. 2375–2385, Sept. 2015, doi: 10.1109/TSG.2014.2387381.

[7] R. V. Yohanandhan, R. M. Elavarasan, P. Manoharan, and L. Mihet-Popa, "Cyber-Physical Power System (CPPS): A Review on Modeling, Simulation, and Analysis with Cyber Security Applications," in IEEE Access, vol. 8, pp. 151019–151064, 2020, doi: 10.1109/ACCESS.2020.3016826.

[8] L. Ribeiro and M. Björkman, "Transitioning from Standard Automation Solutions to Cyber-Physical Production Systems: An Assessment of Critical Conceptual and Technical Challenges," in IEEE Systems Journal, vol. 12, no. 4, pp. 3816–3827, Dec. 2018, doi: 10.1109/JSYST.2017.2771139.

[9] P. Lau, L. Wang, Z. Liu, W. Wei, and C. -W. Ten, "A Coalitional Cyber-Insurance Design Considering Power System Reliability and Cyber Vulnerability," in IEEE Transactions on Power Systems, vol. 36, no. 6, pp. 5512–5524, Nov. 2021, doi: 10.1109/TPWRS.2021.3078730.

[10] R. Altawy and A. M. Youssef, "Security Tradeoffs in Cyber Physical Systems: A Case Study Survey on Implantable Medical Devices," in IEEE Access, vol. 4, pp. 959–979, 2016, doi: 10.1109/ACCESS.2016.2521727.

[11] J. -R. Jiang, "An improved Cyber-Physical Systems architecture for Industry 4.0 Smart Factories," 2017 International Conference on Applied System Innovation (ICASI), 2017, pp. 918–920, doi: 10.1109/ICASI.2017.7988589.

[12] Lee, Jay, Behrad Bagheri, and Hung-An Kao, "A Cyber-Physical Systems Architecture for Industry 4.0-based Manufacturing Systems", Manufacturing Letters, vol. 3, pp. 18–23, 2015.

[13] Wei Wang, Lei Fan, Pu Huang, and Hai Li, "A New Data Processing Architecture for Multi-Scenario Applications in Aviation Manufacturing", IEEE Access, vol. 7, pp. 83637–83650, 2019.

[14] Vítor Alcácer, Carolina Rodrigues, Helena Carvalho, and Virgilio Cruz-Machado, "Tracking the Maturity of Industry 4.0: The Perspective of a Real Scenario", The International Journal of Advanced Manufacturing Technology, vol. 116, no. 7–8, p. 2161, 2021.

[15] Diego G.S. Pivoto, Luiz F.F. de Almeida, Rodrigo da Rosa Righi, Joel J.P.C. Rodrigues, Alexandre Baratella Lugli, and Antonio M. Alberti, "Cyber-physical Systems Architectures for Industrial Internet of Things Applications in Industry 4.0: A Literature Review", Journal of Manufacturing Systems, vol. 58, p. 176, 2021.

[16] Ahmadzai Ahmadi, Ali Hassan Sodhro, Chantal Cherifi, Vincent Cheutet, and Yacine Ouzrout, "Evolution of 3C Cyber-Physical Systems Architecture for Industry 4.0", Service Orientation in Holonic and Multi-Agent Manufacturing, vol. 803, p. 448, 2019.

[17] L. Yongfu, S. Dihua, L. Weining, and Z. Xuebo, "A Service-oriented Architecture for the Transportation Cyber-Physical Systems," Proceedings of the 31st Chinese Control Conference, 2012, pp. 7674–7678.
[18] S. Gautham, A. V. Jayakumar, A. Rajagopala and C. Elks, "Realization of a Model-Based DevOps Process for Industrial Safety Critical Cyber Physical Systems," 2021 4th IEEE International Conference on Industrial Cyber-Physical Systems (ICPS), 2021, pp. 597–604, doi: 10.1109/ICPS49255.2021.9468213.
[19] M. Li, Y. Xue, M. Ni, and X. Li, "Modeling and Hybrid Calculation Architecture for Cyber Physical Power Systems," in IEEE Access, vol. 8, pp. 138251–138263, 2020, doi: 10.1109/ACCESS.2020.3011213.
[20] J. Wan, A. Canedo, and M. A. Al Faruque, "Cyber–Physical Codesign at the Functional Level for Multidomain Automotive Systems," in IEEE Systems Journal, vol. 11, no. 4, pp. 2949–2959, Dec. 2017, doi: 10.1109/JSYST.2015.2472495.
[21] R. S. Ransing and M. Rajput, "Smart Home for Elderly Care, Based on Wireless Sensor Network," 2015 International Conference on Nascent Technologies in the Engineering Field (ICNTE), Navi Mumbai, 2015, pp. 1–5.
[22] S. Kumar and S. R. Lee, "Android-based Smart Home System with Control via Bluetooth and Internet Connectivity," The 18th IEEE International Symposium on Consumer Electronics (ISCE 2014), JeJu Island, 2014, pp. 1–2.
[23] H. Young-Guk, "Dynamic Integration of Zigbee Home Networks into Home Gateways Using OSGI Service Registry," IEEE Transactions on Consumer Electronics., vol. 55, Issue. 2, pp. 470–476, May. 2009.
[24] J. Stankovic, Q. Cao, T. Doan, L. Fang, Z. He, R. Kiran, "Wireless Sensor Networks for In-Home Healthcare: Potential and Challenges," Proc. of the Workshop on High Confidence Medical Devices Software and Systems, June 2005.
[25] W. Zhao, C. Jiang, H. Gao, S. Yang, and X. Luo, "Blockchain-Enabled Cyber–Physical Systems: A Review," in IEEE Internet of Things Journal, vol. 8, no. 6, pp. 4023–4034, March 15, 2021, DOI: 10.1109/JIOT.2020.3014864.
[26] Jay Lee, Moslem Azamfar, and Jaskaran Singh, "A Blockchain Enabled Cyber-Physical System Architecture for Industry", Elsevier Manufacturing Letters, Volume 20, 2019, pp. 34–39, https://doi.org/10.1016/j.mfglet.2019.05.003.
[27] H. K. Singh, S. Verma, S. Pal, and K. Pandey, "A Step Towards Home Automation Using IOT," 2019 Twelfth International Conference on Contemporary Computing (IC3), 2019, pp. 1–5, doi: 10.1109/IC3.2019.8844945.

Index

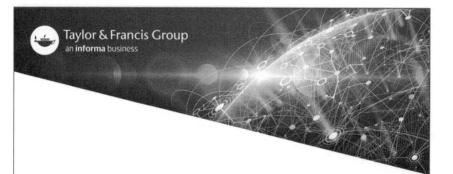